3D Digital Geological Models

3D Digital Geological Models

From Terrestrial Outcrops to Planetary Surfaces

Edited by

Andrea Bistacchi
Department of Environmental and Earth Sciences
University of Milano-Bicocca
Milan, Italy

Matteo Massironi
Department of Geosciences
University of Padua
Padua, Italy

Sophie Viseur
Aix-Marseille University
Marseille, France

WILEY

This edition first published 2022

Registered Office
John Wiley & Sons, Inc., 111 River Street, Hoboken, NJ 07030, USA

Editorial Office
111 River Street, Hoboken, NJ 07030, USA

For details of our global editorial offices, customer services, and more information about Wiley products visit us at www.wiley.com.

Wiley also publishes its books in a variety of electronic formats and by print-on-demand. Some content that appears in standard print versions of this book may not be available in other formats.

Library of Congress Cataloging-in-Publication Data

Names: Bistacchi, Andrea, editor. | Massironi, M. (Matteo), editor. |
Viseur, Sophie, editor.
Title: 3D digital geological models : from terrestrial outcrops to
planetary surfaces / edited by Andrea Bistacchi, Matteo Massironi,
Sophie Viseur.
Description: Hoboken, NJ : Wiley, 2022. | Includes bibliographical
references and index.
Identifiers: LCCN 2021050370 (print) | LCCN 2021050371 (ebook) | ISBN
9781119313885 (hardback) | ISBN 9781119313915 (adobe pdf) | ISBN
9781119313892 (epub)
Subjects: LCSH: Three-dimensional imaging in geology.
Classification: LCC QE26.3 .T15 2022 (print) | LCC QE26.3 (ebook) | DDC
550.28/4–dc23
LC record available at https://lccn.loc.gov/2021050370
LC ebook record available at https://lccn.loc.gov/2021050371

Cover Design: Wiley
Cover Image: Courtesy of Riccardo Pozzobon (front cover); Courtesy of Andrea Bistacchi (back cover)

Set in 9.5/12.5pt STIXTwoText by Straive, Chennai, India

10 9 8 7 6 5 4 3 2 1

Contents

List of Contributors

Sarah Bali
Aix-Marseille University
Marseille
France

Robert Barnes
Imperial College London
London
UK

Roland Baumberger
Institute of Geological Sciences
University of Bern
Bern
Switzerland

and

Federal Office of Topography
Swiss Geological Survey
Wabern
Switzerland

Florian Beguet
Aix-Marseille University
Marseille
France

Ivano Bertini
Department of Physics and Astronomy "Galileo Galilei"
University of Padova
Padova
Italy

Andrea Bistacchi
Dipartimento di Scienze dell'Ambiente e della Terra
Università degli Studi di Milano - Bicocca
Milano
Italy

Sylvain Bouley
Université Paris-Saclay
Orsay
France

Simon J. Buckley
NORCE Norwegian Research Centre AS
Bergen
Norway

and

Department of Earth Science
University of Bergen
Bergen
Norway

Pamela Cambianica
INAF
Astronomical Observatory of Padova
Padova
Italy

Thomas J. B. Dewez
BRGM
Marseille
France

Sabrina Ferrari
Centre of Studies and Activities for Space "G. Colombo"
University of Padua
Padua
Italy

Marco Franceschi
Department of Geosciences
University of Padua
Padua
Italy

Elisa Frattin
INAF
Astronomical Observatory of Padova
Padova
Italy

and

Department of Physics and Astronomy "Galileo Galilei"
University of Padova
Padova
Italy

Robert Gaskell
Planetary Science Institute
Tucson
Arizona
USA

Robert L. Gawthorpe
Department of Earth Science
University of Bergen
Bergen
Norway

Jérémy Giuliano
GeoConseil
Risk and Geological Consulting
Le Val
France

and

Université de Nice Sophia Antipolis
Valbone
France

and

Aix-Marseille Université
Marseille
France

Vincent Godard
Aix-Marseille Université
Marseille
France

Sanjeev Gupta
Imperial College London
London
UK

Marco Herwegh
Institute of Geological Sciences
University of Bern
Bern
Switzerland

Gerd Hesina
Zentrum für Virtual Reality und Visualisierung (VRVis)
Forschungs-GmbH
Vienna
Austria

John A. Howell
Department of Geology and Petroleum Geology
University of Aberdeen
Aberdeen
UK

Laurent Jorda
Aix-Marseille University
Marseille
France

Christian Kehl
Uni Research AS
Bergen
Norway

and

Department of Earth Science
University of Bergen
Bergen
Norway

Edi Kissling
Institute of Geophysics
ETH Zürich
Zürich
Switzerland

Fiorangela La Forgia
Department of Physics and Astronomy "Galileo Galilei"
University of Padova
Padova
Italy

Thomas Lebourg
University of Nice Sophia Antipolis,
Valbone
France

Alice Luccetti
INAF
Astronomical Observatory of Padova
Padova
Italy

Nathalie Marçot
BRGM
Marseille
France

Jean-Luc Mari
Aix-Marseille University
Marseille
France

Mattia Martinelli
Dipartimento di Scienze dell'Ambiente e della Terra
Universita' degli Studi di Milano - Bicocca
Milano
Italy

Matteo Massironi
Dipartimento di Geoscienze
Università degli Studi di Padova
Padua
Italy

Claudio Mazzoli
Department of Geosciences
University of Padua
Padua
Italy

Silvia Mittempergher
Dipartimento di Scienze dell'Ambiente e della Terra
Universita' degli Studi di Milano - Bicocca
Milano
Italy

and

Dipartimento di Scienze Chimiche e Geologiche
Università degli Studi di Modena e Reggio Emilia
Modena
Italy

Jan-Peter Muller
University College London
London
UK

James R. Mullins
Department of Geology and Petroleum Geology
University of Aberdeen
Aberdeen
UK

Jennifer Muscato
Aix-Marseille University
Marseille
France

Giampiero Naletto
Centre of Studies and Activities for Space "G. Colombo"
University of Padua
Padua
Italy

and

CNR-IFN UOS Padua LUXOR
Padua
Italy

and

Department of Physics and Astronomy "Galileo Galilei"
University of Padua
Padua
Italy

Thomas Ortner
Zentrum für Virtual Reality und Visualisierung (VRVis)
Forschungs-GmbH
Vienna
Austria

Gerhard Paar
Joanneum Research
Graz
Austria

Maurizio Pajola
INAF
Astronomical Observatory of Padova
Padova
Italy

Luca Penasa
Centre of Studies and Activities for Space "G. Colombo"
University of Padua
Padua
Italy

and

Dipartimento di Geoscienze
University of Padua
Padua
Italy

Riccardo Pozzobon
Department of Geosciences
University of Padua
Padua
Italy

Mélody Prémaillon
BRGM
Montpellier
France

Nereo Preto
Dipartimento di Geoscienze
University of Padua
Padua
Italy

Frank Preusker
Institute of Planetary Research
German Aerospace Center
Berlin
Germany

Silvia Salvini
Department of Geosciences
University of Padua
Padua
Italy

Tommaso Santagata
Vigea
Reggio Emilia
Italy

Francesco Sauro
Department of Biological, Geological, and Environmental Sciences
University of Bologna
Bologna
Italy

Martin-Pierre Schmidt
Aix-Marseille University
Marseille
France

Frank Scholten
Institute of Planetary Research
German Aerospace Center
Berlin
Germany

Holger Sierks
Max Planck Institute for Solar System Research
Göttingen
Germany

Emanuele Simioni
Astronomical Observatory of Padua
National Institute of Astrophysics
Padua
Italy

and

CNR-IFN UOS Padua LUXOR
Padua
Italy

Yu Tao
University College London
London
UK

Christoph Traxler
Zentrum für Virtual Reality und Visualisierung (VRVis)
Forschungs-GmbH
Vienna
Austria

Sophie Viseur
Aix-Marseille University
Marseille
France

Konrad Willner
Institute of Planetary Research
German Aerospace Center
Berlin
Germany

Preface

A series of sessions was hosted at General Assemblies of the European Geosciences Union in Vienna, focused on techniques allowing geoscientists with various backgrounds to collect 3D geological and geomorphological quantitative data on Digital Outcrop Models (DOMs), Digital Elevation Models (DEMs), or Shape Models (SMs): digital representations of outcrops, topographic surfaces, or whole small bodies of the Solar System (asteroid or comet nuclei) respectively. These sessions were based on the assumption that the two scientific communities, working on the Earth and on planetary bodies of the Solar System, are indeed using very similar datasets and techniques for studying 3D models from the geological point of view. During those meetings the members of the two communities, often unaware of each other's work, had the chance to share their knowhow, procedures, and methodologies, and this volume represents a natural prosecution of that experience. In particular, the volume has two main aims: disclosing the numerous points that geological disciplines have in common, when applied on the Earth and on planetary bodies, and favoring the communication and collaboration between different scientific communities.

The first chapters focus on techniques used to reconstruct DOMs in diverse environments (with examples on the Earth and on Mars) and with different remote sensing techniques (i.e. laser scanning vs. photogrammetry), and propose modern approaches for DOM analysis and interpretation, including semi-automatic image and mesh processing techniques. The second block of chapters presents case studies of quantitative geomorphological analysis on the Earth, Mars, and the Moon. In the final block, some examples on how data collected at the surface can be used to reconstruct 3D subsurface models are discussed.

Reading these chapters, authored by experts in different fields, it will become apparent that (i) the fundamental techniques allowing the production of DEMs, DOMs, and SMs (e.g. photogrammetry, laser scanning, radar interferometry) are well consolidated, and are almost seamlessly shared between the communities of scientists working on the Earth and on other bodies of the Solar System; (ii) the way these techniques are applied in different geological environments may change and, in some cases, can influence the quality of the results; (iii) the development of new techniques for DOM, DEM, and SM processing, elaboration, and analysis is under way and thus highly subject to continuous improvements; and (iv) the production of subsurface geological models based (also) on surface data will be an active field of research in the years to come.

Further challenges will arise with the increase of DOM, DEM, and SM acquisitions, such as: (i) developing integrated routines for the automated analysis and interpretation of topographic datasets (meshes or point-clouds) or imagery (e.g. high-resolution orthophotos); (ii) managing the numerous and huge datasets acquired, for instance, as time series of successive acquisitions for geological risk monitoring; or (iii) providing access to web-platforms for sharing outcrop datasets in geosciences, as already commonly done in astronomy.

We hope that this volume will foster collaborative efforts towards 3D data exploitation and will be able to inspire, with its introductive and general chapters, young researchers interested in 3D data analysis and senior scientists with the more advanced case studies.

Andrea Bistacchi
Dipartimento di Scienze dell'Ambiente e della Terra
Università degli Studi di Milano Bicocca

Matteo Massironi
Dipartimento di Geoscienze
Università degli Studi di Padova

Sophie Viseur
Centre Européen de Recherche et d'Enseignement des Géosciences de l'Environnement (CEREGE)
Aix-Marseille Université

and

Centre National de la Recherche Scientifique (CNRS)
Institut de Recherche pour le Développement (IRD)
Institut National de Recherche pour l'Agriculture l'alimentation et l'Environnement (INRAE)
Collège de France

1

3D Digital Geological Models: From Terrestrial Outcrops to Planetary Surfaces

Andrea Bistacchi[1], Matteo Massironi[2], and Sophie Viseur[3]

[1] Dipartimento di Scienze dell'Ambiente e della Terra, Università degli Studi di Milano - Bicocca, Piazza della Scienza, 4, 20126 Milano
[2] Dipartimento di Geoscienze, Università degli Studi di Padova, Via Gradenigo 6, 35131 Padova
[3] Aix Marseille Univ, CNRS, IRD, INRAE, Coll France, CEREGE., Case 67, 3 place Victor Hugo, 13331 Marseille CEDEX 03, France

Abstract

Collecting quantitative data to support geological analysis and modelling is nowadays a fundamental requirement in all geology disciplines, including structural geology, stratigraphy, and geomorphology, on the Earth and on planetary bodies of the Solar System. In many cases the answer to this need is a Digital Outcrop Model (DOM), a Digital Elevation Model (DEM), or a Shape Model (SM): this can be a digital representation of an outcrop or topographic surface, or of a whole small body (asteroid or comet nucleus) for an SM, generally combined with imagery, that can be quantitatively visualized and studied in 3D, with the goal of obtaining quantitative measurements.

3D datasets and models for geological purposes include different complementary products: DEMs, DOMs, SMs, and subsurface models. The main differences among these different products are: (i) their nature, since DEMs, DOMs, and SMs represent relief surfaces showing outcropping geological structures that are completely accessible to characterization (up to some precision/resolution), while subsurface models reproduce inaccessible subsurface geological structures with some unavoidable level of uncertainty (hence they are *models*); and (ii) their topology/dimensionality, as DEMs are actually 2.5D surfaces, generally covering large areas, DOMs are truly 3D surfaces, including multivalued reliefs (e.g. complex or overhanging reliefs, cliffs, caves, etc.), but are generally limited to smaller-scale outcrops, and SMs are closed surfaces covering a whole small body, where subsurface models are essentially volumetric.

In this volume we collect various examples of methods and techniques used to collect, analyze, and model 3D datasets, based on one or more supports (DEM, DOM, SM, subsurface model), and on different software tools, remote sensing, and modelling techniques. Reading the chapters authored by experts in different fields, it will become apparent that (i) the fundamental techniques allowing the production of DEMs, DOMs, and SMs through photogrammetry, laser scanning devices, and radar interferometry are well consolidated, and are almost seamlessly shared between the community of scientists working on the Earth and on planetary bodies of the Solar System; (ii) the particular way these techniques are applied in specific geological environments may change and, for instance, acquisition schemes in photogrammetry still represent a potentially critical issue; (iii) DOM, DEM, and SM processing, elaboration and analysis, including the analysis of image data associated with these surfaces, are active fields of research that are subject to continuous improvements; and (iv) the production of subsurface geological models based (also) on surface data is still not very common, particularly in planetary geology contexts.

One of the aims of this volume is to disclose the numerous points that geological disciplines have in common in applications on the Earth and on planetary bodies of the Solar System, and to favor the communication and collaboration between different scientific communities.

1.1 Introduction

Collecting 3D quantitative data is a fundamental requirement in many structural geology, stratigraphy, sedimentology, geomorphology, and engineering geology projects both on the Earth and on planetary bodies of the Solar System (e.g. Bistacchi et al., 2011; Simioni et al., 2015; Jones et al., 2016; Tavani et al., 2016; Martinelli et al., 2017, 2020; Penasa et al., 2017; Triantafyllou et al., 2019; Siddiqui et al., 2019; Caravaca et al., 2020; Crane, 2020; De Toffoli et al., 2020; Le Mouélic et al., 2020). This can be achieved using different and complementary 3D datasets:

3D Digital Geological Models: From Terrestrial Outcrops to Planetary Surfaces, First Edition.
Edited by Andrea Bistacchi, Matteo Massironi, and Sophie Viseur.

Digital Elevation Models (DEMs), Digital Outcrop Models (DOMs), and Shape Models (SMs).

DEMs are 2.5D representations of topographic surfaces (e.g. Jones et al., 2008), generally with regional to global extension, produced through consolidated approaches from photogrammetry, laser scanning devices, and radar interferometry. DEMs are generally stored as 2D regular grids, but triangulated surfaces can be used in some applications, and are called Triangulated Irregular Networks (TINs) in GIS systems.

DOMs (e.g. Bellian et al., 2005) are instead digital high-resolution representations of outcrops, or of the topographic surface at smaller scale. DOMs, represented as triangulated surfaces or point clouds, can represent multivalued reliefs (e.g. cliffs, caves, highly rough or overhanging reliefs) and can be really considered as 3D geometrical representations (Jones et al., 2008). DOMs can be textured if stored as triangulated surfaces (Catmull, 1974) and colored in case of point clouds (e.g. with RGB, LiDAR intensity). The textures mapped onto the surfaces may be single photos or, more recently, texture atlases (Lévy et al., 2002). DOMs can be visualized and studied, with dedicated software, with the final goal of obtaining quantitative measurements of sedimentary, stratigraphic, intrusive, tectonic, or geomorphological structures, or mapping lithology or alteration halos, etc.

SMs are produced for irregular small bodies of the Solar System, mainly using photogrammetric approaches (e.g. Carry et al., 2012; Preusker et al., 2012, 2015; Willner et al., 2014). Their particularity is that they are closed surfaces since they represent a whole small body. Apart from this, they share most other properties with DOMs.

3D geological models are 3D reconstructions of the subsurface geology and, historically, they have been produced based on geophysical and borehole/well datasets, mainly in oil and mining exploration contexts (e.g. Mallet, 2002, and references therein). More recently, they have been used also in other contexts, such as academic research projects or engineering geology, and, thanks to the emergence of high-resolution DEMs and DOMs, they are also based on surface datasets. Subsurface geological models are often represented as a set of interface surfaces (e.g. faults, stratigraphic surfaces), termed as 3D Structural Models (Caumon et al., 2009), and sometimes as 3D structured or unstructured grids or meshes (Mallet, 2002) for specific applications (e.g. flow simulation, mechanical modelling, geostatistics). On planetary bodies of the Solar System, subsurface models have been reconstructed using surface data collected on DEMs, DOMs, or SMs (Penasa et al., 2017; Pozzobon et al., 2020; Franceschi et al., 2020), and also on subsurface geophysical datasets such as radargrams (Yuan et al., 2017).

Using the same techniques when reconstructing and analyzing DOMs, DEMs, SMs, and 3D subsurface geological models on the Earth and planets can be considered a fair example of replicable science, which helps to reduce the uncertainty when interpreting geological features on planetary surfaces without any help from manned surveys.

The goal of this volume is to review and discuss a collection of techniques and workflows that can be applied to study DEMs and DOMs and to retrieve 3D geomodels on the Earth and on planetary bodies, starting from data collected with different instruments and platforms (e.g. laser scanning vs. photogrammetry, aerial and orbital vs. terrestrial), in different environmental and logistic conditions, at different scales, and for different purposes. The volume is organized in three sections. The five chapters of the first section focus on techniques used to reconstruct DOMs in different environments and with different remote sensing techniques (i.e. laser scanning vs. photogrammetry) and propose modern approaches for DOM analysis and interpretation. The three chapters of the second section propose examples of morphometric analysis at different scales on the Earth, on Mars, and on the Moon. Finally, the two chapters of the final section show how data collected at the surface can be used to reconstruct 3D models of the subsurface.

1.2 DOM/SM Reconstruction and Interpretation Workflows

In the last 10 years, many papers have been published based on the analysis of DOM and SM datasets, with interesting results in many fields of the Earth and planetary sciences. Considering structural analysis, many contributions appeared where the authors use DOMs to collect large datasets on fracture networks (e.g. Martinelli et al., 2020, and references therein) and the increase in dataset size was so huge that it resulted in a renewed interest in techniques used to characterize fracture statistics (e.g. Guerriero et al., 2011; Marrett et al., 2018; Bistacchi et al., 2020). Other applications include the larger-scale characterization of fault zones (e.g. Tavani et al., 2016) and folds (e.g. Vollgger and Cruden, 2016). The opportunity to collect huge datasets for fracture network characterization is appealing also in rock mechanics and engineering projects, both on natural slopes and rock faces (e.g. Jaboyedoff et al., 2007) and on man-made mine faces and roadcuts (e.g. Sturzenegger and Stead, 2009).

In stratigraphy and sedimentology the main applications of DOMs are those where quantitative data are collected to provide data for facies distribution models, particularly in clastic systems (e.g. Buckley et al., 2013; Siddiqui et al., 2019). Special applications of SM analyses allowed Massironi et al. (2015) to reconstruct the stratigraphy of the 67P Churyumov–Gerasimenko comet, with implications concerning mechanisms of comet accretion and evolution, as well as the works of Matonti et al. (2019) concerning fault interpretations in the neck of the comet and Simioni et al. (2015) retrieving the 3D fracture pattern of Phobos grooves.

Another growing field of application for DOMs is the high-resolution mapping of compositional features, either in terms of lithology, mineralogy, or chemical alteration. Typical examples on the Earth deal with mapping of hydrothermal dolostone bodies (Kurz et al., 2013; Bistacchi et al., 2015).

In addition to allowing collecting quantitative data, DOMs also allow these measurements to be performed remotely. This is an advantage in situations where accessing an outcrop can be dangerous (e.g. steep rock walls liable to rockfall) or the fieldwork would be very time-consuming (e.g. large outcrops exposing thousands of joints) or even impossible (e.g. on planetary bodies). Virtual reality environments called Virtual Outcrops (VOs) represent the only way to perform some sort of fieldwork on planetary bodies and, in fact, an important impetus to develop advanced applications in this field came from recent rover missions on Mars (e.g. Barnes et al., 2018; Caravaca et al., 2020).

The efficient, time- and cost-effective, and precise reconstruction and analysis of DOMs does not result from a single and simple receipt but is a combination of multiple ingredients that can be mixed in different ways, depending on the goals of the study, on logistical and environmental conditions, on the availability of instruments and software, and on the expertise and personal inclination of the geologist.

Chapter 2 by Bistacchi et al. (2021) discusses photogrammetric and laser-scanning techniques used to reconstruct DOMs on the Earth, both with terrestrial and aerial drone platforms, particularly for outcrops ranging from a few square meters to about 1 km^2. These authors suggest a best-practice workflow in photogrammetric projects and provide a review of common pitfalls, which can make the difference between successful and unreliable photogrammetric processing. Luckily, it turns out that obtaining a successful photogrammetric reconstruction, with high accuracy and low noise, is more a matter of using proper acquisition schemes and software than costly cameras, lenses, drones, and hardware. Also, free software is very competitive with respect to commercial competitors. This means that well-trained geologists could collect high-quality photogrammetric DOMs with limited expense, and this is one key to understand the rapid growth of DOM projects in many geological disciplines.

Chapter 3 by Traxler et al. (2021) provides a top-notch example of the integration of rover and orbiter images of the surface of Mars, aimed at reconstructing multiresolution DOMs to be inserted in Virtual Reality environments. The availability of this kind of data is seeding a revolution in planetary geology, allowing planetary geologists to perform multiscale analyses that are the standard on the Earth, but were not possible on terrestrial planets just a few years ago. The chapter itself shows examples of geological data extraction on sites explored by the Opportunity and Curiosity rovers.

In any case, these two chapters highlight how, in completely different environments, the fundamental starting point in any DOM project is to implement a correct acquisition scheme, which must be at the same time cost- and time-effective, prone to generate accurate reconstructions, and tailored to the requirements of the study in terms of resolution, imagery products, etc.

If the acquisition and reconstruction of a DOM, either with laser scanning or photogrammetry, is a task shared with many other disciplines (e.g. architecture, archeology, civil engineering, environmental sciences, etc.), what is specific to geology, and to particular disciplines in geology (e.g. structural geology, stratigraphy, sedimentology, geomorphology, etc.), is DOM interpretation. Probably the most important obstacle that has slowed down the growth of quantitative analysis of DOM datasets is the almost complete lack of dedicated software. The solution to this problem consists in either borrowing, in some "creative" way, software originally developed for other tasks, or developing new dedicated software. In the first case the price to pay is, almost always, to have not optimal or missing functions. In the second case, relevant time (and money) must be invested, but in the end software tools with perfectly tailored functions could be developed.

Chapter 4 by Penasa et al. (2021) discusses the Vombat plugin (github.com/luca-penasa/vombat) that adds tools for geological interpretation and measurement of stratigraphic logs to the well-known and very efficient CloudCompare 3D point cloud processing software (cloudcompare.org). The development of a plugin (i.e. a software component that extends the capabilities of a computer program) is an interesting shortcut that allows developing specific functions with a limited time (and economic) effort if a base program exists, which implements a useful set of base functions. In this case, CloudCompare is amongst the best and more efficient programs to visualize and analyze 3D point clouds and Vombat adds specific geological and stratigraphic interpretation functions.

Chapter 5 by Kehl et al. (2021) discusses a software tool that is somehow similar to Vombat, but runs on a handheld device (Android tablet). This means that the analysis can be performed in the field in a sort of augmented reality way, which is very interesting since it allows the breadth of the remote-sensing based DOM dataset to combine with detailed and focused observations or cross-checks that can be performed just in person, directly on the outcrop.

Finally, Chapter 6 by Mittempergher and Bistacchi (2021) addresses what is becoming a key issue due to the growing size and resolution of DOM datasets: (semi-)automatic interpretation. It is nowadays a common practice to collect photogrammetric DOMs covering a few hundred square meters with submillimeter/pixel ground resolution, or up to 1–2 km^2 with 1–5 cm/pixel resolution. This corresponds to some gigabytes of image data, and the manual interpretation of these images could be extremely time-consuming or sometimes unfeasible, hindering some important advantages of DOM analysis. Even if the problem is far from being solved in all geological situations and outcrop conditions, Mittempergher and Bistacchi (2021) propose a choice of algorithms that could help the analyst at least in some standard situations (e.g. fracturing in sedimentary rocks).

1.3 Morphometric Analysis Across Different Scales and Planets

DEMs and DOMs are fundamental datasets for geomorphological interpretations used to analyze and model surface phenomena (crater detection for relative dating, landslides, erosional features, etc.). Quite often these interpretations are still performed manually, which becomes very time-consuming or even impossible when dealing with large areas including many geological and geomorphological structures at different scales. Automatic approaches were proposed to extract features from 2D images (Csillag, 1982; Blondel et al., 1992; Yésou et al., 1993; Koike et al., 1995; Mugglestone and Renshaw, 1998; Reid and Harrison, 2000), but in the case of geological applications, this often leads to distortions of the interpreted structures and morphologies that are 3D in nature rather than 2D. With the emergence of high-resolution DEMs and DOMs, automatic approaches were proposed to help solve these problems. The implementation of these approaches depends on three factors: (i) the nature of the structures to be interpreted, (ii) their surface expressions, and (iii) the dataset that is available for the analysis.

On the one hand, geological structures include surfaces such as faults or stratigraphic surfaces but also volumes such as layers or sedimentary bodies. Their intersections with the topographic surface leads respectively to: (i) lineaments (e.g. fracture or stratigraphy traces) or surfaces (e.g. fracture planes or structural surfaces) belonging to the relief; (ii) partitions of the topographic surfaces, which correspond to a geological mapping onto the numerical outcrop. On the other hand, geomorphological features such as impact craters, lava flows, volcanic cones, fluvial terraces, landslides, and glacial/periglacial forms are expressed on the topographic surface and their interpretation leads to a geomorphological mapping (i.e. a surface partition in terms of geomorphological units). In geomorphology, a particular attention is also devoted to quantitatively estimate the volume and morphometric parameters (i.e. slope and curvatures) of geological and geomorphological features as well as quantitatively assess their evolution in the time frame of successive acquisitions (4D data analysis).

Most approaches used to extract morphometric parameters from DEMs or DOMs rely on differential geometry and especially on the computation of the different normal curvatures (minimum, maximum, mean, and Gaussian). In Kudelski et al. (2011) approaches are proposed for extracting continuous lineaments (fracture traces or stratigraphic planes) despite the roughness of the outcrop surfaces. A similar approach has been used for extracting the impacts of crater rims by Mari et al. (2021) and many other different algorithms have been proposed for the same aim (Bandeira et al., 2012; Cohen and Ding, 2014; Salamunićcar et al., 2014; Christoff et al., 2020) in order to find the crater size frequency distribution in the most reliable way to indirectly date planetary surfaces (e.g. Neukum et al., 2001; Marchi et al., 2009; Le Feuvre and Wieczorek, 2011). Morphometric approaches for geomorphological mapping on extra-terrestrial surfaces also deal with other structures such as lava tube pits and skylights (e.g. Sauro et al., 2020), mounds and mud volcanoes (Pozzobon et al., 2020), volcanic rootless cones, and transverse aeolian ridges (Palafox et al., 2017).

The use of triangulated surfaces for representing DOMs generally leads to decimation of the dataset due to performance issues. Even if algorithms have been proposed to optimize triangle resolution according to surface curvature (Nivoliers et al., 2015), applying them on DOMs still remains tricky as features associated to subtle surface roughness, such as fractures or bedding, may be obliterated. Therefore, detection algorithms have been developed to directly deal with point clouds. Among them, many approaches applied to laser scanning data use the LiDAR intensity as a proxy for lithology and many corrected estimations have been established for enhancing the resulting facies or lithological mapping (Franceschi et al., 2009; Burton et al., 2011; Penasa et al., 2014; Carrea et al., 2016). The same goals have been achieved by combining DOM acquisition with hyperspectral data (Hartzell et al., 2014).

In this volume, a sample of up-to-date techniques for automatic feature extraction and/or quantitative data analysis on DEMs are presented.

Chapter 9 by Mari et al. (2021) presents an approach for extracting morphometric parameters of impact craters from extra-terrestrial surfaces represented as full 3D meshes. In this work, the algorithms are based on vertex labelling using mean and Gaussian curvature thresholds for automatically extracting crater rims and floors. In particular, two techniques have been proposed and applied on 3D meshes representing asteroid reliefs (e.g. Lutetia observed by the ROSETTA space probe and Vesta imaged by the DAWN mission). The results have afterwards been compared and validated.

Chapter 7 by Giuliano et al. (2021) proposes an approach based on laser scanning acquisitions for estimating erosion rates through time in a coastal environment. The authors discuss the use of repeated boat-borne laser scanning surveys to quantify cliff erosion in micro-tidal environments as well as the performance and resolution of this proposed processing. The huge dataset was projected onto a series of vertical planes and cylinder arcs in order to process the point clouds efficiently into a 2.5D GIS software. Thanks to this workflow, the average cliff recession rate was estimated with a good confidence level as well as erosion rates in the function of rock types.

Chapter 8 by Pozzobon et al. (2021) presents a multiscale quantitative approach for estimating volume variations from the microscopic scale of rock samples to the kilometer scale of volcanic features using the same approach. The microscopic analysis is performed on the carbonate rock samples analogue to historical buildings materials. Stone surface models were successively acquired from a confocal laser scanning microscope between series of immersion cycles simulating erosion and recession due to rainwater aggression. The key point of this study is to consider this microscopic surface model as a DEM in a GIS software and to use a reference surface for calibrating the different acquisitions of the same sample, retrieving the lost volumes and the related recession rates. The same strategy was used at a kilometer scale for quantifying the real volume of collapsed sections of lava tubes from DEMs of Earth, Moon, and Mars.

1.4 3D Modelling of the Subsurface from Surface Data

One of the goals of DOMs, DEMs, and SMs is to provide surface geological data that can be used to reconstruct 3D models of the subsurface (e.g. Bistacchi et al., 2010, 2015; Penasa et al., 2017; Franceschi et al., 2020; Pozzobon et al., 2020). Since reconstructing 3D models of the subsurface from surface data always requires some sort of extrapolation (e.g. Bistacchi et al., 2008), at least the quality of the input data must be verified very strictly, and for this reason a quantitative and high-resolution topography such as that provided by a DOM, a high-resolution DEM, or an SM is invaluable. This requirement is even more pronounced when the goal is to model complicated geological structures.

In Chapter 10 by Baumberger et al. (2021), a model of fault zones in the Aar Massif (Central Alps) is discussed. Noteworthy is the fact that the authors developed a methodology allowing an estimation to be made of the uncertainty in the modelling, demonstrating that, even if extrapolation from a DEM or DOM is not completely free from uncertainty, the modelling is in any case some orders of magnitude more reliable than the extrapolation of structural measurements collected using traditional methods in the field (e.g. compass/clinometer). In addition, the uncertainty can be further reduced, or adapted to the goals of the study, by tuning the resolution of the topographic and surface-geology dataset. As recently pointed out by Bistacchi et al. (2020), Fondriest et al. (2020), and Martinelli et al. (2020), and also Baumberger et al. (2021), show how DOM analysis allows many aspects of a fault and fracture network, such as fault and fracture connectivity, to be quantitatively characterized in a way that cannot be attained with traditional surveys (e.g. Viseur et al., 2020).

Chapter 11 by Penasa et al. (2021) provides a detailed explanation of how the 3D geological model of the lobes of the 67P Churyumov–Gerasimenko comet has been reconstructed from an SM dataset. This is a recent and major achievement in the geology of small bodies of the Solar System, since it allowed some important constraints to be placed on the accretion and evolution of a comet (Massironi et al., 2015; Penasa et al., 2017) as well as on its overall mechanical behavior (Franceschi et al., 2020). In this case, the subellipsoidal geometry of the lobes of the comet and their complete accessibility (in remote sensing terms) reduces the uncertainty in 3D modelling, since the authors are interpolating the internal structure based on a complete mapping of the outer surface on an SM. On the other hand, the authors describe how they used an implicit surface modelling algorithm to tackle the curvilinear geometries of the comet layering.

1.5 Summary and Perspectives

To our knowledge this volume represents the first attempt at collecting contribution from Earth and planetary science communities working on 3D datasets in diverse contexts, scales, and for different purposes. It is high time that different techniques, software, workflows, and methodologies are shared among these communities. We hope our volume will foster collaborative efforts towards 3D data exploitation, inspiring young researchers interested in 3D data analysis with its introductive and general chapters and senior scientists with the advanced case studies presented here.

Acknowledgments

First of all, we would like to warmly thank all the authors of the very interesting chapters in this volume. The volume was inspired by sessions at the EGU General Meetings, and all contributors to these sessions are also acknowledged. We would like to thank particularly Claudio Rosenberg, who encouraged us to convene these sessions. Finally, we would like to acknowledge the editorial staff at Wiley for assisting us in this effort.

References

Bandeira, L., Ding, W., and Stepinski, T.F. (2012). Detection of sub-kilometer craters in high resolution planetary images using shape and texture features. *Advances in Space Research* 49: 64–74. https://doi.org/10.1016/j.asr.2011.08.021.

Barnes, R., Gupta, S., Traxler, C. et al. (2018). Geological analysis of Martian Rover-derived digital outcrop models using the 3-D visualization tool, Planetary Robotics 3-D Viewer-PRo3D. *Earth and Space Science* 5: 285–307. https://doi.org/10.1002/2018EA000374.

Baumberger, R., Herwegh, M., and Kissling, E. (2021). Remote sensing and field data based structural 3D modelling (Haslital, Switzerland) in combination with uncertainty estimation and verification by underground data. In: *3D Digital Geological Models: From Terrestrial Outcrops to Planetary Surfaces*. Wiley this volume.

Bellian, J.A.J.A., Kerans, C., and Jennette, D.C. (2005). Digital outcrop models: Applications of terrestrial scanning LiDAR technology in stratigraphic modeling. *Journal of Sedimentary Research* 75: 166–176. https://doi.org/10.2110/jsr.2005.013.

Bistacchi, A., Massironi, M., Dal Piaz, G.V. et al. (2008). 3D fold and fault reconstruction with an uncertainty model: An example from an Alpine tunnel case study. *Computers and Geosciences* 34: 351–372. https://doi.org/10.1016/j.cageo.2007.04.002.

Bistacchi, A., Massironi, M., and Menegon, L. (2010). Three-dimensional characterization of a crustal-scale fault zone: The Pusteria and Sprechenstein fault system (Eastern Alps). *Journal of Structural Geology* 32: 2022–2041. https://doi.org/10.1016/j.jsg.2010.06.003.

Bistacchi, A., Griffith, W.A., Smith, S.A.F. et al. (2011). Fault roughness at seismogenic depths from LIDAR and photogrammetric analysis. *Pure and Applied Geophysics* 168: 2345–2363. https://doi.org/10.1007/s00024-011-0301-7.

Bistacchi, A., Balsamo, F., Storti, F. et al. (2015). Photogrammetric digital outcrop reconstruction, visualization with textured surfaces, and three-dimensional structural analysis and modeling: Innovative methodologies applied to fault-related dolomitization (Vajont Limestone, Southern Alps, Italy). *Geosphere* 11: 2031–2048. https://doi.org/10.1130/GES01005.1.

Bistacchi, A., Mittempergher, S., Martinelli, M., and Storti, F. (2020). On a new robust workflow for the statistical and spatial analysis of fracture data collected with scanlines (or the importance of stationarity). *Solid Earth* 11: 2535–2547. https://doi.org/10.5194/se-2020-83.

Bistacchi, A., Mittempergher, S., and Martinelli, M. (2021). Digital outcrop model reconstruction and interpretation. In: *3D Digital Geological Models: From Terrestrial Outcrops to Planetary Surfaces*. Wiley this volume.

Blondel, P., Sotin, C., and Masson, P. (1992). Adaptive filtering and structure-tracking for statistical analysis of geological features in radar images. *Computers and Geosciences* 18: 1169–1184. https://doi.org/10.1016/0098-3004(92)90038-S.

Buckley, S.J., Kurz, T.H., Howell, J.A., and Schneider, D. (2013). Terrestrial LiDAR and hyperspectral data fusion products for geological outcrop analysis. *Computers and Geosciences* 54: 249–258. https://doi.org/10.1016/j.cageo.2013.01.018.

Burton, D., Dunlap, D.B., Wood, L.J., and Flaig, P.P. (2011). LiDAR intensity as a remote sensor of rock Properties. *Journal of Sedimentary Research* 81: 339–347. https://doi.org/10.2110/jsr.2011.31.

Caravaca, G., Le Mouélic, S., Mangold, N. et al. (2020). 3D digital outcrop model reconstruction of the Kimberley outcrop (Gale crater, Mars) and its integration into Virtual Reality for simulated geological analysis. *Planetary and*

Space Science 182: 104808. https://doi.org/10.1016/j.pss.2019.104808.

Carrea, D., Abellan, A., Humair, F. et al. (2016). Correction of terrestrial LiDAR intensity channel using Oren–Nayar reflectance model: An application to lithological differentiation. *ISPRS Journal of Photogrammetry and Remote Sensing* 113: 17–29. https://doi.org/10.1016/j.isprsjprs.2015.12.004.

Carry, B., Kaasalainen, M., Merline, W.J. et al. (2012). Shape modeling technique KOALA validated by ESA Rosetta at (21) Lutetia. *Planetary and Space Science* 66: 200–212. https://doi.org/10.1016/j.pss.2011.12.018.

Catmull, E.E. (1974). A subdivision algorithm for computer display of curved surfaces. PhD Thesis, University of Utah.

Caumon, G., Collon-Drouaillet, P., Le Carlier De Veslud, C. et al. (2009). Surface-based 3D modeling of geological structures. *Mathematical Geosciences* 41: 927–945. https://doi.org/10.1007/s11004-009-9244-2.

Christoff, N., Jorda, L., Viseur, S. et al. (2020). Automated extraction of crater rims on 3D meshes combining artificial neural network and discrete curvature labeling. *Earth, Moon, and Planets* 124: 51–72. https://doi.org/10.1007/s11038-020-09535-7.

Cohen, J.P. and Ding, W. (2014). Crater detection via genetic search methods to reduce image features. *Advances in Space Research* 53: 1768–1782. https://doi.org/10.1016/j.asr.2013.05.010.

Crane, K. (2020). Structural interpretation of thrust fault-related landforms on Mercury using Earth analogue fault models. *Geomorphology* 369: 107366. https://doi.org/10.1016/j.geomorph.2020.107366.

Csillag, F. (1982). Significance of tectonics in linear feature detection and interpretation on satellite images. *Remote Sensing of Environment* 12: 235–245. https://doi.org/10.1016/0034-4257(82)90056-6.

De Toffoli, B., Mangold, N., Massironi, M. et al. (2020). Structural analysis of sulfate vein networks in Gale crater (Mars). *Journal of Structural Geology* 137: 104083. https://doi.org/10.1016/j.jsg.2020.104083.

Fondriest, M., Balsamo, F., Bistacchi, A. et al. (2020). Structural complexity and mechanics of a shallow crustal seismogenic source (Vado di Corno Fault Zone, Italy). *Journal of Geophysical Research: Solid Earth* 125. https://doi.org/10.1029/2019JB018926.

Franceschi, M., Penasa, L., Massironi, M. et al. (2020). Global-scale brittle plastic rheology at the cometesimals merging of comet 67P/Churyumov–Gerasimenko. *Proceedings of the National Academy of Sciences* 117: 10181–10187. https://doi.org/10.1073/pnas.1914552117.

Franceschi, M., Teza, G., Preto, N. et al. (2009). Discrimination between marls and limestones using intensity data from terrestrial laser scanner. *ISPRS Journal of Photogrammetry and Remote Sensing* 64: 522–528. https://doi.org/10.1016/j.isprsjprs.2009.03.003.

Giuliano, J., Dewez, T., Lebourg, T. et al. (2021). Mapping coastal erosion of a Mediterranean cliff with a boat-borne laser scanner: Performance, processing and cliff erosion rate. In: *3D Digital Geological Models: From Terrestrial Outcrops to Planetary Surfaces*. Wiley this volume.

Guerriero, V., Vitale, S., Ciarcia, S., and Mazzoli, S. (2011). Improved statistical multi-scale analysis of fractured reservoir analogues. *Tectonophysics* 504: 14–24. https://doi.org/10.1016/j.tecto.2011.01.003.

Hartzell, P., Glennie, C., Biber, K., and Khan, S. (2014). Application of multispectral LiDAR to automated virtual outcrop geology. *ISPRS Journal of Photogrammetry and Remote Sensing* 88: 147–155. https://doi.org/10.1016/j.isprsjprs.2013.12.004.

Jaboyedoff, M., Metzger, R., Oppikofer, T. et al. (2007). New insight techniques to analyze rock-slope relief using DEM and 3D-imaging cloud points. In: *Rock Mechanics: Meeting Society's Challenges and Demands*, 61–68. Taylor and Francis: https://doi.org/10.1201/NOE0415444019-c8.

Jones, R.R., Pearce, M.A., Jacquemyn, C., and Watson, F.E. (2016). Robust best-fit planes from geospatial data. *Geosphere* 12: 196–202. https://doi.org/10.1130/GES01247.1.

Jones, R.R., Wawrzyniec, T.F., Holliman, N.S. et al. (2008). Describing the dimensionality of geospatial data in the earth sciences—Recommendations for nomenclature. *Geosphere* 4: 354. https://doi.org/10.1130/GES00158.1.

Kehl, C., Mullins, J., Buckley, S.J. et al. (2021). Interpretation and mapping of geological features using mobile devices in outcrop geology—A case study of the Saltwick Formation, North Yorkshire, UK. In: *3D Digital Geological Models: From Terrestrial Outcrops to Planetary Surfaces*. Wiley this volume.

Koike, K., Nagano, S., and Ohmi, M. (1995). Lineament analysis of satellite images using a Segment Tracing Algorithm (STA). *Computers and Geosciences* 21: 1091–1104. https://doi.org/10.1016/0098-3004(95)00042-7.

Kudelski, D., Viseur, S., Scrofani, G., and Mari, J.-L. (2011). Feature line extraction on meshes through vertex marking and 2D topological operators. *International Journal of Image and Graphics* 11: 531–548. https://doi.org/10.1142/S0219467811004226.

Kurz, T.H., Buckley, S.J., and Howell, J.A. (2013). Close-range hyperspectral imaging for geological field studies: workflow and methods. *International Journal of Remote Sensing* 34: 1798–1822. https://doi.org/10.1080/01431161.2012.727039.

Le Feuvre, M. and Wieczorek, M.A. (2011). Nonuniform cratering of the Moon and a revised crater chronology of the inner Solar System. *Icarus* 214: 1–20. https://doi.org/10.1016/j.icarus.2011.03.010.

Le Mouélic, S., Enguehard, P., Schmitt, H.H. et al. (2020). Investigating lunar boulders at the Apollo 17 landing site using photogrammetry and virtual reality. *Remote Sensing* 12: 1900. https://doi.org/10.3390/rs12111900.

Lévy, B., Petitjean, S., Ray, N., and Maillot, J. (2002). Least squares conformal maps for automatic texture atlas generation. *ACM Transactions on Graphics* 21: 362–371. https://doi.org/10.1145/566654.566590.

Mallet, J.-L. (2002). *Geomodeling*. New York: Oxford University Press.

Marchi, S., Mottola, S., Cremonese, G. et al. (2009). A new chronology for the Moon and Mercury. *The Astronomical Journal* 137: 4936–4948. https://doi.org/10.1088/0004-6256/137/6/4936.

Mari, J.-L., Viseur, S., Bouley, S. et al. (2021). Robust detection of circular shapes on 3D meshes based on discrete curvature—Application to impact craters recognition. In: *3D Digital Geological Models: From Terrestrial Outcrops to Planetary Surfaces*. Wiley this volume.

Marrett, R.A., Gale, J.F.W., Gómez, L.A., and Laubach, S.E. (2018). Correlation analysis of fracture arrangement in space. *Journal of Structural Geology* 108: 16–33. https://doi.org/10.1016/j.jsg.2017.06.012.

Martinelli, M., Franceschi, M., Massironi, M. et al. (2017). An extensional syn-sedimentary structure in the Early Jurassic Trento Platform (Southern Alps, Italy) as analogue of potential hydrocarbon reservoirs developing in rifting-affected carbonate platforms. *Marine and Petroleum Geology* 79: 360–371. https://doi.org/10.1016/j.marpetgeo.2016.11.002.

Martinelli, M., Bistacchi, A., Mittempergher, S. et al. (2020). Damage zone characterization combining scan-line and scan-area analysis on a km-scale Digital Outcrop Model: The Qala Fault (Gozo). *Journal of Structural Geology* 140: 104144. https://doi.org/10.1016/j.jsg.2020.104144.

Massironi, M., Simioni, E., Marzari, F. et al. (2015). Two independent and primitive envelopes of the bilobate nucleus of Comet 67P. *Nature* 526: 402–405. https://doi.org/10.1038/nature15511.

Matonti, C., Attree, N., Groussin, O. et al. (2019). Bilobate comet morphology and internal structure controlled by shear deformation. *Nature Geoscience* 12: 157–162. https://doi.org/10.1038/s41561-019-0307-9.

Mittempergher, S. and Bistacchi, A. (2021). Image analysis algorithms for semi-automatic lineament detection on digital outcrop models. In: *3D Digital Geological Models: From Terrestrial Outcrops to Planetary Surfaces*. Wiley, this volume.

Mugglestone, M.A. and Renshaw, E. (1998). Detection of geological lineations on aerial photographs using two-dimensional spectral analysis. *Computers and Geosciences* 24: 771–784. https://doi.org/10.1016/S0098-3004(98)00065-X.

Neukum, G., Ivanov, B.A., and Hartmann, W.K. (2001). Cratering records in the inner solar system in relation to the Lunar Reference System. *Space Science Reviews* 96: 55–86. https://doi.org/10.1023/A:1011989004263.

Nivoliers, V., Lévy, B., and Geuzaine, C. (2015). Anisotropic and feature sensitive triangular remeshing using normal lifting. *Journal of Computational and Applied Mathematics* 289: 225–240. https://doi.org/10.1016/j.cam.2015.01.041.

Palafox, L.F., Hamilton, C.W., Scheidt, S.P., and Alvarez, A.M. (2017). Automated detection of geological landforms on Mars using Convolutional Neural Networks. *Computers and Geosciences* 101: 48–56. https://doi.org/10.1016/j.cageo.2016.12.015.

Penasa, L., Franceschi, M., and Preto, N. (2021). Vombat: an open source tool for creating stratigraphic logs from virtual outcrops. In: *3D Digital Geological Models: From Terrestrial Outcrops to Planetary Surfaces*. Wiley, this volume.

Penasa, L., Franceschi, M., Preto, N. et al. (2014). Integration of intensity textures and local geometry descriptors from terrestrial laser scanning to map chart in outcrops. *ISPRS Journal of Photogrammetry and Remote Sensing* 93: 88–97. https://doi.org/10.1016/j.isprsjprs.2014.04.003.

Penasa, L., Massironi, M., Naletto, G. et al. (2017). A three-dimensional modelling of the layered structure of comet 67P/Churyumov-Gerasimenko. *Monthly Notices of the Royal Astronomical Society* 469: S741–S754. https://doi.org/10.1093/mnras/stx2899.

Penasa, L., Massironi, M., Simioni, E. et al. (2021). Application of implicit 3D modeling to reconstruct the layered structure of the comet 67P. In: *3D Digital Geological Models: From Terrestrial Outcrops to Planetary Surfaces*. Wiley, this volume.

Pozzobon, R., Mazzoli, C., Salvini, S. et al. (2021). A DTM-based volume extraction approach: from micro-scale weathering forms to planetary lava tubes. In: *3D Digital Geological Models: From Terrestrial Outcrops to Planetary Surfaces*. Wiley, this volume.

Pozzobon, R., Pesce, D., and Massironi, M. (2020). 3D geomodel of the deformed deposits in Crommelin Crater (Mars). EPSC Abstracts, vol. 14. https://doi.org/10.5194/epsc2020-770.

Preusker, F., Scholten, F., Knollenberg, J. et al. (2012). The northern hemisphere of asteroid (21) Lutetia—topography and orthoimages from Rosetta OSIRIS NAC image data. *Planetary and Space Science* 66: 54–63. https://doi.org/10.1016/j.pss.2012.01.008.

Preusker, F., Scholten, F., Matz, K.-D. et al. (2015). Shape model, reference system definition, and cartographic mapping standards for Comet 67P/Churyumov-Gerasimenko—Stereo-photogrammetric analysis of

Rosetta/OSIRIS image data. *Astronomy and Astrophysics* 583: A33. https://doi.org/10.1051/0004-6361/201526349.

Reid, T.R. and Harrison, J.P. (2000). A semi-automated methodology for discontinuity trace detection in digital images of rock mass exposures. *International Journal of Rock Mechanics and Mining Sciences* 37: 1073–1089. https://doi.org/10.1016/S1365-1609(00)00041-1.

Salamunićcar, G., Lončarić, S., Grumpe, A., and Wöhler, C. (2014). Hybrid method for crater detection based on topography reconstruction from optical images and the new LU78287GT catalogue of Lunar impact craters. *Advances in Space Research* 53: 1783–1797. https://doi.org/10.1016/j.asr.2013.06.024.

Sauro, F., Pozzobon, R., Massironi, M. et al. (2020). Lava tubes on Earth, Moon and Mars: A review on their size and morphology revealed by comparative planetology. *Earth-Science Reviews* 209: 103288. https://doi.org/10.1016/j.earscirev.2020.103288.

Siddiqui, N.A., Ramkumar, M., Rahman, A.H.A. et al. (2019). High resolution facies architecture and digital outcrop modeling of the Sandakan formation sandstone reservoir, Borneo: Implications for reservoir characterization and flow simulation. *Geoscience Frontiers* 10: 957–971. https://doi.org/10.1016/j.gsf.2018.04.008.

Simioni, E., Pajola, M., Massironi, M., and Cremonese, G. (2015). Phobos grooves and impact craters: A stereographic analysis. *Icarus* 256: 90–100. https://doi.org/10.1016/j.icarus.2015.04.009.

Sturzenegger, M. and Stead, D. (2009). Close-range terrestrial digital photogrammetry and terrestrial laser scanning for discontinuity characterization on rock cuts. *Engineering Geology* 106: 163–182. https://doi.org/10.1016/j.enggeo.2009.03.004.

Tavani, S., Corradetti, A., and Billi, A. (2016). High precision analysis of an embryonic extensional fault-related fold using 3D orthorectified virtual outcrops: The viewpoint importance in structural geology. *Journal of Structural Geology* 86: 200–210. https://doi.org/10.1016/j.jsg.2016.03.009.

Traxler, C., Ortner, T., Hesina, G. et al. (2021). The PRoViDE Framework: Accurate 3D geological models for virtual exploration of the Martian surface from rover and orbital imagery. In: *3D Digital Geological Models: From Terrestrial Outcrops to Planetary Surfaces*. Wiley, this volume.

Triantafyllou, A., Watlet, A., Le Mouélic, S. et al. (2019). 3-D digital outcrop model for analysis of brittle deformation and lithological mapping (Lorette Cave, Belgium). *Journal of Structural Geology* 120: 55–66. https://doi.org/10.1016/j.jsg.2019.01.001.

Viseur, S., Lamarche, J., and Gauthier, B. (2020). Integrated analysis of fracture corridors. Fourth Naturally Fractured Reservoir Workshop. Presented at the Fourth Naturally Fractured Reservoir Workshop, European Association of Geoscientists and Engineers, Ras Al Khaimah, United Arab Emirates, 1–5. https://doi.org/10.3997/2214-4609.2020622027.

Vollgger, S.A. and Cruden, A.R. (2016). Mapping folds and fractures in basement and cover rocks using UAV photogrammetry, Cape Liptrap and Cape Paterson, Victoria, Australia. *Journal of Structural Geology* 85: 168–187. https://doi.org/10.1016/j.jsg.2016.02.012.

Willner, K., Shi, X., and Oberst, J. (2014). Phobos' shape and topography models. *Planetary and Space Science* 102: 51–59. https://doi.org/10.1016/j.pss.2013.12.006.

Yésou, H., Besnus, Y., and Rolet, J. (1993). Extraction of spectral information from Landsat TM data and merger with SPOT panchromatic imagery—a contribution to the study of geological structures. *ISPRS Journal of Photogrammetry and Remote Sensing* 48: 23–36. https://doi.org/10.1016/0924-2716(93)90069-Y.

Yuan, Y., Zhu, P., Zhao, N. et al. (2017). The 3-D geological model around Chang'E-3 landing site based on lunar penetrating radar Channel 1 data: Three-dimensional geological model of CE-3 landing site. *Geophysical Research Letters* 44: 6553–6561. https://doi.org/10.1002/2017GL073589.

Part I

DOM and SM Reconstruction and Interpretation Workflows

2

Digital Outcrop Model Reconstruction and Interpretation

Andrea Bistacchi[1], Silvia Mittempergher[1,2], and Mattia Martinelli[1]

[1] *Dipartimento di Scienze dell'Ambiente e della Terra, Universita' degli Studi di Milano - Bicocca, Piazza della Scienza, 4, 20126 Milano*
[2] *Dipartimento di Scienze Chimiche e Geologiche, Università degli Studi di Modena e Reggio Emilia, Via G. Campi 106, 41125 Modena*

Abstract

Collecting quantitative and extensive datasets in the field is fundamental in structural geology, stratigraphy, and sedimentology, rock mechanics, and in other fields of the Earth and planetary sciences. Digital Outcrop Models (DOMs) provide a 3D framework for collecting these large datasets and can be obtained from laser scanning or photogrammetric surveys, carried out either with an avionic platform (airplane, helicopter, drone) or with terrestrial methods. In this chapter we review best-practice methods for collecting DOMs, focusing particularly on terrestrial and drone photogrammetric surveys and on critical issues that determine their efficiency, reliability, and accuracy. Then we compare the two main formats for DOMs: point clouds (PC-DOMs) and textured surfaces (TS-DOMs). Finally, we outline typical goals and workflows for the geological interpretation of DOMs on PC- and TS-DOMs, either from laser scanning or photogrammetric surveys.

2.1 Introduction

Collecting quantitative and extensive datasets in the field is becoming more and more important in projects of structural geology (e.g. McCaffrey et al., 2005; Bistacchi et al., 2008, 2011, 2015, 2020; Smith et al., 2013;Tavani et al., 2016; Triantafyllou et al., 2019; Vollgger and Cruden, 2016; Martinelli et al., 2017, 2020; De Toffoli et al., 2020), stratigraphy and sedimentology (e.g. Bellian et al., 2005, Fabuel-Perez et al., 2010; Penasa et al., 2014; Franceschi et al., 2015; Tomassetti et al., 2018; Dujoncquoy et al., 2019; Siddiqui et al., 2019; Qu et al., 2021), rock mechanics and hydrogeology (e.g. Jaboyedoff et al., 2007; Francese et al., 2009; Massironi et al., 2013; Bistacchi et al., 2013; Riquelme et al., 2017), and in other fields of the Earth and planetary sciences (e.g. Silva et al., 2017; Barnes et al., 2018). In many cases, as in the ones cited above, the answer to this need is a 3D Digital Outcrop Model (DOM): a digital 3D model of an outcrop (also called Virtual Outcrop), based on remote-sensing data, that provides a quantitative reference for collecting and interpreting data with a combination of high-resolution topographic and imaging data.

DOMs can be obtained using two techniques: laser scanning or photogrammetry (e.g. Hodgetts, 2013; Tavani et al., 2014). In the first case an airborne or terrestrial laser scanner is used to collect a topographic dataset, and a camera (sometimes multi- or hyper-spectral), attached to the laser scanner, is used to collect imaging data to be fused with the topographic data. In photogrammetric surveys, large collections of images are used, obtained from an airborne platform (both drone and manned aerial photogrammetry) or directly with a camera located on the surface (close-range photogrammetry). In this case, both the topography and imaging data are obtained from images and this makes photogrammetry a simpler and more economic technique in most scenarios.

Once a DOM has been obtained, with either photogrammetry or laser scanning, the interpretation can be carried out on a workstation or even a laptop with various techniques and software, using manual, semiautomatic, or completely automatic workflows (e.g. Vasuki et al., 2017), which will be briefly discussed in the second part of this contribution and also in Chapter 6 of this volume by Mittempergher and Bistacchi (2021).

Many recent contributions highlight the fact that remote sensing techniques based on DOMs are particularly useful when working on large inaccessible or unsafe outcrops (e.g. Hodgetts, 2013, and references therein). This is true, obviously, but we would like to stress that the best results can be achieved by combining a DOM approach with traditional analysis on the outcrop, which allows defining

3D Digital Geological Models: From Terrestrial Outcrops to Planetary Surfaces, First Edition.
Edited by Andrea Bistacchi, Matteo Massironi, and Sophie Viseur.

or "ground-truthing" details that cannot be solved with remote sensing.

Two fields of geological analysis have mainly benefitted from DOM studies: structural geology (including rock mechanics studies) and stratigraphy/sedimentology. In structural geology studies, DOMs are used to extract geometrical and topological data of structures that intersect the outcrop. We will call them collectively "discontinuities", but they are fractures, joints, foliations, stylolites, faults, bedding, etc. Sometimes these discontinuities appear as "facets:" planar portions of the outcrop surface that correspond to a portion of a certain discontinuity (Figure 2.1). In other situations, depending on the outcrop morphology and on the relative orientation of the outcrop surface and discontinuities, the discontinuities appear as traces: linear intersections of a geological subplanar feature and the outcrop surface (Figure 2.1). Many authors have developed workflows to extract information from facets (e.g. Jaboyedoff et al., 2007; Chen et al., 2017), other authors have concentrated on traces (e.g. Sturzenegger et al., 2011; Vasuki et al., 2014), but not so many workflows are able to consider both facets and traces (Thiele et al., 2017). This can be a problem since considering both could be important in order to be able to obtain a complete picture of the outcrop and the underlying geological structure.

In sedimentary geology, the aim of DOM studies is mainly to map in details the stratigraphic sequence and/or the occurrence and geometries of particular facies or sedimentary bodies in siliciclastic deposits (e.g. Siddiqui et al., 2019), platform carbonates (e.g. Martinelli et al., 2017), and other kind of sediments. Other studies are devoted to mapping lithology, with applications in diagenesis or hydrothermal alteration studies (e.g. Bistacchi et al., 2015; Jacquemyn et al., 2015; Madjid et al., 2018). In all these applications, the DOM is used to recognize and map different bodies with particular sedimentological or compositional characteristics, and also in this case the interpretation can be completely manual, semi-automatic, or completely automatic.

In this contribution, we cover some key steps of a workflow leading from data collection to quantitative interpretation of DOMs at different scales. Some relatively standardized techniques like TLS and LiDAR acquisition and base data processing are not described in detail, since a wide literature (e.g. Buckley et al., 2008; Hodgetts, 2013) and instrument manuals already cover them. For the same reason, we do not indulge in technical details on software tools that are already discussed in their manuals. On the other hand, we dedicate space to non-standard techniques that are particularly important in geological applications (some very important software tools are not specifically developed for geological studies), and on advice aimed at ensuring a high quality and reliability of datasets and interpretations.

2.2 Photogrammetric Surveys and Processing for DOMs

Photogrammetric surveys have been performed for decades, since when photography has been used to quantitatively reconstruct the geometry of objects and scenes in the mid-nineteenth century (Mikhail et al., 2001).

Figure 2.1 Structural interpretation on a DOM: comparison of manually digitized facets (in green) and traces (in red). Interpretation was carried out on a PC-DOM in CloudCompare with the Compass plugin. The outcrop exposes turbidities of the Marnoso-Arenacea Formation with joints related to folding and thrusting (Val Santerno, Italy, 44°09'27"N−11°27'28"E).

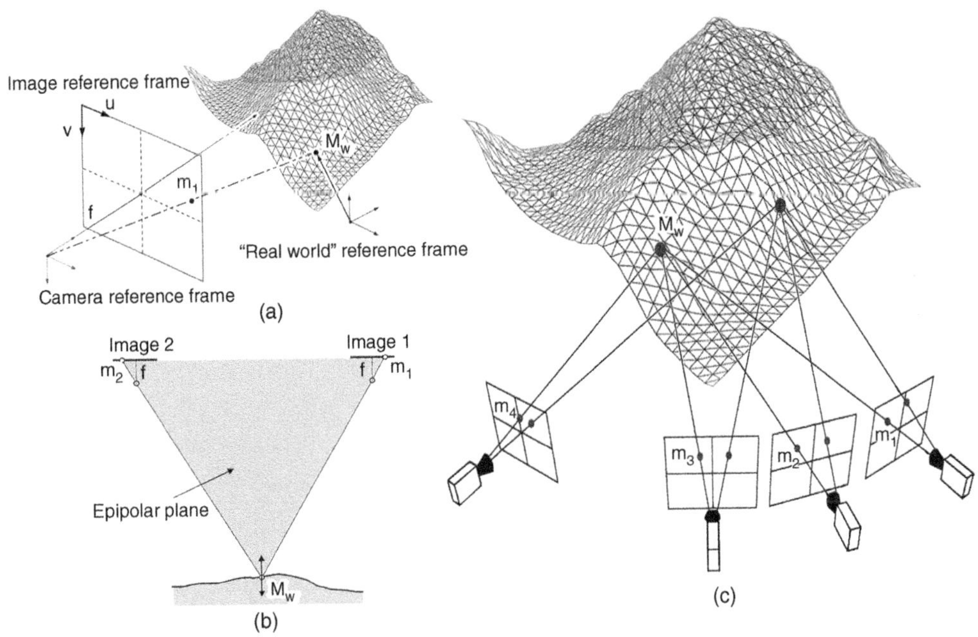

Figure 2.2 Principles of SFM photogrammetry: (a) Image projection as a function of relationships between the real world, camera and image reference frames (extrinsic orientation), and the camera optical properties (intrinsic orientation); (b) Principle of traditional stereophotogrammetry with two images (stereo-couple); (c) Principle of multiview stereophotogrammetry as applied in SFM, with several redundant cameras viewing the same feature on the outcrop. A feature M_W on the outcrop is seen as m_1, m_2, etc., in the images.

Photogrammetry is based on the parallax principle: when a scene is viewed from at least two different viewpoints, objects at different distances from the observer appear shifted in different ways with respect to a background, and closer objects are more shifted than more distant ones (Figure 2.2). This allows reconstructing the 3D position of objects or "features" in a 3D scene with a high accuracy (e.g. Mikhail et al., 2001).

Probably the most important revolution in these techniques, since the introduction of digital photography, was represented by the introduction of Structure From Motion (SFM) and Multi-View Stereo (MVS) (Lowe, 2004; Snavely et al., 2006; Furukawa and Hernández, 2015). While the traditional photogrammetric surveys (e.g. classical aerial surveys) were based on the analysis of pairs of photos (stereo couples), SFM/MVS methods allow extracting information from large redundant collections of images that are processed altogether (e.g. Furukawa and Ponce, 2010). This is possible thanks to advanced and highly optimized algorithms for automatic feature detection (e.g. Lowe, 2004), feature matching and bundle adjustment (e.g. Wu, 2013), and multiview stereo (e.g. Furukawa and Ponce, 2010).

Automatic feature detection allows recognizing a large number of characteristic "features" in each image taken from a large collection. The most efficient feature detection algorithms are able to recognize features irrespective of scale, rotation, and (moderate) perspective distortion (due to the different viewpoints from which each image is taken). One of the most efficient, and more commonly used, of these algorithms is SIFT—Scale Invariant Feature Transform (Lowe, 2004).

Features detected in each photo are then matched between different photos, in order to reconstruct which images view the same part of a given scene (e.g. the same part of a rock outcrop). This can be a rather time-consuming processing step, with time complexity $O(n^4)$ in standard SFM (where n is the number of images), but recent technical advancement allows (for some implementations) reducing this time up to $O(n)$ (Wu, 2013), enabling processing of very large image collections (e.g. thousands of images).

Then, in the bundle adjustment step of SFM (e.g. Wu et al., 2011), the "structure" of the 3D scene is reconstructed (Figure 2.2), yielding: (1) the relative position of points recognized as features in the images ("sparse point cloud"); (2) the relative position and orientation of cameras that have taken the images (external or extrinsic orientation in photogrammetry terms); and (3) optical parameters of the cameras (focal length, lens distortion, etc., i.e. internal or intrinsic orientation). Since camera parameters are also obtained from inversion, thanks to the large and redundant number of images, an accurate camera calibration is no more necessary as it was in traditional methods based on pairs of images (stereo-couples). This has opened the possibility to use, in photogrammetric surveys, common

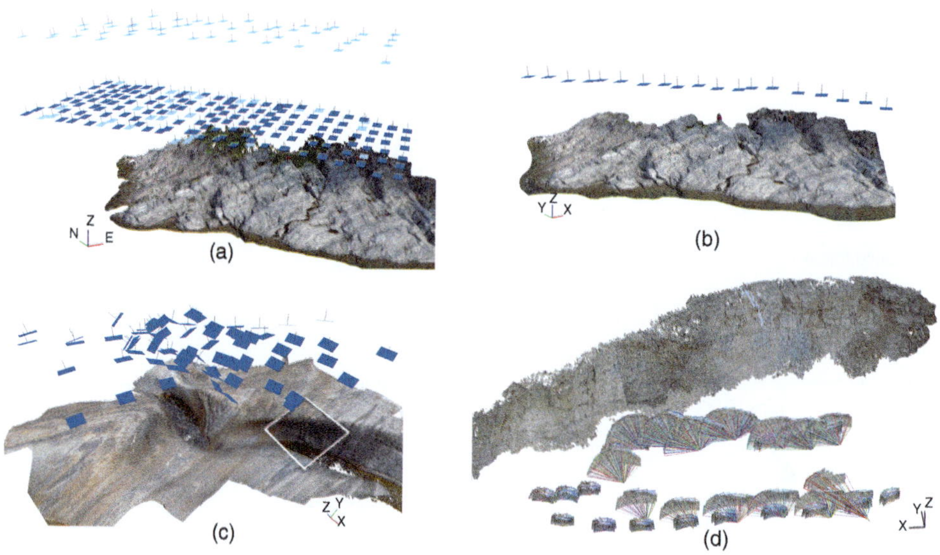

Figure 2.3 Photogrammetric image collection schemes: (a) Aerial survey with multiple passes at variable altitude; (b) Classical aerial photo strip; (c) Small scale terrestrial survey with multiple camera orientations, adapted to complex outcrop morphology; (d) Terrestrial image fans.

consumer-grade cameras and small cameras carried by drones, and resulted in an unprecedented popularization of photogrammetry.

Finally, a second phase of image correlation is performed by MVS algorithms (Furukawa and Hernández, 2015), but in this case image extrinsic and intrinsic orientations are already known, so image correlation can be carried out at the level of the single pixel, greatly increasing the spatial resolution, up to the point that the output point cloud could almost reach the resolution of the input images ("dense point cloud"). In close range surveys of DOMs, it is nowadays common to obtain point clouds with a point density of 5,000,000 points/m^2 for DOMs covering around 50 m^2 (e.g. a typical DOM obtained from terrestrial photogrammetry) and point densities of between 500 and 50,000 points/m^2 for DOMs obtained from drone images, which can cover areas up to some km^2.

In the SFM/MVS approach, a higher detail and accuracy in the reconstruction can be obtained by increasing the number of images that describe a scene in a very redundant way (multiview reconstruction). This means processing large datasets of thousands of images, with relevant computational challenges. A fundamental improvement of SFM techniques has been obtained in the last 5–6 years thanks to the availability of parallel algorithms implemented on CUDA GPUs (Graphical Processing Units): high-end video cards combining thousands of processing cores and RAM exceeding some gigabytes (https://developer.nvidia.com/cuda-zone). All the relevant steps of an SFM/MVS package have been ported on this architecture, resulting in 10× to 100× improvements in processing time and model size that is manageable on a workstation (e.g. Wu et al., 2011).

The advent of SFM techniques also resulted in a radical change in the way a survey is planned and carried out in order to achieve the best quality and accuracy (Wu, 2014). The classical aerial photogrammetry survey scheme, characterized by parallel "strips" of partly overlapping nadiral photos (Mikhail et al., 2001)(Figure 2.3), is no longer the best to ensure high accuracy and distortion-free results from SFM algorithms. Much better results, which leverage on the multiview reconstruction principle, are obtained if oblique photos and images taken from different distances are combined in a survey (James and Robson, 2014; Wu, 2014; Jaud et al., 2018) (Figure 2.3). This is particularly critical to control large-scale distortion ("dishing" or "bowl" effect), which can arise due to error propagation in a simple strip of photos (Wu, 2014).

In the next three sections, we discuss survey schemes that have been successfully applied to terrestrial "close range photogrammetry" surveys, and to surveys carried out on larger areas with drones, as well as general requirements about photo quality.

2.2.1 Calculating Ground Resolution for Photogrammetric Surveys

A fundamental parameter that we must consider when designing a photogrammetric survey is the ground resolution (e.g. in *pixel/m*) that can be achieved with a certain shooting scheme. The ground resolution should be high enough to enable accomplishing the goals of the analysis (i.e. recognizing the smallest features that must be

Table 2.1 Ground resolution and image calculation for a Nikon D700 SLR camera as a function of focal length and distance to the outcrop. In the lower box, the suggested distance between shooting stations for a fan survey is indicated.

		distance to outcrop d [m]										
		1.25	2.5	5	10	20	40	80	160	320	640	
		ground resolution GR [mm/pixel]										
focal length f [mm]	14	0.75	1.50	3.01	6.03	12.08	24.16	48.33	96.66	193.33	386.67	
	24	0.43	0.87	1.75	3.52	7.04	14.09	28.19	56.38	112.77	225.56	
	50	0.20	0.41	0.84	1.68	3.38	6.76	13.53	27.06	54.13	108.26	
	70	0.14	0.29	0.60	1.20	2.41	4.83	9.66	19.33	38.66	77.33	
	100	0.10	0.20	0.41	0.84	1.68	3.38	6.76	13.53	27.06	54.13	
	120	0.08	0.17	0.34	0.70	1.40	2.81	5.63	11.27	22.55	45.10	
	200	0.04	0.10	0.20	0.41	0.84	1.68	3.38	6.76	13.53	27.06	
	300	0.03	0.06	0.13	0.27	0.56	1.12	2.25	4.50	9.01	18.04	
		image width [m]										
focal length f [mm]	14	3.18	6.39	12.82	25.68	51.39	102.82	205.68	411.39	822.82	1645.68	
	24	1.84	3.71	7.46	14.96	29.96	59.96	119.96	239.96	479.96	959.96	
	50	0.86	1.76	3.56	7.16	14.36	28.76	57.56	115.16	230.36	460.76	
	70	0.61	1.25	2.54	5.11	10.25	20.54	41.11	82.25	164.54	329.11	
	100	0.41	0.86	1.76	3.56	7.16	14.36	28.76	57.56	115.16	230.36	
	120	0.34	0.71	1.46	2.96	5.96	11.96	23.96	47.96	95.96	191.96	
	200	0.19	0.41	0.86	1.76	3.56	7.16	14.36	28.76	57.56	115.16	
	300	0.11	0.26	0.56	1.16	2.36	4.76	9.56	19.16	38.36	76.76	
		image height [m]										
focal length f [mm]	14	2.11	4.25	8.53	17.09	34.20	68.42	136.86	273.75	547.52	1095.06	
	24	1.22	2.47	4.97	9.96	19.94	39.90	79.83	159.68	319.37	638.77	
	50	0.57	1.17	2.37	4.77	9.56	19.14	38.30	76.63	153.29	306.60	
	70	0.40	0.83	1.69	3.40	6.82	13.66	27.35	54.73	109.48	218.99	
	100	0.28	0.57	1.17	2.37	4.77	9.56	19.14	38.30	76.63	153.29	
	120	0.23	0.48	0.97	1.97	3.97	7.96	15.95	31.92	63.86	127.74	
	200	0.13	0.28	0.57	1.17	2.37	4.77	9.56	19.14	38.30	76.63	
	300	0.08	0.18	0.38	0.77	1.57	3.17	6.36	12.75	25.53	51.08	
		distance between fan shooting stations [m]										**angle [°]**
		0.33	0.67	1.34	2.68	5.36	10.72	21.44	42.87	85.74	171.49	15
		0.22	0.44	0.88	1.76	3.53	7.05	14.11	28.21	56.42	112.85	10

Nikon D700 12 Mpixel	
width [mm]	**width pixels**
36.00	4256
height [mm]	**height pixels**
23.95	2832

mapped), but at the same time are not too high, in order to minimize the survey and processing time. Considering a simple pinhole camera model (Tsai, 1987), ground resolution GR can be calculated as:

$$GR = p\frac{d - f}{f}$$

where d is the distance between the camera and outcrop, f is the focal length, and p is the pixel size of the sensor, obtained by dividing the width of the sensor by the number of pixels along the width. In this equation, units are considered homogeneous, so, for instance, if all linear units are expressed in meters, GR will be expressed in *pixel/m*. Then, knowing the image width and height in pixels (wp and hp respectively), it is also possible to calculate the width W and height H of the area covered by the image as:

$$W = \frac{wp}{GR}$$

$$H = \frac{hp}{GR}$$

When carrying out surveys in the field, it is very handy to have tables with pre-calculated values of GR, W, H as a function of f and d. An example of these tables for a DSLR camera is shown in Table 2.1. Similar values are provided by applications used to design autonomous drone flight paths (e.g. DJI Ground Station Pro, UgCS, and DroneDeploy, available on the Apple or Android app stores).

2.2.2 Terrestrial Surveys for SFM

Typical scenarios for a close-range photogrammetric survey in geology would be aimed at reconstructing a DOM of a vertical cliff (Figure 2.3b and Figure 2.4) or of a small-size "pavement" outcrop (an almost flat or gently sloping outcrop surface). In this case the best survey scheme would be the "multiple fans" scheme, where several shooting "stations" are selected, allowing a complete view of the outcrop under different angles, and several photos are shot from each station, covering the largest part of the outcrop from each location. Even if a random shooting scheme from each station could work, in order to limit the number of photos needed to cover the whole scene, a regular rotation pattern should be implemented, e.g. with the help of a tripod with a graduated head (Figure 2.4).

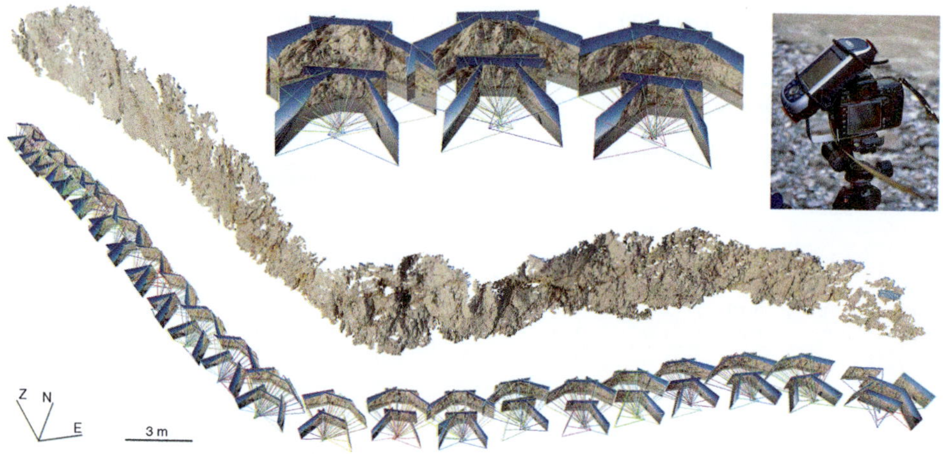

Figure 2.4 Terrestrial image fans shooting scheme for a DOM of a subvertical cliff. Note that image overlap within the same fan is limited (ca. 5–10%) and that the change in view angle between different shooting stations should not exceed 10–15°. Insets show a detail of three image fans, captured from three shooting stations, and a DSLR camera mounted on a tripod with a graduated head, equipped with a decimeter-grade GPS. The outcrop exposes fractured and faulted Eocene foraminiferal limestones (Island of Pag, Croatia, 44°19'24"N–15°15'16"E).

It is important to point out that, in this shooting scheme, photo overlap is not achieved at the level of a single station (where overlap is kept to ca. 5%, just to avoid "holes" in the coverage), but thanks to the fact that the whole outcrop (or the largest part of the outcrop) is repeatedly imaged from each shooting station (Figure 2.4). In this way, if the shooting stations are, for example, 10, each portion of the outcrop is imaged in at least 10 photos, from different points of view, ensuring a highly redundant and robust reconstruction.

In order to allow the automatic feature detection and matching algorithms to find common features in images, the images must not appear to be too different, so when moving from one station to the next, the viewing angle should not change by more than 10–15° and the area covered by each image (or vice versa for the magnification factor) should change gradually (Figure 2.4). Distances between shooting stations corresponding to angles of 10° and 15° are reported for different configurations in Table 2.1.

An even greater redundancy can be achieved by collecting photos with different lenses, ranging from wide-angle (e.g. 12 mm) to tele (e.g. 200 mm). If this strategy is used, wide-angle photos will cover a large area of the outcrop, reducing the risk of large-scale distortion (James and Robson, 2014; Wu, 2014), and at the same time tele-photos will provide a high spatial resolution. In this case, to ensure that the automatic feature detection and matching algorithms are able to find common features in images taken with different lenses (so with different magnification), the images must not be too different, so the change in focal length should not be larger than a factor 2 at every step (e.g. focal lengths of 12 mm, 25 mm, 50 mm, 100 mm, and 200 mm can be used in sequence).

Other survey schemes have been proposed, such as shooting a sequence of photos almost perpendicular to the outcrop face with just one or two photos per shooting station (Bilmes et al., 2019). We strongly warn against this method, as it results in poor correlation between photos and in a poor reconstruction of facets at high angles to the average outcrop surface.

2.2.3 Drone Surveys for SFM

Drone surveys can be carried out in two ways: with a fan scheme as for terrestrial surveys or with a "modified aerial-photo strip" scheme. In the first case, the discussion in the previous paragraph applies. One problem with the fan scheme is that to-date no software is available to automatically design and run an appropriate autonomous flight plan, and this is an important limitation since flying with autonomous control by the drone autopilot is very efficient and greatly reduces battery consumption.

When mapping large areas with a large horizontal extension with respect to the vertical difference in elevation, it is more efficient to use a modified aerial-photo strip scheme (Figure 2.5). The advantage here is that most flight planning software (e.g. DJI Ground Station Pro, UgCS, and DroneDeploy, available on the Apple or Android app stores) are able to create a similar efficient automatic flight plan. In order to avoid large-scale distortion problems in the reconstruction, the traditional aerial-photo strip scheme must be modified as follows (Figure 2.5): (1) the photos are shot with Nadir orientation, or pointing perpendicular to

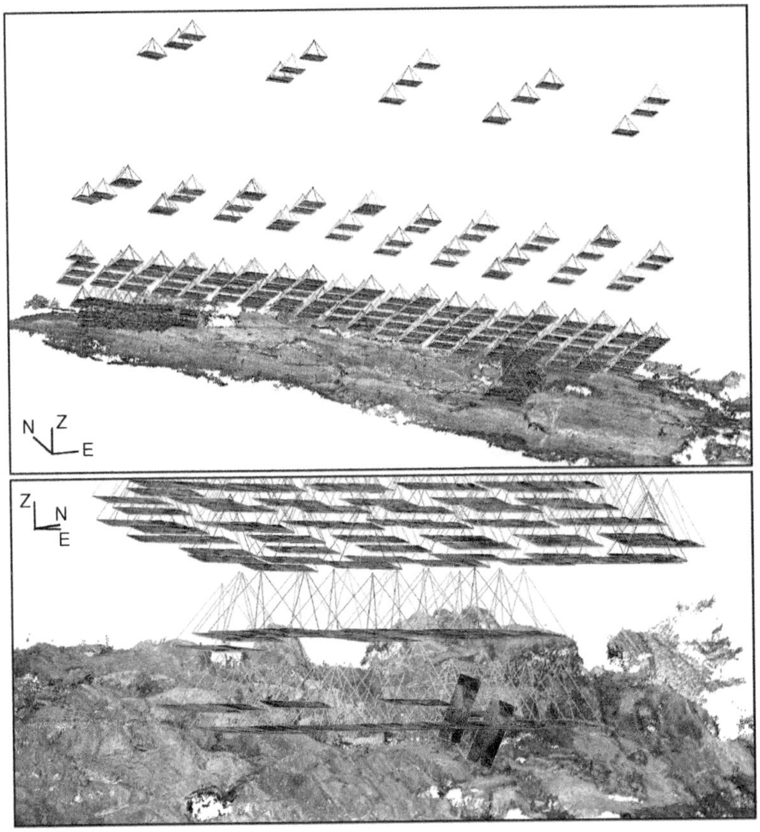

Figure 2.5 Drone survey shooting scheme with multiple photo strips at variable altitude, processed in VSFM. Cameras and dense point cloud are shown. The outcrop, c. 140 m × 40 m, exposes amphibolite facies gneisses with folds and shear zones in Lavertezzo (Valle Verzasca, CH, 46°15'33"N–8°50'16"E).

the regional slope, as in the traditional scheme; (2) a large overlap, between 65 and 70%, is necessary, not only along each photo strip, but also between adjoining photo strips (in this way each portion of the scene is imaged at least four times); and (3) to avoid large scale distortion, multiple passes are performed, each time increasing the altitude by a factor 2 (e.g. 30 m, 60 m, 120 m, 240 m), which results in a similar effect, such as changing the focal length (generally not possible on a drone camera) in terrestrial surveys. In some cases, a limited number of oblique photos can also be shot from the higher altitude (Figure 2.5). This strategy, suggested by James and Robson (2014), Wu (2014), and Jaud et al. (2018), has been tested several times by the authors and is the only way to be sure to avoid large-scale distortion ("dishing" or "bowl" effect) in all conditions. Other solutions, such as using a large number of ground control points (GCPs), as suggested in the Agisoft® manual (https://www.agisoft.com/), are not so reliable.

2.2.4 Image Quality and Pre-processing

Image quality is fundamental in a photogrammetric survey, but understanding which are the most critical parameters is not straightforward. Probably the most critical concern is for noise. In fact, noise is randomly distributed in images and results in incorrect feature detection and matching, so, if noise is too much, a collection of images could be completely useless. Noise in images can be measured with the noise/signal ratio and results from three factors: the sensor quality, acquisition conditions, and image compression. Illumination conditions have a great influence since more illumination means more energy reaching the sensor and a lower noise/signal ratio (Mather and Koch, 2011). In any case, in most outdoor situations, illumination is not a problem, and a more significant noise can result from image compression. For instance, JPEG image compression (https://jpeg.org/jpeg/) works by eliminating higher frequencies after applying a cosine transform to the image, and could introduce block and grid artefacts if the compression ratio is too high (Foi et al., 2006). For this reason, we suggest that non-compressed image formats, such as RAW, TIF, etc., should always be used or to limit the compression ratio at a minimum (e.g. with JPG images).

Other problems might arise from image distortion due to a lens geometric aberration. A moderate spherical aberration can be managed by SFM algorithms as the aberration is considered in the internal or intrinsic orientation parameters (Tsai, 1987). However, when the aberration is too pronounced or follows a very complicate non-linear pattern, significant problems in image matching and even inaccuracy in the photogrammetric reconstruction can occur. A general guideline is to use high-quality low-distortion lenses, but luckily this does not always mean to necessarily use heavy and expensive lenses. For instance, some lightweight consumer-grade drones are equipped with fairly low distortion lenses, and have a higher flight autonomy and range than heavy drones equipped with large cameras.

Correct and balanced exposure is possibly the most important concern when designing a photogrammetric survey. All images should have a balanced exposure, allowing all geological features in the scene to be properly recognized, and very deep shadows or very bright lights must be avoided. Moreover, the exposure should be the same for all images, to facilitate feature matching. This means that a photogrammetric dataset must be collected in a short time (similar light conditions), with a high angle of incidence of sunlight on the outcrop (to avoid shadows). Quite often, an optimal image acquisition can be carried out with a slightly cloudy sky, resulting in diffuse lighting with gentle shadows. If some photos in a dataset show a markedly different exposure, they can be adjusted, e.g. with histogram matching, before photogrammetric processing. Under other conditions, e.g. when working on low contrast outcrops, contrast enhancement algorithms (Mather and Koch, 2011) can also be applied to all images in the dataset before photogrammetric reconstruction.

Finally, image resolution is not as critical as one might imagine. In fact, the overall resolution of a photogrammetric reconstruction results from the sum of all images in the dataset. Hence, if only a moderate resolution camera is available, to obtain a high-resolution DOM one just needs to shoot more photos from a closer distance. This strategy can be very effective where different constraints limit the resolution of a single image, e.g. when collecting photos with a lightweight drone.

2.2.5 Photogrammetric Processing with SFM Software Packages

In this section, we highlight practical issues and the main differences between two software packages that are commonly used for SFM/MVS reconstruction in geological applications: VSFM (http://ccwu.me/vsfm/) and Agisoft Metashape® (previously called Photoscan®, https://www.agisoft.com/). Many other SFM software tools are available, both commercial and open source (e.g. AliceVision Meshroom, https://alicevision.org), but unfortunately we are not able to discuss all of them here. We will discuss VSFM and Agisoft® because, to our knowledge, they are some of the most widely used open and commercial packages and because they are both mature packages that enable large projects to be managed with thousands of photos and have good control on outputs. In any case, the following discussion covers fundamental aspects of any SFM/MVS software; hence it can be taken as a useful base to also evaluate other software that is already available, or that might become available in the future.

The main difference between Agisoft® and VSFM (apart from the first being a commercial software and the second being free for personal, non-profit, or academic use, and based on open-source libraries) are (1) a more or less user-friendly GUI (Graphical User Interface); (2) a very different way of dealing with georeferencing data; (3) partly different lens distortion models; (4) a more or less optimized usage of GPU (Graphical Processing Unit); and (5) a more or less transparent way of dealing with accuracy and noise in the reconstruction.

2.2.5.1 Graphical User Interface (GUI)

Agisoft® certainly has the most user-friendly GUI, and this could be the main reason for its widespread diffusion. In a few minutes, a non-experienced user can start a project and see the first results. VSFM has a GUI that requires some training and experience before mastering all tools and options, but, on the other hand, using VSFM in command-line mode, or better with batch scripts, is very powerful and probably the best option for large projects that require several hours or days of processing. Another advantage of the VSFM interface is that, partly hidden in initially not-so-intuitive tools, much more information is available about the quality and accuracy of the photogrammetric reconstruction.

2.2.5.2 Usage of Georeferencing Data

VSFM builds a reconstruction in relative coordinates (image pixels) in an arbitrary model-centric reference frame. Only at the end of the processing is the user able to specify real-word coordinates for some target-points in the scene, or for cameras (i.e. positions from where photos were shot), and apply a rigid-body rotation, scaling and translation transformation to output a georeferenced model. In other words, georeferencing information is not used at all by VSFM in the reconstruction, and SFM equations are solved just based on imaging data. In practical applications, this means that VSFM can either output a reconstruction with a very high-quality level, with small errors at the pixel scale, or not yield a reconstruction at all when large errors are envisaged. We judge this behavior very reliable, since with VSFM there is no possibility of obtaining a 3D model from an ill-posed problem.

Agisoft® can use georeferencing information as an initial guess in model reconstruction, using algorithms that unfortunately are not completely documented. This means that in some borderline conditions, if the survey was carried out with a poor error-prone geometry (see Sections 2.2.2 and 2.2.3), Agisoft® can still output a reconstruction.

However, it is not possible for the user to evaluate if the reconstruction is reliable or not, particularly regarding large-scale distortion (see the discussions in James and Robson, 2014; Wu, 2014; Jaud et al., 2018). We judge this behavior to be somehow unreliable, and we suggest always using proper acquisition strategies, with oblique views and images taken at different distances from the outcrop, even if apparently Agisoft® allows the use of more simple schemes (that unfortunately are also suggested in the software manual, such as the classical photo strip scheme).

2.2.5.3 Lens Distortion Models

The lens distortion model of VSM is simpler than that of Agisoft®. In VSFM an eight-parameter camera model is used, with position (XYZ coordinates), rotation (three angles), focal length and radial distortion (a simple one-parameter radial distortion is used). In Agisoft®, additional parameters are allowed for principal point coordinates, affinity and skew coefficients, four different radial distortion coefficients, and four tangential distortion coefficients (see Tsai, 1987, for details on these parameters). In practical applications, the larger number of parameters in Agisoft® can either lead to a successful calibration of almost pathologic cameras (such as cameras with fisheye lenses) or to relevant errors in calibration due to a too large number of parameters to be calibrated with non-sufficient data and non-unique solutions.

Based on several years of experience, we tend to use good cameras with high-quality and low-distortion lenses. These can be successfully calibrated with the VSFM simplified model, and when we obtain a reliable calibration from VSFM, we use the same calibration for processing in Agisoft® in "fixed calibration" mode. The practice of calibrating a camera with a full-size survey processed with VSFM, and then using the same calibration also in Agisoft®, has also been successfully experimented with consumer-grade drone cameras.

2.2.5.4 GPU (Graphical Processing Unit) Computation

Both Agisoft® and VSFM rely heavily on GPU computation carried out with CUDA® libraries on Nvidia® high-end graphic cards (https://developer.nvidia.com/cuda-zone). These graphic cards have several gigabytes of dedicated RAM and several thousand parallel processors, and all key algorithms in Agisoft® and VSFM are highly optimized for massively parallel computation. VSFM does not run at all without a GPU, while Agisoft® can be run, but only on small models and with very poor performance.

In a few words, only with a GPU workstation is it possible to process very large collections of photos (some thousand to some tens of thousands of photos), in order to obtain high resolution and accuracy, at levels comparable to or even better than a laser scanning survey. For these reasons, GPU computing has been the game-changer in photogrammetry, and it is not possible nowadays to use a software not based on the CUDA parallel programming technique.

2.2.5.5 Control on Accuracy and Noise

An area where VSFM and Agisoft® are very different is in the control on accuracy of the reconstruction and noise of the sparse and dense point clouds. Agisoft® allows setting a "quality level" for the reconstruction using qualitative terms only ("high quality," "low quality," etc.) and does not output a complete report of estimated errors after the reconstruction is completed. In VSFM the quality and accuracy of the reconstruction can be fine-tuned by editing a configuration file, where many parameters are set quantitatively. Amongst the more important parameters, it is possible to define (i) the minimum number of projections and matched features needed to orient a camera (influences errors in image orientation and sparse point clouds) and (ii) the minimum number of cameras required to recognize a point with subpixel accuracy, in order to include the point in the dense point cloud. This latter parameter is fundamental to control errors and even more to reduce noise in the dense point cloud. In many situations, dense point clouds obtained from VSFM, with restrictive processing parameters, include a smaller number of points than those obtained from Agisoft®, but are also characterized by a significantly lower noise.

2.3 Point-Cloud vs. Textured-Surface DOMs

The datasets that can be obtained from either a photogrammetric VSFM or a laser scanning workflow are not as different as one might expect. Differences are evidenced in the discussion paragraph, but, in most cases, the data include a georeferenced dense point cloud and a set of oriented photos, and the dense point cloud is normally "photorealistic" in the sense that each point also carries RGB values obtained from the images. This similarity means that visualization, interpretation, and analysis can be carried out with similar tools irrespective of the technique used to collect the data. In the next two paragraphs we will see how a DOM can be represented directly with a photorealistic dense point cloud or with a textured triangulated surface obtained from the point cloud and images. Strictly speaking, a third option exists for DOMs where the horizontal extension is much larger than the vertical one. In this case, the DOM can be represented as an orthophoto (orthorectified image) and a DEM (Digital Elevation Model). However, we will explain why this must be considered just as a subclass of a textured surface DOM.

2.3.1 Point-Cloud DOMs

When using point-cloud DOMs (in the following, PC-DOMs), the outcrop surface is represented by a dense set of points, characterized by X,Y,Z coordinates and additional properties, including RGB values in a "photorealistic" DOM. The RGB properties are obtained by projecting on each point the RGB value obtained by photos that are associated with the point cloud. This association is obvious in a photogrammetric workflow, and all photogrammetric point clouds are natively colorized.

On the other hand, in laser scanning point clouds the projection of RGB values onto points must be performed a-posteriori, using images from a camera attached to the laser scanner head. In this case, the perfect alignment of the camera with respect to the laser scanner head is fundamental to obtain a perfect correspondence of the image and laser data (White and Jones, 2008). In addition, the camera can have a different resolution with respect to the laser scanner, and in common setups one pixel in the image is associated with several points in the point cloud, i.e. the imaging data have a lower resolution than the topographic data, and thus several neighboring points will share the same RGB information (White and Jones, 2008).

In any case, point clouds obtained from laser scanners or photogrammetry must be processed to eliminate noise and non-geological features before a proper DOM interpretation can be carried out. When a good-quality point cloud has been obtained, areas covered by vegetation, soil, man-made artefacts or debris must be removed, in order to highlight areas of the outcrop where solid rocks are exposed (e.g. Jones et al., 2011; Kromer et al., 2019). This task can be easily performed using the manual segmentation tools in CloudCompare (https://www.danielgm.net/cc/), a very powerful opensource tool for point-cloud processing. Also, semiautomatic segmentation tools implemented in CloudCompare can be used for this purpose. For instance, the "Connected Components" tool allows labelling subsets of the point cloud that are disconnected from the main set and are generally associated to vegetation or man-made artefacts. In addition, a segmentation based on a roughness or curvature filter can help to select areas where the point cloud shows high roughness related to vegetation or debris.

Once the bare rock surface is segmented and "cleaned", a PC-DOM can be interpreted with techniques described in Section 2.4.1.

2.3.2 Textured-Surface DOMs

Textured-surface DOMs (in the following TS-DOMs) require some additional processing with respect to PC-DOMs. Once a clean point cloud and a set of oriented images are available, a surface must be generated from the point cloud, and then images are projected onto this surface in order to provide a realistic texture. This kind of visualization is very common in computer graphics (Catmull, 1974), e.g. in architectural applications or videogames, but less common in geology (Tavani et al., 2014; Bistacchi et al., 2015).

The first step to obtain a TS-DOM is meshing of the point cloud data to obtain a discrete surface representation of the outcrop. A triangulated surface (a mesh composed of triangles; see Figure 2.6) can be generated, for instance, in CloudCompare (https://www.danielgm.net/cc/) or Meshlab (http://www.meshlab.net/) using the Poisson surface reconstruction algorithm (Kazhdan and Hoppe, 2013). It must be noted that the quality (e.g. low noise) and resolution of the point cloud is critical to obtain a clean and reasonably smooth surface.

In order to texture the DOM surface with the associated images, image coordinates (locally defined in each image plane) must be projected onto each vertex of the triangulated surface (Figure 2.6). This projection is performed using the camera intrinsic and extrinsic orientation, or with a simplified pin-hole projection transformation (Figure 2.6). The simplified pin-hole projection is valid if the input images have been corrected to remove spherical aberration (undistorted in SFM jargon) in the photogrammetric workflow. For instance, this correction is implemented in VSFM, at the step where the dense point cloud is generated, but not in Agisoft®. In any case, the internal consistency of the resulting DOM is guaranteed since the same projection transformation, calculated by the SFM package to generate the 3D model, is used.

When modelling a complex multifaceted 3D DOM, different portions of its surface are best represented in different images. Therefore, the DOM triangulated surface must be subdivided into a mosaic of different portions, each one associated with a particular image (Figure 2.6). The selection of the set of images used for texturing can be a problem since the images used in an SFM project are generally between hundreds and thousands (Sima et al., 2013). In the workflow that we generally use, a first selection of this subset is performed manually; then we select automatically the triangles to be textured using each image, based on two criteria (Figure 2.6): (1) to minimize the angle between the normal to the triangle and the normal to the image plane and (2) to minimize the distance of the triangle with respect to the projection of the image optical axis (symmetry axis of the image).

Texturing can be performed in a completely automatic way in Agisoft®, but we strongly suggest to always select manually the subset of images to be used for texturing, since this greatly improves the quality of the results (and the processing time).

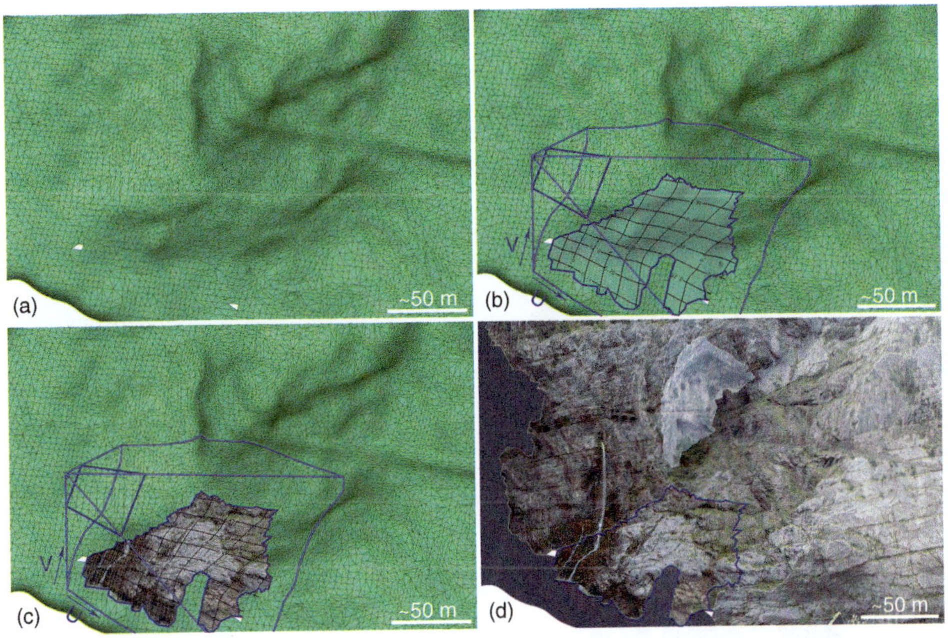

Figure 2.6 Construction of a TS-DOM by image projection onto a triangulated surface: (A) Triangulated surface obtained with interpolation from point cloud; (B) Projection of image coordinates (UV) onto a triangulated surface and selection of area to be textured with this image; (C) Texturing with a single image; (D) Texturing of the whole triangulated surface with multiple images.

Once each portion of the outcrop has been associated with one of the selected images, the textured surface can be saved in suitable formats (e.g. Gocad.TS or .OBJ) for visualization and interpretation, which will be discussed in Section 2.4.2.

It is important to note that, if we consider the imaging data, the TS-DOM workflow generally results in a much higher resolution dataset with respect to PC-DOM, since images can be projected onto the triangulated surface, maintaining their original resolution (Bistacchi et al., 2015).

2.4 Geological Interpretation of DOMs

We have seen in the previous paragraphs how to obtain a 3D model of an outcrop (a 3D DOM). This is not very different from obtaining a 3D model that can be used in engineering, architectural, or archeology applications, but very important differences arise when we would like to extract geological information, i.e. to carry out a geological interpretation on a DOM.

The general goal of software tools aimed at geological interpretation on DOMs is to allow a geologist to carry out on a DOM the same measurements that are usually carried out in the field on a physical outcrop, e.g. (1) measuring the attitude of a surface or lineation with a compass, (2) measuring the position, spacing, thickness, and size of structures, and (3) mapping lithology or boundaries, etc.

All these measurements could be carried out on a DOM "manually", e.g. digitizing with mouse clicks a polyline representing a boundary, or a small planar surface that allows measuring an attitude (Figure 2.1), or with some (semi-)automatic algorithms, allowing a large number of measurements to be carried quickly and in a reproducible manner. Some approaches have also been proposed to perform this digital interpretation directly in the field with a mobile device (also see Chapter 5 by Kehl et al., 2021). In the next two paragraphs we will outline different interpretation methodologies as they are carried out in the DOMstudio workflow on PC-DOMs and TS-DOMs.

2.4.1 Interpretation on Point-Cloud DOMs

We mainly use PC-DOMs for the structural characterization of fractured rock masses, based on the assumption that small subplanar facets on an outcrop are the morphological expression of planar structures such as fractures, joints, foliation, or bedding surfaces (Figure 2.7; in the following we will use the collective term "discontinuity" for all these structural features). The main reason why PC-DOMs are particularly suitable for this kind of interpretation is that the local attitude of the DOM can be easily computed by fitting a small planar surface to points in a small neighborhood of each point ("kernel"; e.g. see Jaboyedoff et al., 2007). The attitude obtained by applying the kernel to each point in the point cloud is attributed to the point itself, as a normal unit vector (three components)

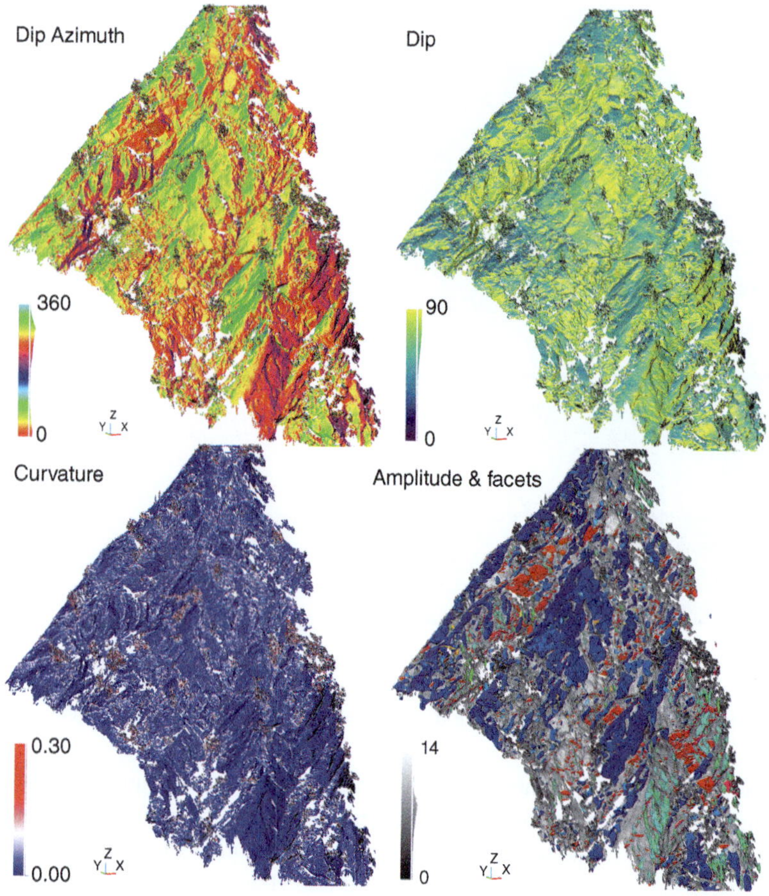

Figure 2.7 Example of a PC-DOM (ca. 100 m × 300 m) with color representation of dip azimuth, dip and curvature properties, and structural interpretation with facets superposed on a point cloud colorized with amplitude. Interpretation carried out on a PC-DOM in CloudCompare. The outcrop exposes gneisses of the Sesia-Lanzo Zone (Valle del Lys, Italy, 45°43'50"N–7°51'50"E).

or as dip/azimuth properties (two angles). Once the local attitude has been calculated, dip and azimuth or normal properties can be used as a color scale to highlight facets that share a common attitude and could represent different discontinuity sets (Figure 2.7). Using a similar processing, with different kernel functions, additional morphometric parameters can be calculated, as for instance different curvature or roughness indexes (Figure 2.7). In all these algorithms, properly selecting the radius of the kernel, considering both the DOM morphology and the point-cloud spatial resolution, is fundamental to obtain meaningful results. All these algorithms are available in CloudCompare (https://www.danielgm.net/cc/), a highly optimized open-source package used for the analysis of point clouds.

A point-cloud dataset enhanced with morphometric and attitude properties, can be analyzed in different ways that roughly fall into three categories: manual, semiautomatic (e.g. Riquelme et al., 2014), and automatic methods. In the workflow that we generally apply, we first clean the point cloud from outcrop portions that show a high roughness or curvature. This can be done in CloudCompare or in Skua/Gocad® via an automatic segmentation, based on a threshold roughness or curvature value. The result is a dataset where only subplanar facets, really associated with planar structural features (i.e. discontinuities), are retained for subsequent analysis (Figure 2.7).

Then we can choose between two main options: manual extraction of single facets or semiautomatic segmentation. In the first case, the interpreter can manually select points on a facet and fit a small surface to them. The points are selected by digitizing a loop, either in CloudCompare or in Skua/Gocad®, or with the very convenient Compass plugin in CloudCompare (Thiele et al., 2017). This is the DOM counterpart of placing a compass on an outcrop surface and measuring the dip and azimuth (Figure 2.7). The accuracy of the measurement is similar or better than that of a measurement performed with a compass in the field, and the main advantage is the possibility to collect measurements in places that are not accessible. However, collecting a large number of data in this way can be relatively time-consuming (even if not as in the field), and generally no more than some hundred measurements can be collected in this way.

A faster alternative is to apply a threshold segmentation on dip/azimuth values to extract subset point clouds representing only one discontinuity set each. The dip and azimuth thresholds for each discontinuity set can be defined based on preliminary field measurements (this is the preferred option) or based on visual inspection and orientation analysis performed on the point cloud colored with dip/azimuth attributes. Once the dip/azimuth segmentation has been performed, automatic algorithms can be used to fit small surfaces to each facet in CloudCompare (Facet plugin, discussed in Dewez et al., 2016) or in Skua/Gocad® (macros custom-developed by the authors), and eventually the attitude, length, height, and area of each small surface can be calculated and extracted for structural analysis (see below at the end of this subsection the structural significance of length, height, and area). The same automatic algorithms can be potentially applied to the whole point cloud (e.g. the original workflow by Dewez et al., 2016), without extracting the subset point clouds associated with individual discontinuity sets, but in our experience this invariably leads to errors and noise.

This workflow is semiautomatic since the selection of orientation thresholds for each discontinuity set is done explicitly, using proven structural criteria, and then the extraction of facets is performed automatically. In this way, large datasets can be obtained quickly, but still applying proper structural analysis criteria. Other completely automatic workflows have been proposed in the literature (e.g. Zhang et al., 2018; Ge et al., 2018), but we do not completely trust their "black box" approach.

We emphasize that in both the manual and semiautomatic workflows, we do not use the local orientation defined at each single point with the kernel function as the direct source of structural information, since the orientation measured in this way is affected by various noise sources. This local orientation is used only as a first approximation, to apply a color scale to the point cloud and to define the number and mean attitude of each discontinuity set. Then larger surfaces are fit to facets to obtain better structural measurements (Figure 2.7). In approaches where the point cloud is used directly to obtain orientation statistics (e.g. Jaboyedoff et al., 2007), it is very likely that the scatter in dip and azimuth represents the noise in the point cloud, rather than real variability in discontinuity attitude.

The small facets interpolating each discontinuity can be used to extract other information, in addition to attitude, such as spacing or fracture intensity. However, it is important to note that facets only represent the portion of a discontinuity that intersects the outcrop, so they always underestimate length, height, and area, and any attempt to extract these parameters from a facets dataset is irremediably flawed (Figure 2.7).

2.4.2 Interpretation on Textured-Surface DOMs

We use TS-DOMs for mapping linear features (polylines) or areal features (polygons) (Figure 2.8). Polylines can be the expression of discontinuities that cross the outcrop surface; hence they represent traces of fractures, joints, foliation or bedding surfaces, or lithological boundaries. Closed polygons are generally used to represent the intersection with the outcrop surface of lithological or other compositional boundaries (e.g. reaction fronts), but in any case they are obtained exactly as polylines, so the following discussion applies to both kinds of features.

Manual digitization of polylines can be carried out in CloudCompare or in geomodelling packages like Skua/Gocad® or Move® with simple "digitize polyline" tools (Figure 2.8). The advantage of performing the digitization in Skua/Gocad® or Move® is that typical geological and structural analysis tools are already available in these packages (e.g. it is possible to assign a polyline to a certain "stratigraphic horizon" or "fault set").

If the DOM is represented by an orthophoto and a DEM (a special case of TS-DOM representation), manual digitization of polylines can also be performed in a 2D GIS package. However, visualizing the results in 3D is always suggested, as in any other geological interpretation workflow.

Figure 2.8 Example of a TS-DOM (ca. 5 m × 15 m) textured with the original high-resolution images, with interpretation carried out as traces (see also Chapter 6 by Mittempergher and Bistacchi, 2021). The outcrop exposes tonalites of the Adamello Batholith with pseudotachylyte-bearing faults of the Gole Larghe Fault Zone (Val di Genova, Italy, 46°10'18"N–10°34'48"E).

Gaining inspiration from Vasuki et al. (2014), we also developed a semiautomatic feature extraction workflow, using advanced image processing tools, which is discussed in Chapter 6 by Mittempergher and Bistacchi (2021). In this workflow (called DOMstudio Image) we perform image processing in 2D, starting from the original images collected in the field. In this way we minimize data loss and noise due to resampling of images. When the feature extraction is completed, we use the internal and external orientation parameters associated with each image to project the polylines on the TS-DOM surface for subsequent analysis (Figure 2.8), which can then be carried out, for example, in Skua/Gocad® with the same methods used for polylines obtained from manual digitization.

The main difference between the manual digitization and the automatic feature extraction methods is that the first is easily applied for a small number of features, but could become very time-consuming if thousands of features have to be digitized. On the other hand, automatic feature extraction methods require tuning various parameters to be effective, and this requires some trial-and-error, but, in favorable situations where they can be calibrated, they are able to yield thousands of features automatically (see Chapter 6 by Mittempergher and Bistacchi, 2021).

Every geological data that is represented by a polyline can be collected in this way: fault or fracture traces, foliation traces, bedding intersections, compositional or reaction fronts, to name the more common types. The following analysis depends on several factors. For instance, if fracture traces have been digitized, small fracture surfaces can be fit to the traces, in order to measure the attitude of fractures and perform other analyses already discussed in the paragraph on PC-DOM interpretation. If bedding intersections have been digitized, a stratigraphic model can be reconstructed (as in Bistacchi et al., 2015), and virtual wells can be also extracted to perform analyses generally used in subsurface projects (e.g. in hydrocarbon reservoirs, Siddiqui et al., 2019). In other situations, reaction fronts that crosscut the stratigraphy unconformably can be mapped and used to model complex 3D bodies (e.g. hydrothermal dolostone bodies, as in Bistacchi et al., 2015).

Finally, we would like to emphasize here that we generally do not use the outcrop surface of a TS-DOMs (represented by a triangulated surface) to directly collect orientation data as in the PC-DOM workflow described in the previous paragraph. This approach has been used by some authors (e.g. Lato and Vöge, 2012; Zhang et al., 2018), but we do not consider it very reliable. In fact, the surface of the outcrop is always slightly smoothed by the meshing algorithms used to obtain the triangulated surface, to an extent that is difficult to assess, and this might influence orientation statistics.

2.5 Discussion and Conclusion

We have reviewed best-practices to reconstruct 3D DOMs with laser scanning techniques or with SFM photogrammetry, and then we have discussed how PC- and TS-DOMs are obtained and the different kinds of analyses that can be performed on these different datasets. In the following we discuss pros and cons of different data acquisition and processing techniques, based on our practical experience in several years of carrying out DOM studies.

2.5.1 Data Acquisition: Platform

The choice of a platform should be mainly based on logistics and the geometry and size of the outcrop to be mapped. Terrestrial acquisition (both TLS—Terrestrial Laser Scanning—and close-range photogrammetry) is suggested for small outcrops to be imaged at very high resolution and for vertical walls or cliffs, which are better imaged standing in front of them with a horizontal line of sight. When planning a photogrammetric survey, also the possibility to find several shooting stations with an unobstructed view of the area to be surveyed, must be considered and carefully planned (e.g. using a DEM or simply Google Earth).

On the other hand, drone surveys are best suited for larger areas, between a few hundred square meters and a few square kilometers, particularly if the outcrops are characterized by a limited difference in elevation (see the discussion on the limitations of software used for autonomous flight planning in Section 2.2.3). According to our experience, it is relatively easy to map areas of several hundred square meters with a lightweight drone like a DJI Mavic Pro 2®, and with some planning and several batteries we have successfully mapped areas of a few square kilometers always with the same drone (Martinelli et al., 2020). By the way, it is not necessarily true (as it was just five years ago) that larger and heavier drones are required for larger areas, since their autonomy and maximum velocity are not much higher than those of a good lightweight drone. Nowadays, large and heavy drones are required only when the payload is heavy (e.g. hyperspectral spectrometers), and in many cases their autonomy and range are smaller than those of lightweight drones carrying simple optical cameras.

When the area to be mapped is even larger, it is possible do use a fixed wing drone or a manned aircraft, either an airplane, best suited for very large areas, or a helicopter, which is still unrivalled for complex surveys in mountain areas with large differences in elevation of maybe more than one thousand meters. In this case it is very convenient to use an airborne Lidar mounted on a "helicopter pod":

an inertial platform that can be easily mounted on any helicopter available in the survey area (e.g. RIEGL VP-1®).

2.5.2 Data Acquisition: Laser Scanning vs. Photogrammetry

If we compare the pros and cons of laser scanning versus photogrammetry in practical DOM studies, according to our experience we should say that in general laser scanning is more "standardized" while photogrammetry is more flexible. As summarized in Table 2.2, the ground spatial resolution of a laser scanning survey is controlled, within a certain range, by technical details of the laser scanner and the distance to the outcrop (which for a certain laser scanner is limited to the instrument maximum range). On the other hand, in photogrammetry the ground spatial resolution is a function of the many parameters discussed in Section 2.2.1. These parameters can be easily adjusted (e.g. by using a different lens); hence the spatial resolution can be easily adapted to the needs of each particular study.

One advantage of laser scanning is that the point cloud can be visualized and quality checked directly in the field during the survey. On the other hand, the results of a photogrammetric survey can be quality checked only after a preliminary processing, and this can be a problem particularly when working in very remote areas. If this is the case, our suggestion for a successful field campaign is to collect very redundant datasets, with the best light conditions, and to perform a preliminary processing at the first occasion, e.g. back in the camp at night, to assess the quality of the survey.

Another advantage of laser scanning is that it returns a signal on a wide variety of surfaces, while photogrammetry does not yield any data when working on featureless surfaces (e.g. an homogeneous white wall) or on changing surfaces (e.g. reflective surfaces that yield a different image according to the viewing angle or moving objects like trees or water in windy conditions). However, this limitation does not affect surveys of rock outcrops, which are generally very rich in features useful for photogrammetric reconstruction.

Laser scanning has a very good reputation for accuracy and repeatability, while some authors tend to consider photogrammetry, and particularly close-range photogrammetry, to be less reliable. However, in our experience, and based on a few tests where the two approaches have been compared, photogrammetry is at least as accurate as laser scanning, and in some conditions yields point clouds that are less noisy, if a proper survey strategy and processing pipeline (as discussed above) are used (e.g. James and Robson, 2014; Wu, 2014; Wilkinson et al., 2016).

One situation where laser scanning is unsurpassed is when we would like to survey an outcrop that is partly covered by vegetation. In this case the laser scanner signal is composed by a first arrival that represents the reflection from the vegetation and by a last arrival that represents the reflection from the outcrop, and it is possible to select just the latter for the analysis. Under these conditions, photogrammetry can only yield a 3D model of the first object in the line of sight (e.g. the treetops).

The advantages of laser scanning are obtained at the cost of a heavier and more complex instrumentation to be carried and used in the field, and by a much higher cost, so photogrammetry should be favored, for instance, when working in the field in remote areas accessible only on foot. Nowadays, a lightweight drone (ca. 750 g) with a very good camera can be easily carried in a backpack just in case an interesting outcrop is found. The weight and increased complexity of a laser scanner with respect to photogrammetry, however, is partly counterbalanced by the possibility to complete a survey from a limited number of survey stations, whereas in a photogrammetric survey a large number of shooting stations is needed to obtain a high-quality survey.

Table 2.2 A comparison of laser scanning vs. photogrammetric data acquisition.

Lidar/TLS	SFM-MVS photogrammetry
More standard	**More flexible**
Limited resolution range	**Arbitrary resolution**
Easy quality check in the field	Quality check after preliminary processing
Granted accuracy on several kinds of surfaces	Granted accuracy on textured/patterned surfaces (rocks)
People say it is more reliable	**With proper software and acquisition, possibly even more accurate**
"Sees" through vegetation	Do not "see" through vegetation
Less measurement stations	Needs several shooting stations
Heavier and more complex in the field	**Lighter and faster in the field**
Possible errors in pointcloud/images alignment	**Granted point cloud/images alignment**

Finally, laser scanning systems are potentially affected by inaccuracies in the relative orientation of the laser scanner and the camera used to collect the photos (Jones et al., 2009).

2.5.3 Pointcloud vs. Textured Surface DOMs

Pointcloud vs. textured surface (or orthophoto and DEM) DOMs provide rather different support for the interpretation and should be used in different situations. A schematic comparison is provided in Table 2.3. Pointcloud DOMs are obtained directly at the end of a laser scanning or photogrammetry acquisition workflow; hence there is a slight processing advantage with respect to textured-surface (or orthophoto-DEM) DOMs that require an additional processing step. However, this additional step is completely automatic and does not require much interaction with the interpreter, and quite often is counterbalanced by the lighter weight (i.e. for rendering) of a textured-surface DOM at the same ground resolution. If the DOM is rather flat, carrying out the interpretation on a 2D orthophoto results in an even lighter weight dataset.

In any case, the choice between a point-set and a textured-surface DOM must be based mainly on the geological features that we are going to extract from the DOM and on how they appear in a particular outcrop. If we are interpreting fractures, bedding surfaces, or other discontinuities that correspond to planar facets on the outcrop surface, a point-cloud DOM must be used. In fact, a textured surface is always a more or less smoothed version of the original point cloud, so the edges at the transition between different facets could be slightly smoothed and the orientation of smaller facets can be slightly altered. The best support to study the orientation and roughness of discontinuities is always provided by a PC-DOM.

On the other hand, a textured-surface DOM must be preferred when the image carries more information than the 3D geometry, for instance when the discontinuities intersect the outcrop at a high angle and appear as fracture or bedding traces (the line corresponding to the intersection of a geological surface—fracture, bedding, etc.—and the outcrop surface). Working on imaging data also allows applying spectral analysis and image processing and analysis algorithms, such as those used for automatic feature extraction in structural geology (e.g. see Chapter 6 by Mittempergher and Bistacchi, 2021) or in mineral and lithological mapping (e.g. Massironi et al., 2008; van der Meer et al., 2012; Cudahy, 2016).

2.6 Summary and Perspectives

To summarize, using Lidar/TLS vs. SFM photogrammetry is mainly a matter of logistics, outcrop geometry, and morphology, and since in the absence of other important constraints using photogrammetry is so simple and so much less expensive than laser scanning, this is becoming the obvious choice in many situations. This explains the rapid and steady growth of photogrammetry over laser scanning in the last few years, and in most cases laser scanning is justified only to overcome problems related to vegetation hiding the outcrop.

A partial bottleneck in a complete workflow is still represented by the software used for interpretation. CloudCompare is probably the best solution for point-cloud DOMs, but tools developed specifically for the interpretation of geological structures are still underdeveloped, even if a few very interesting ones have been introduced in the last few years (Dewez et al., 2016; Thiele et al., 2017). When considering the interpretation and analysis of textured-surface DOMs, we see that CloudCompare can be used for visualization, but no specific tools are available. On the other hand, the geomodelling packages Move and Skua/Gocad can be used and the latter can be extended with custom-developed scripts, but they show limitations since DOM interpretation is not their primary goal. An approach using custom-developed tools for automatic image segmentation tools and Skua/Gocad just for the final structural interpretation is presented by Mittempergher and Bistacchi (2021) in Chapter 6.

Table 2.3 A comparison of point-cloud DOMs (PC-DOMs) vs. textured surface DOMs (TS-DOMs).

Point clouds	Textured surfaces/Orthophoto & DTM
Direct output from TLS/LIDAR and photogrammetry (needs some cleaning)	Processing to obtain textured surface or orthophoto (automatic)
Computationally heavier at the same image resolution	**Computationally lighter-weight at the same image resolution**
Perfect to study surface orientation from facets of DOM (e.g. joints)	Perfect to study structures as traces (e.g. faults and joints)
Spectral analysis more difficult (e.g. Compass plugin)	**Perfect for spectral analysis (e.g. automatic segmentation in Mittempergher & Bistacchi, this volume)**

Acknowledgments

We warmly acknowledge Changchang Wu for developing and distributing VSFM, one of the best structures from motion packages available. We are all indebted to Daniel Girardeau-Montaut for developing, maintaining, and distributing CloudCompare, a really powerful point-cloud analysis and processing software. We acknowledge Agisoft LLC for providing an educational license for Photoscan/Metashape at a reduced cost.

References

Barnes, R., Gupta, S., Traxler, C. et al. (2018). Geological analysis of Martian Rover-derived digital outcrop models using the 3-D visualization tool, Planetary Robotics 3-D Viewer-PRo3D. *Earth and Space Science* 5: 285–307. https://doi.org/10.1002/2018EA000374.

Bellian, J.A.J.A., Kerans, C., and Jennette, D.C. (2005). Digital outcrop models: Applications of terrestrial scanning LiDAR technology in stratigraphic modeling. *Journal of Sedimentary Research* 75: 166–176. https://doi.org/10.2110/jsr.2005.013.

Bilmes, A., D'Elia, L., Lopez, L. et al. (2019). Digital outcrop modelling using "structure-from-motion" photogrammetry: Acquisition strategies, validation and interpretations to different sedimentary environments. *Journal of South American Earth Sciences* 96, 102325: https://doi.org/10.1016/j.jsames.2019.102325.

Bistacchi, A., Massironi, M., Dal Piaz, G.V. et al. (2008). 3D fold and fault reconstruction with an uncertainty model: An example from an Alpine tunnel case study. *Computers and Geosciences* 34: 351–372. https://doi.org/10.1016/j.cageo.2007.04.002.

Bistacchi, A., Griffith, W.A., Smith, S.A.F. et al. (2011). Fault roughness at seismogenic depths from LiDAR and photogrammetric analysis. *Pure and Applied Geophysics* 168: 2345–2363. https://doi.org/10.1007/s00024‐011‐0301‐7.

Bistacchi, A., Massironi, M., Superchi, L. et al. (2013). A 3D geological model of the 1963 vajont landslide. *Italian Journal of Engineering Geology and Environment* 2013: 531–539. https://doi.org/10.4408/IJEGE.2013-06.B-51.

Bistacchi, A., Balsamo, F., Storti, F. et al. (2015). Photogrammetric digital outcrop reconstruction, visualization with textured surfaces, and three-dimensional structural analysis and modeling: Innovative methodologies applied to fault-related dolomitization (Vajont Limestone, Southern Alps, Italy). *Geosphere* 11: 2031–2048. https://doi.org/10.1130/GES01005.1.

Bistacchi, A., Mittempergher, S., Martinelli, M., and Storti, F. (2020). On a new robust workflow for the statistical and spatial analysis of fracture data collected with scanlines (or the importance of stationarity). *Solid Earth* 11: 2535–2547. https://doi.org/10.5194/se-2020-83.

Buckley, S.J., Howell, J., Enge, H.D.D., and Kurz, T.H.H. (2008). Terrestrial laser scanning in geology: Data acquisition, processing and accuracy considerations. *Journal of the Geological Society* 165: 625–638. https://doi.org/10.1144/0016-76492007-100.

Catmull, E.E. (1974). *A subdivision algorithm for computer display of curved surfaces. PhD Thesis*. University of Utah.

Chen, N., Kemeny, J., Jiang, Q., and Pan, Z. (2017). Automatic extraction of blocks from 3D point clouds of fractured rock. *Computers and Geosciences* 109: 149–161. https://doi.org/10.1016/j.cageo.2017.08.013.

Cudahy, T. (2016). Mineral mapping for exploration: An Australian journey of evolving spectral sensing technologies and industry collaboration. *Geosciences* 6: 52. https://doi.org/10.3390/geosciences6040052.

De Toffoli, B., Mangold, N., Massironi, M. et al. (2020). Structural analysis of sulfate vein networks in Gale crater (Mars). *Journal of Structural Geology* 137: 104083. https://doi.org/10.1016/j.jsg.2020.104083.

Dewez, T.J.B., Girardeau-Montaut, D., Allanic, C., and Rohmer, J. (2016). Facets : A CloudCompare plugin to extract geological planes from unstructured 3d point clouds. In: *ISPRS—International Archives of the Photogrammetry, Remote Sensing and Spatial Information Sciences*, vol. XLI-B5, 799–804. https://doi.org/10.5194/isprsarchives-XLI-B5-799-2016.

Dujoncquoy, E., Masse, P., Nicol, Y. et al. (2019). UAV-based 3d outcrop analog models for oil and gas exploration and production. In: *IGARSS 2019 -—2019 IEEE International Geoscience and Remote Sensing Symposium*, 6791–6794. IEEE https://doi.org/10.1109/IGARSS.2019.8900176.

Fabuel-Perez, I., Hodgetts, D., and Redfern, J. (2010). Integration of digital outcrop models (DOMs) and high resolution sedimentology—workflow and implications for geological modelling: Oukaimeden Sandstone Formation, High Atlas (Morocco). *Petroleum Geoscience* 16: 133–154. https://doi.org/10.1144/1354-079309-820.

Foi, A., Katkovnik, V., and Egiazarian, K. (2006). Pointwise shape-adaptive DCT for high-quality deblocking of compressed color images. *European Signal Processing Conference* 16: 1–17. https://doi.org/10.1.1.75.416.

Franceschi, M., Martinelli, M., Gislimberti, L. et al. (2015). Integration of 3D modeling, aerial LiDAR and photogrammetry to study a synsedimentary structure in the Early Jurassic Calcari Grigi (Southern Alps, Italy). *European Journal of Remote Sensing* 48: 527–539. https://doi.org/10.5721/EuJRS20154830.

Francese, R., Mazzarini, F., Bistacchi, A. et al. (2009). A structural and geophysical approach to the study of fractured aquifers in the Scansano-Magliano in Toscana Ridge, southern Tuscany, Italy. *Hydrogeology Journal* 17: 1233–1246. https://doi.org/10.1007/s10040-009-0435-1.

Furukawa, Y. and Hernández, C. (2015). Multi-view stereo: A Tutorial. *Foundations and Trends® in Computer Graphics and Vision* 9: 1–148. https://doi.org/10.1561/0600000052.

Furukawa, Y. and Ponce, J. (2010). Accurate, dense, and robust multiview stereopsis. *IEEE Transactions on Pattern Analysis and Machine Intelligence* 32: 1362–1376. https://doi.org/10.1109/TPAMI.2009.161.

Ge, Y., Tang, H., Xia, D. et al. (2018). Automated measurements of discontinuity geometric properties from a 3D-point cloud based on a modified region growing algorithm. *Engineering Geology* 242: 44–54. https://doi.org/10.1016/j.enggeo.2018.05.007.

Hodgetts, D. (2013). Laser scanning and digital outcrop geology in the petroleum industry: A review. *Marine and Petroleum Geology* 46: 335–354. https://doi.org/10.1016/j.marpetgeo.2013.02.014.

Jaboyedoff, M., Metzger, R., Oppikofer, T. et al. (2007). New insight techniques to analyze rock-slope relief using DEM and 3D-imaging cloud points. In: *Rock Mechanics: Meeting Society's Challenges and Demands*, 61–68. Taylor and Francis https://doi.org/10.1201/NOE0415444019-c8.

Jacquemyn, C., Huysmans, M., Hunt, D. et al. (2015). Multi-scale three-dimensional distribution of fracture- and igneous intrusion-controlled hydrothermal dolomite from digital outcrop model, Latemar platform, Dolomites, northern Italy. *AAPG Bulletin* 99: 957–984. https://doi.org/10.1306/10231414089.

James, M.R. and Robson, S. (2014). Mitigating systematic error in topographic models derived from UAV and ground-based image networks. *Earth Surface Processes and Landforms* 39: 1413–1420. https://doi.org/10.1002/esp.3609.

Jaud, M., Passot, S., Allemand, P. et al. (2018). Suggestions to limit geometric distortions in the reconstruction of linear coastal landforms by SfM Photogrammetry with PhotoScan® and MicMac® for UAV surveys with restricted GCPs pattern. *Drones* 3: 2. https://doi.org/10.3390/drones3010002.

Jones, R.R., Mccaffrey, K.J.W., Clegg, P. et al. (2009). Integration of regional to outcrop digital data: 3D visualisation of multi-scale geological models. *Computers and Geosciences* 35: 4–18. https://doi.org/10.1016/j.cageo.2007.09.007.

Jones, R.R., Pringle, J.K., Mccaffrey, K.J.W. et al. (2011). *Extending Digital Outcrop Geology into the Subsurface. Outcrops Revitalized. SEPM (Society for Sedimentary Geology)* 31–50. https://doi.org/10.2110/sepmcsp.10.031.

Kazhdan, M. and Hoppe, H. (2013). Screened poisson surface reconstruction. *ACM Transactions on Graphics* 32: 1–13. https://doi.org/10.1145/2487228.2487237.

Kehl, C., Mullins, J.R., Buckley, S.J. et al. (2021). Interpretation and mapping of geological features using mobile devices in outcrop geology—A case study of the Saltwick Formation, North Yorkshire, UK. In: *3D Digital Geological Models: From Terrestrial Outcrops to Planetary Surfaces*. Wiley, this volume.

Kromer, R., Walton, G., Gray, B. et al. (2019). Development and optimization of an automated fixed-location time lapse photogrammetric rock slope monitoring system. *Remote Sensing* 11: 1890. https://doi.org/10.3390/rs11161890.

Lato, M.J. and Vöge, M. (2012). Automated mapping of rock discontinuities in 3D LiDAR and photogrammetry models. *International Journal of Rock Mechanics and Mining Sciences* 54: 150–158. https://doi.org/10.1016/j.ijrmms.2012.06.003.

Lowe, D.G. (2004). Distinctive image features from scale-invariant keypoints. *International Journal of Computer Vision* 60: 91–110. https://doi.org/10.1023/B:VISI.00000.

Madjid, M.Y.A., Vandeginste, V., Hampson, G. et al. (2018). Drones in carbonate geology: Opportunities and challenges, and application in diagenetic dolomite geobody mapping. *Marine and Petroleum Geology* 91: 723–734. https://doi.org/10.1016/j.marpetgeo.2018.02.002.

Martinelli, M., Franceschi, M., Massironi, M. et al. (2017). An extensional syn-sedimentary structure in the Early Jurassic Trento Platform (Southern Alps, Italy) as analogue of potential hydrocarbon reservoirs developing in rifting-affected carbonate platforms. *Marine and Petroleum Geology* 79: 360–371. https://doi.org/10.1016/j.marpetgeo.2016.11.002.

Martinelli, M., Bistacchi, A., Mittempergher, S. et al. (2020). Damage zone characterization combining scan-line and scan-area analysis on a km-scale Digital Outcrop Model: The Qala Fault (Gozo). *Journal of Structural Geology* 140, 104144: https://doi.org/10.1016/j.jsg.2020.104144.

Massironi, M., Bertoldi, L., Calafa, P. et al. (2008). Interpretation and processing of ASTER data for geological mapping and granitoids detection in the Saghro massif (eastern Anti-Atlas, Morocco). *Geosphere* 4: 736. https://doi.org/10.1130/GES00161.1.

Massironi, M., Zampieri, D., Superchi, L. et al. (2013). Geological structures of the vajont landslide. *Italian Journal of Engineering Geology and Environment* 2013: 573–582. https://doi.org/10.4408/IJEGE.2013-06.B-55.

Mather, P.M. and Koch, M. (2011). *Computer Processing of Remotely-Sensed Images: An Introduction*, 4the. Wiley-Blackwell.

McCaffrey, K.J.W., Jones, R.R., Holdsworth, R.R.E. et al. (2005). Unlocking the spatial dimension: Digital technologies and the future of geoscience fieldwork. *Journal of the Geological Society* 162: 927–938. https://doi.org/10.1144/0016-764905-017.

Mikhail, E.M., Bethel, J.S., and McGlone, J.C. (2001). *Introduction to Modern Photogrammetry*. John Wiley & Sons.

Mittempergher, S. and Bistacchi, A. (2021). Image analysis algorithms for semi-automatic lineament detection on Digital Outcrop Models. In: *3D Digital Geological Models: From Terrestrial Outcrops to Planetary Surfaces*. Wiley, this volume.

Penasa, L., Franceschi, M., Preto, N. et al. (2014). Integration of intensity textures and local geometry descriptors from Terrestrial Laser Scanning to map chert in outcrops. *ISPRS Journal of Photogrammetry and Remote Sensing* 93: 88–97. https://doi.org/10.1016/j.isprsjprs.2014.04.003.

Qu, D., Frykman, P., Stemmerik, L. et al. (2021). Upscaling of outcrop information for improved reservoir modelling—exemplified by a case study on chalk. *Petroleum Geoscience* 27: petgeo2020-126. https://doi.org/10.1144/petgeo2020-126.

Riquelme, A.J., Abellán, A., Tomás, R., and Jaboyedoff, M. (2014). A new approach for semi-automatic rock mass joints recognition from 3D point clouds. *Computers and Geosciences* 68: 38–52. https://doi.org/10.1016/j.cageo.2014.03.014.

Riquelme, A., Cano, M., Tomás, R., and Abellán, A. (2017). Identification of rock slope discontinuity sets from laser Ssanner and photogrammetric point clouds: A comparative analysis. *Procedia Engineering* 191: 838–845. https://doi.org/10.1016/j.proeng.2017.05.251.

Siddiqui, N.A., Ramkumar, M., Rahman, A.H.A. et al. (2019). High resolution faces architecture and digital outcrop modeling of the Sandakan formation sandstone reservoir, Borneo: Implications for reservoir characterization and flow simulation. *Geoscience Frontiers* 10: 957–971. https://doi.org/10.1016/j.gsf.2018.04.008.

Silva, O.L., Bezerra, F.H.R., Maia, R.P., and Cazarin, C.L. (2017). Karst landforms revealed at various scales using LiDAR and UAV in semi-arid Brazil: Consideration on karstification processes and methodological constraints. *Geomorphology* 295: 611–630. https://doi.org/10.1016/j.geomorph.2017.07.025.

Sima, A.A., Bonaventura, X., Feixas, M. et al. (2013). Computer-aided image geometry analysis and subset selection for optimizing texture quality in photorealistic models. *Computers and Geosciences* 52: 281–291. https://doi.org/10.1016/j.cageo.2012.11.004.

Smith, S.A.F., Bistacchi, A., Mitchell, T.M. et al. (2013). The structure of an exhumed intraplate seismogenic fault in crystalline basement. *Tectonophysics* 599: 29–44. https://doi.org/10.1016/j.tecto.2013.03.031.

Snavely, N., Seitz, S., and Szeliski, R. (2006). PhotoTourism: Exploring photo collections in 3D. In: *SIGGRAPH Conference Proceedings*, 835–846. https://doi.org/10.1145/1141911.1141964.

Sturzenegger, M., Stead, D., and Elmo, D. (2011). Terrestrial remote sensing-based estimation of mean trace length, trace intensity and block size/shape. *Engineering Geology* 119: 96–111. https://doi.org/10.1016/j.enggeo.2011.02.005.

Tavani, S., Corradetti, A., and Billi, A. (2016). High precision analysis of an embryonic extensional fault-related fold using 3D orthorectified virtual outcrops: The viewpoint importance in structural geology. *Journal of Structural Geology* 86: 200–210. https://doi.org/10.1016/j.jsg.2016.03.009.

Tavani, S., Granado, P., Corradetti, A. et al. (2014). Building a virtual outcrop, extracting geological information from it, and sharing the results in Google Earth via OpenPlot and Photoscan: An example from the Khaviz Anticline (Iran). *Computers and Geosciences* 63: 44–53. https://doi.org/10.1016/j.cageo.2013.10.013.

Thiele, S.T., Grose, L., Samsu, A. et al. (2017). Rapid, semi-automatic fracture and contact mapping for point clouds, images and geophysical data. *Solid Earth* 8: 1241–1253. https://doi.org/10.5194/se-8-1241-2017.

Tomassetti, L., Petracchini, L., Brandano, M. et al. (2018). Modeling lateral facies heterogeneity of an upper Oligocene carbonate ramp (Salento, southern Italy). *Marine and Petroleum Geology* 96: 254–270. https://doi.org/10.1016/j.marpetgeo.2018.06.004.

Triantafyllou, A., Watlet, A., Le Mouélic, S. et al. (2019). 3-D digital outcrop model for analysis of brittle deformation and lithological mapping (Lorette cave, Belgium). *Journal of Structural Geology* 120: 55–66. https://doi.org/10.1016/j.jsg.2019.01.001.

Tsai, R.Y. (1987). A versatile camera calibration technique for high-accuracy 3D machine vision metrology using off-the-shelf TV cameras and lenses. *IEEE Journal on Robotics and Automation* 3: 323–344. https://doi.org/10.1109/JRA.1987.1087109.

van der Meer, F.D., van der Werff, H.M.A., van Ruitenbeek, F.J.A. et al. (2012). Multi- and hyperspectral geologic remote sensing: A review. *International Journal of Applied Earth Observation and Geoinformation* 14: 112–128. https://doi.org/10.1016/j.jag.2011.08.002.

Vasuki, Y., Holden, E.-J.J., Kovesi, P., and Micklethwaite, S. (2014). Semi-automatic mapping of geological structures using UAV-based photogrammetric data: An image analysis approach. *Computers and Geosciences* 69: 22–32. https://doi.org/10.1016/j.cageo.2014.04.012.

Vasuki, Y., Holden, E.-J., Kovesi, P., and Micklethwaite, S. (2017). An interactive image segmentation method for lithological boundary detection: A rapid mapping tool for geologists. *Computers and Geosciences* 100: 27–40. https://doi.org/10.1016/j.cageo.2016.12.001.

Vollgger, S.A. and Cruden, A.R. (2016). Mapping folds and fractures in basement and cover rocks using UAV photogrammetry, Cape Liptrap and Cape Paterson, Victoria, Australia. *Journal of Structural Geology* 85: 168–187. https://doi.org/10.1016/j.jsg.2016.02.012.

White, P.D. and Jones, R.R. (2008). A cost-efficient solution to true color terrestrial laser scanning. *Geosphere* 4: 564. https://doi.org/10.1130/GES00155.1.

Wilkinson, M.W., Jones, R.R., Woods, C.E. et al. (2016). A comparison of terrestrial laser scanning and structure-from-motion photogrammetry as methods for digital outcrop acquisition. *Geosphere* 12: 1865–1880. https://doi.org/10.1130/GES01342.1.

Wu, C. (2013). Towards linear-time incremental structure from motion. In: *3DV. IEEE*, 127–134. https://doi.org/10.1109/3DV.2013.25.

Wu, C. (2014). Critical configurations for radial distortion self-calibration. In: *2014 IEEE Conference on Computer Vision and Pattern Recognition*, 1–16. https://doi.org/10.1109/CVPR.2014.11.

Wu, C., Agarwal, S., Curless, B., and Seitz, S.M. (2011). *Multicore bundle adjustment. CVPR* 2011: 3057–3064. https://doi.org/10.1109/CVPR.2011.5995552.

Zhang, P., Du, K., Tannant, D.D. et al. (2018). Automated method for extracting and analysing the rock discontinuities from point clouds based on digital surface model of rock mass. *Engineering Geology* 239: 109–118. https://doi.org/10.1016/j.enggeo.2018.03.020.

3

The PRoViDE Framework: Accurate 3D Geological Models for Virtual Exploration of the Martian Surface from Rover and Orbital Imagery

Christoph Traxler[1], Thomas Ortner[1], Gerd Hesina[1], Robert Barnes[2], Sanjeev Gupta[2], Gerhard Paar[3], Jan-Peter Muller[4], Yu Tao[4], and Konrad Willner[5]

[1]*Zentrum für Virtual Reality und Visualisierung (VRVis) Forschungs-GmbH, Vienna, Austria*
[2]*Imperial College London, United Kingdom*
[3]*Joanneum Research, Graz, Austria*
[4]*University College London, London, United Kingdom*
[5]*German Aerospace Center (DLR), Berlin, Germany*

Abstract

In this chapter, we describe the PRoViDE (Planetary Robotics Vision Data Exploitation) framework that supports an entire workflow to generate 3D geological models of planetary surfaces from separate or fused rover and orbiter imagery, which are used for an efficient and reliable geologic analysis. PRoViDE provides a comprehensive solution for planetary geological studies including tools and data products from multiresolution co-registered orbital imagery to multiresolution rover derived imagery. A special processing called Super-Resolution Restoration (SRR) is employed to increase the spatial resolution of some orbital imagery. The main components of PRoViDE consist of a geo-database (PRoDB) to gather and manage huge volumes of data, a prototype of a planetary webGIS (PRoGIS) providing interactive map exploration, a processing pipeline producing multiresolution 3D reconstructions of planetary surfaces (PRoViP) and an interactive 3D viewer for virtual exploration and visual analysis (PRo3D). This viewer relies on advanced real-time visualization and interaction methods tailored to geospatial data. Various measurement tools are provided for a quantitative analysis of geological features. In summary, the PRoViDE framework enables an efficient and accurate investigation chain for detailed geologic interpretation of a planetary region, which also serves as decision support for mission planning.

3.1 Introduction

For planetary science, 3D geological models are an essential asset for quantitative analyses. These models come in different data formats such as Digital Terrain Models (DTMs), Digital Surface Models (DSMs), Digital Elevation Models (DEMs), or Digital Outcrop Models (DOMs). The latter are high-resolution 3D representations of rock outcrops and are particularly interesting as they reveal the geological past of a planetary environment. The common practice to solely rely on 2D images for geological analysis has proven to be cumbersome and error prone. The third dimension is critical to fully understand surface structures and past geological processes. While on Earth 3D reconstructions mostly supplement traditional field work and might bring additional insights, they are the only opportunity for planetary scientists to investigate surfaces in their full geospatial context—at least so long as no geologists are being sent to other planets such as Mars.

Here, we present a software framework that supports the entire workflow of generating 3D geological models of planetary surfaces from rover and orbiter imagery and exploit them for an efficient and reliable analysis. Developments took place mainly in the EU-funded "Planetary Robotics Vision Data Exploitation" (PRoViDE) project (Paar et al., 2013, 2015; Traxler et al., 2016) and were extended in several follow-up projects. The framework includes a geo-database (PRoDB) to gather huge volumes of imagery and transform them into a common geospatial context (Willner and Tasdelen, 2015), including fusion between rover and orbital imagery (Tao et al., 2016). A computer vision processing chain (PRoViP) produces multiresolution and fused 3D reconstructions based on this imagery. A webGIS (PRoGIS, Giordano et al., 2015) provides a map view where scientists can browse through various layers showing rover paths, capturing locations, and available 3D surface reconstructions (DOMs) among other data. Although it was successfully demonstrated to

3D Digital Geological Models: From Terrestrial Outcrops to Planetary Surfaces, First Edition.
Edited by Andrea Bistacchi, Matteo Massironi, and Sophie Viseur.

serve for a variety of applications (e.g. Paar et al., 2015; Traxler et al., 2016; Tao et al., 2016; Tao and Muller, 2016), the development of PRoGIS was frozen at the prototype stage in 2015 due to a lack of funding. Finally, scientists can explore 3D products in an interactive 3D viewer (PRo3D, 2021a, 2021b) that provides them with smooth navigation and various annotation tools to digitize features directly on the 3D surface, resembling traditional fieldwork. In recent projects GIS functionality was integrated into PRo3D compensating for the deprecated PRoGIS (Ortner et al., 2019). In future all the required GIS functionality will be available within PRo3D, which evolves into a 3D GIS.

In the following sections, we will describe the components of the PRoViDE framework in more detail and show how they support the whole workflow from image acquisition to an extensive geologic interpretation of 3D reconstructions. The framework was evaluated on two use cases, which are rock outcrops around the rim of the Victoria Crater visited by NASA's Mars Exploration Rover (MER) "Opportunity" and a region known as Yellowknife Bay in Gale Crater, visited by the NASA's Mars Science Laboratory (MSL) rover "Curiosity." These use cases also demonstrate how interactive measurement tools of PRo3D are used for visual analysis.

3.2 Components and Methods

3.2.1 Overview

Figure 3.1 shows the components of the PRoViDE framework and the data flow between them. Next to PRoViP, which processes mainly rover image data, SRR is an additional processing chain providing context images of enhanced resolutions from orbital data. It creates super-resolution restoration products of up to five times the original resolution from stacks of five or more images taken at slightly different view angles irrespective of the time difference between the images. These high-resolution products can then be used by PRoViP for 3D reconstructions or as a raster layer in PRoGIS. PRoDB harvests links to images from PDS sources providing imagery taken by rover cameras and evaluates the meta-data of these images to define certain products to be processed by PRoViP. PRoViP queries PRoDB for products to be processed and saves processed images and products back into the PRoDB database, from where the 2D and 3D products can be accessed by PRoGIS and PRo3D respectively. More details on the interconnections are given by Paar et al. (2016b). Moreover, PRoGIS can be used to select 3D image products for detailed geological assessment in PRo3D.

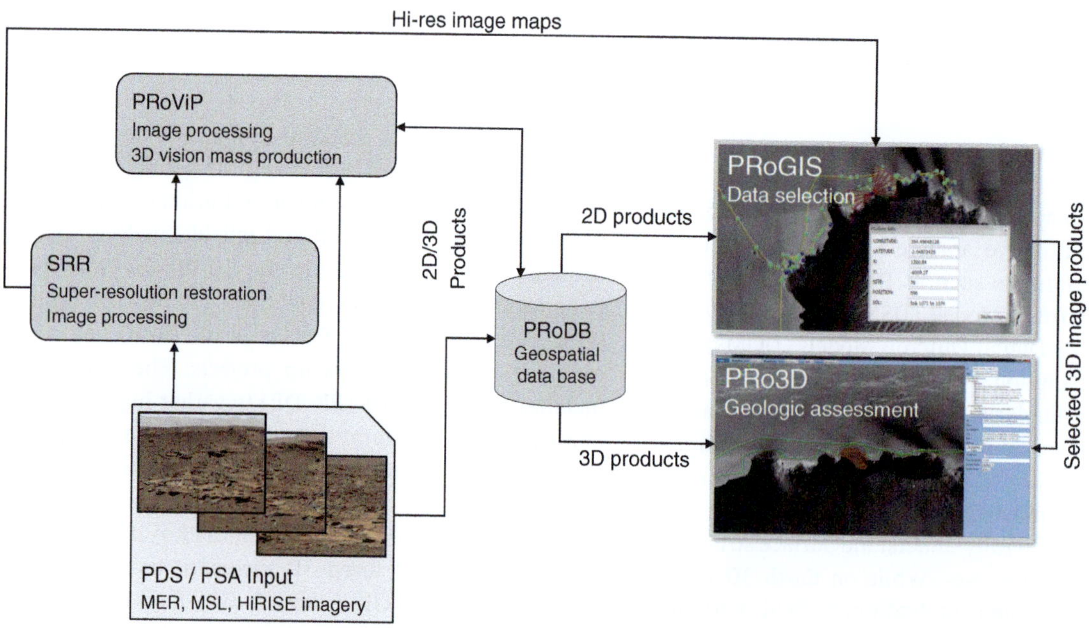

Figure 3.1 Components of the PRoViDE framework and the data flow. PRoGIS is deprecated. In future, PRo3D will provide functionality of PRoGIS.

3.2.2 PRoDB–A Geospatial Data Base for Planetary Data

To keep track of the input data (including metadata) for processing that was/is to be harvested, as well as to define products to be processed and keep track of the processing history, a data base had to be developed. This data base was initially designed and described as a "data catalogue" and was extended into the relational data base termed PRoDB (Planetary Robotics Data Base) over the course of the PRoViDE project. The data catalogue/PRoDB was designed and technically implemented based on PostgreSQL and holds information of relevant source images stemming from robotic missions such as MER-A, MER-B, MSL, Pathfinder, and Phoenix, to be used for 3D reconstruction and panorama generation. PRoDB is a central component that gathers links to image data from either the ESA Planetary Science Archive (PSA) or NASA Planetary Data System (PDS), along with their metadata, and PRoViP processed products, including relevant information such as images contributing to the product, rover location in the global reference frame, etc. PRoDB provides an overview of available data to be considered for processing within the project, easy access to the data itself, and a link between several other components of the PRoViDE overall system. It provides the Batch processing component with relevant input information. Furthermore, PRoDB forwards information about finalized products to a separate data base that can be queried by the viewing components, PRoGIS (Planetary Robotics GIS, Section 2.5) and PRo3D (Planetary Robotics 3D Viewer, Section 2.6).

A list of default products was defined and integrated into PRoDB for automated processing. Interfaces to PRoViP and PRoGIS were defined and verified to run the full processing chain including Data Catalogue readout, scheduling, processing by PRoViP, and insertion of products back into the Data Catalogue. Based on this design and these interfaces, the batch-based processing of all defined products was executed within the project. The complexity of the structure of the database can be seen in the structure chart in Figure 3.2.

Data from multiple PDS releases for MSL and the MER missions was incrementally harvested and inserted into the respective tables of the PRoDB including data, metadata, and product definitions. Figure 3.3 shows an excerpt of a database query of relevant surface images.

PRoDB incorporates automatic link procedures between the harvested PDS datasets (calibrated instrument data) and possible data products to be gained therefrom, making use of specific rules (e.g. stereo pairs, sets of stereo pairs to form unique panoramas, and overlaying sequences of different color filters belonging to the same sequence). Using the candidates established through queries in PRoDB, in the PRoViP built-in scheduler system, image matching is subsequently performed on suitably identified stereo-pairs, followed by 3D reconstruction and grid point interpolation into DTMs in various geometries, generation of an intermediate dataset ("GPC": Generic Point Cloud), combination of the DTMs into unique consistent mosaicked products,

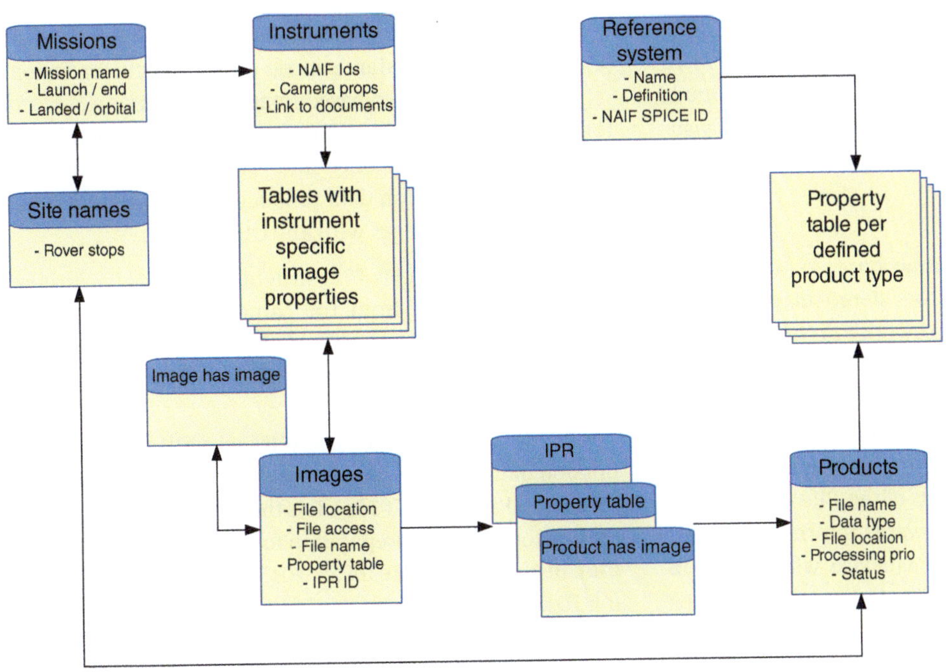

Figure 3.2 Schematic view of the relational database PRoDB.

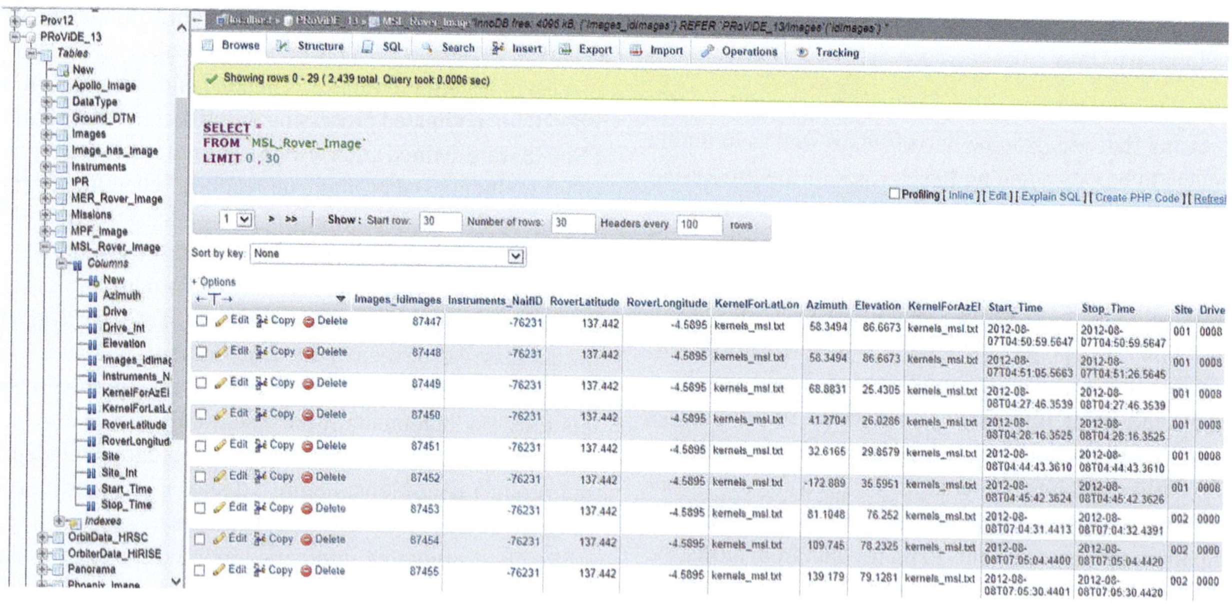

Figure 3.3 PRoDB data query example.

and finally the export into products such as Ordered Point Clouds (see Section 3.2.3.2) to be exploited by scientists and operations personnel.

3.2.3 PRoViP–A Computer Vision Processing Chain to Create 3D Reconstructions

In this section, we first describe the PRoViP processing chain listing which kind of products PRoViP can produce and how the overall processing works. We next describe the proprietary spatial data structure generated by the PRoViP pipeline, called Ordered Point Clouds (OPC) for subsequent use within the interactive 3D visualization tool, PRo3D, which is described later in this chapter.

3.2.3.1 Image-Based 3D Reconstruction

3D Scene reconstruction from Rover stereo imagery is realized through the processing framework *PRoViP* (Planetary Robotics Vision Processing) (see Paar et al., 2015). PRoDB provides an overview of available data to be considered for processing for a specific investigation site (or even a whole mission), easy access to the data itself, and a link between several other components of the PRoViDE overall system. Furthermore, PRoDB forwards information about finalized products to a separate database within PRoDB that can be queried by the visualization components, i.e., PRoGIS (Planetary Robotics GIS) and PRo3D (Planetary Robotics 3D Viewer), as described in Sections 3.2.5 and 3.2.6 respectively.

Examples of PRoViP processing products and capabilities are:

(a) DTMs in spherical, cylindrical and Cartesian coordinate space.
(b) Ortho images.
(c) 3D TIN (Triangulated Irregular Network) textured meshes.
(d) Derived thematic maps of the surrounding area describing reconstruction accuracy, occlusions, solar illumination, slopes, roughness, hazards, etc.
(e) Geographic and Mars-centric data products for fusion of rover and orbiter-based image products.
(f) For the ExoMars PanCam case: fusion between WAC (Wide-Angle Camera) and HRC (High-Resolution Camera) 3D vision data products (e.g. overlay of WAC DTM/Ortho images with an HRC texture).

All 3D data products (3D structures, overlaid with texture either directly from the camera image content, or derived information, see point d above) can be represented as Ordered Point Clouds (OPC) (see Section 3.2.3.2) for further geologic analysis.

PRoViP has been extensively tested with various Mars mission datasets from stereo imaging sensors such as MSL Mastcam (Paar et al., 2016a). It is the primary 3D vision processing tool chain in preparation for the ExoMars 2022 PanCam instrument (Coates et al., 2015), and is currently in use for 3D vision product generation of the Mastcam-Z instrument of the Mars 2020 Rover mission (Bell et al., 2021).

PRoViP processing is based on photogrammetric methods and produces 3D point clouds that are the same as the ones acquired in terrestrial laser scanning (e.g. see Chapter 2 by Bistacchi et al., 2021). The points can be understood as surface samples while the density of these samples depends on the distance to the capturing instrument. When generated from rover imagery, the points are arranged in an irregular grid, which is non-linearly distorted and typically has a fan-shaped appearance. Nevertheless, the point cloud is strictly ordered from the data-capturing characteristics of the initial images (e.g. the pixel grid on a camera sensor) or mosaics therefrom, i.e. the neighbors of each point are unique. This allows them to be stored in OPCs (see Section 3.2.3.2). Exported OPCs contain further information, such as transformation parameter sets for efficient display.

3.2.3.2 Ordered Point Clouds (OPC)

Rendering multiple surface reconstructions simultaneously makes it necessary to employ a *Levels of Detail* (LOD) strategy, which only displays high-resolution data in the vicinity of the viewer, while it gradually omits detail with increasing distance. For our LOD approach, which will be discussed in more detail in Section 3.2.6.1, we developed the OPC data structure, short for Ordered Point Clouds, which provides PRo3D useful meta data and different levels of detail of surface reconstructions.

Output of the image-based 3D reconstruction (Section 3.2.3.1) is a high-resolution DTM and an ortho image, where each quad of the DTM exactly matches a pixel of the image, which we refer to as level 0. First, level 0 is cut into patches of certain dimensions, e.g. 128 × 128. Four of such patches are then grouped together and simplified to a single patch of the same dimensions belonging to the next level. This bottom-up simplification is performed until we arrive at level n at the root patch. Each patch contains geometry and image data and spatially encloses its subpatches, illustrated by Figure 3.4. Since the geometric data originates from image data, the reconstructed vertices are ordered, which makes it trivial for PRo3D to triangulate this data at runtime. For reconstructions based on orbiter images, we get DTM-like surfaces lying on a nearly horizontal plane, similar to reconstructions from terrestrial satellite images. However, the mesh resulting from rover imagery can be seen as a triangulated depth map, which results in artefacts (elongated triangles) where there are sudden jumps in the depth values, e.g. behind the edge of a rock. Furthermore, the detail near the camera is much higher, because more pixels cover an area close by than far away. The resulting artefacts in the outcrop reconstructions are mitigated by clipping large depth values, i.e. spatially large areas only covered by a few pixels, and by discarding elongated or degenerate triangles, i.e. where one edge is considerably smaller than the other two. This

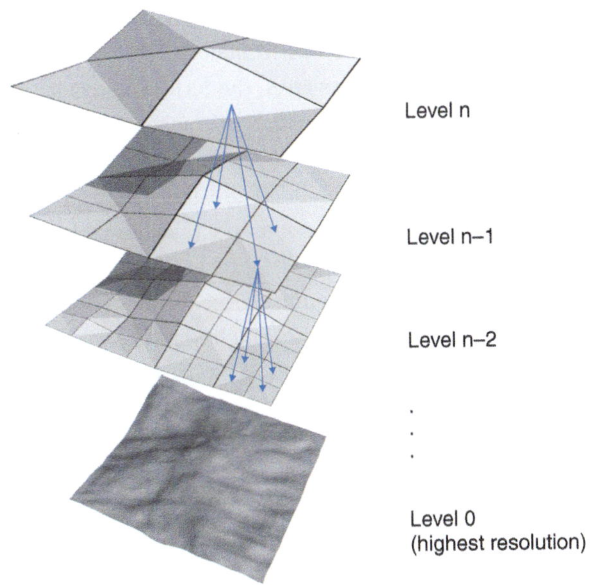

Figure 3.4 Hierarchical organization of OPCs. Patches are further subdivided in each layer to provide more geometric detail.

results in holes in the mesh, which are visually filled by the underlying orbiter DTM.

In essence, the construction and hierarchical organization of our data resembles a quadtree with center splits for the geometry and a mipmap pyramid for the image data. To provide applications with the data at maximum accuracy, the vertices of each patch are saved in double-precision in a local coordinate system with a zero origin, while the transformation to the common global coordinate system is attached in the metadata file associated with each patch. All processing steps discussed happen at the preprocessing stage to keep startup times of the target applications, as for instance PRo3D, at a minimum.

3.2.4 Super-Resolution Restoration (SRR) Processing

In PRoViDE, we developed a novel super-resolution algorithm called Gotcha Partial Differential Equation-Based Total Variation (GPT), which was specifically developed to address the task of enhancing the image effective resolution of orbital optical data using non-redundant information contained from the overlapping multiangle views (Tao and Muller, 2016). The GPT-SRR system (a separate processing chain to PRoViP) is used to generate up to 5 cm super resolution orthorectified base maps from repeat-pass HiRISE input images (25 cm). This enables the retrieval and visualization of very detailed surface features of Mars. HiRISE SRR bridges the resolution gap between rover-derived products and the original HiRISE images in which hazards are not viewable or are insufficiently clear. From the experiments performed in PRoViDE over the MER and MSL rover traverses, the SRR products have demonstrated restoration of surface features including the imaging of individual rocks (diameter ≥ 25 cm) by comparison with both the original HiRISE images and rover Navcam orthorectified image mosaics (Tao and Muller, 2016b).

The first SRR process (Figures 3.5–3.7) was performed using the GPT SRR algorithm from eight repeat-pass 25 cm HiRISE images covering the MER-A Spirit rover traverse in Gusev Crater to resolve a 5 cm SRR image of the area surrounding the MER-A rover traverse. An example of the entire area is shown in Figure 3.5 before and after restoration, a detailed close-up is shown in Figure 3.6 and their use within PRo3D is shown in Figure 3.7 compared against the original 25 cm image.

Owing to the computational demands of the SRR implementation, only images sized up to 2,000 by 2,000 pixels can be currently processed in a sensible time frame on a multicore Linux blade (24 cores, 96GB RAM). SRR has been applied to images centered around the rover track

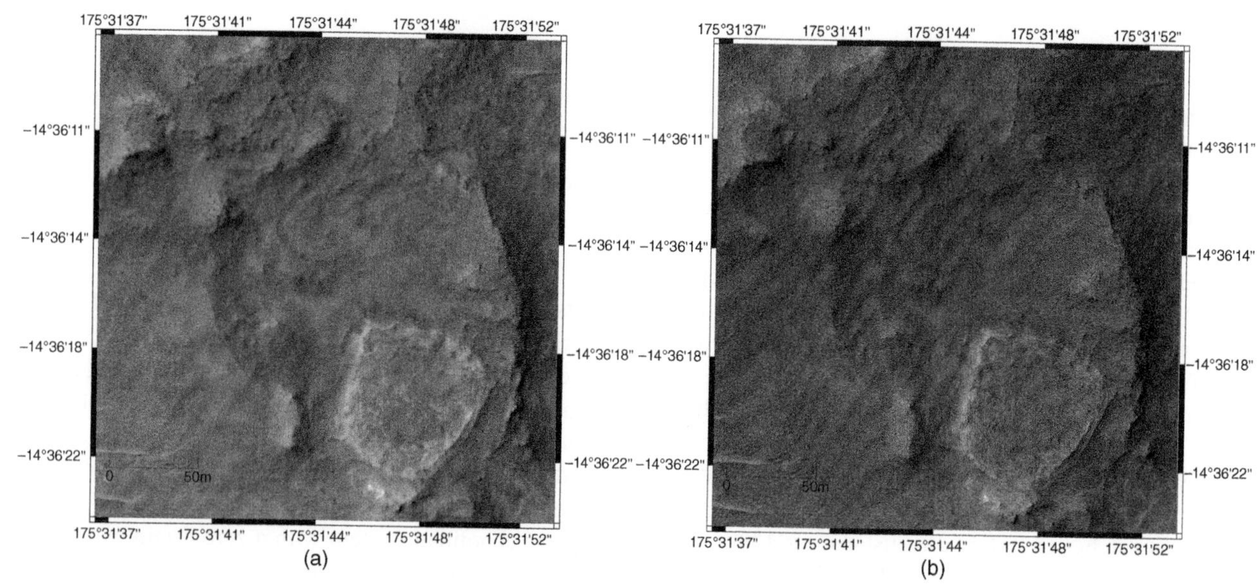

Figure 3.5 Examples of the 25 cm HiRISE image (a) and 5 cm SRR image (b) produced from eight multiangle repeat-pass HiRISE images for the MER-A Homeplate area.

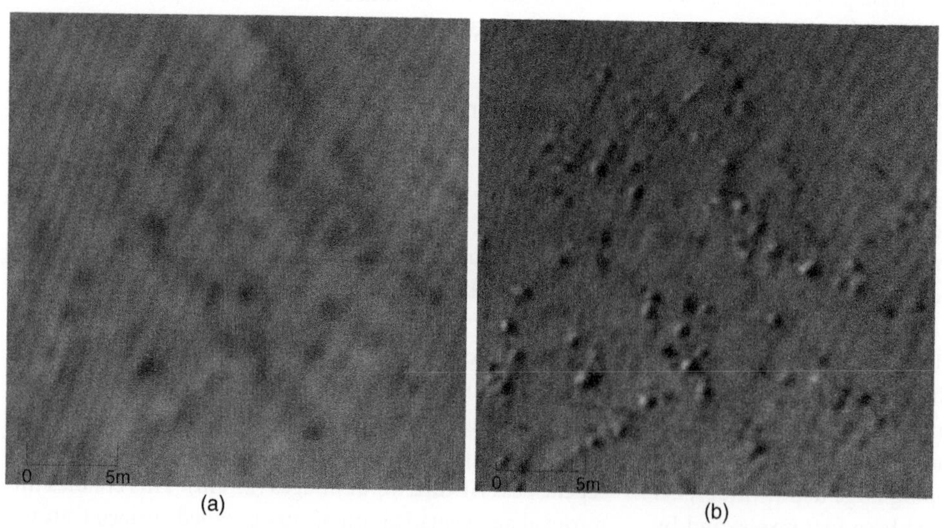

Figure 3.6 A zoom-in view of the MER-A Homeplate 25 cm HiRISE image (a) and 5 cm SRR image (b). These SRR results have revealed new information including the imaging of individual rocks (diameter > 25 cm) (Tao and Muller, 2016).

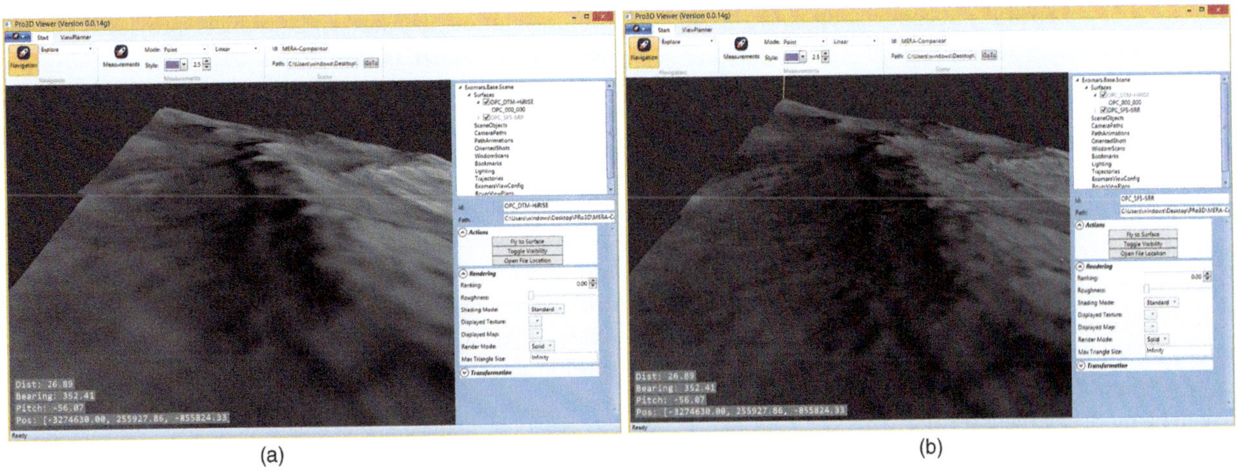

Figure 3.7 Example of the MER-A Homeplate 25 cm HiRISE image (a) and the 5 cm SRR image (b) shown in the PRo3D viewer.

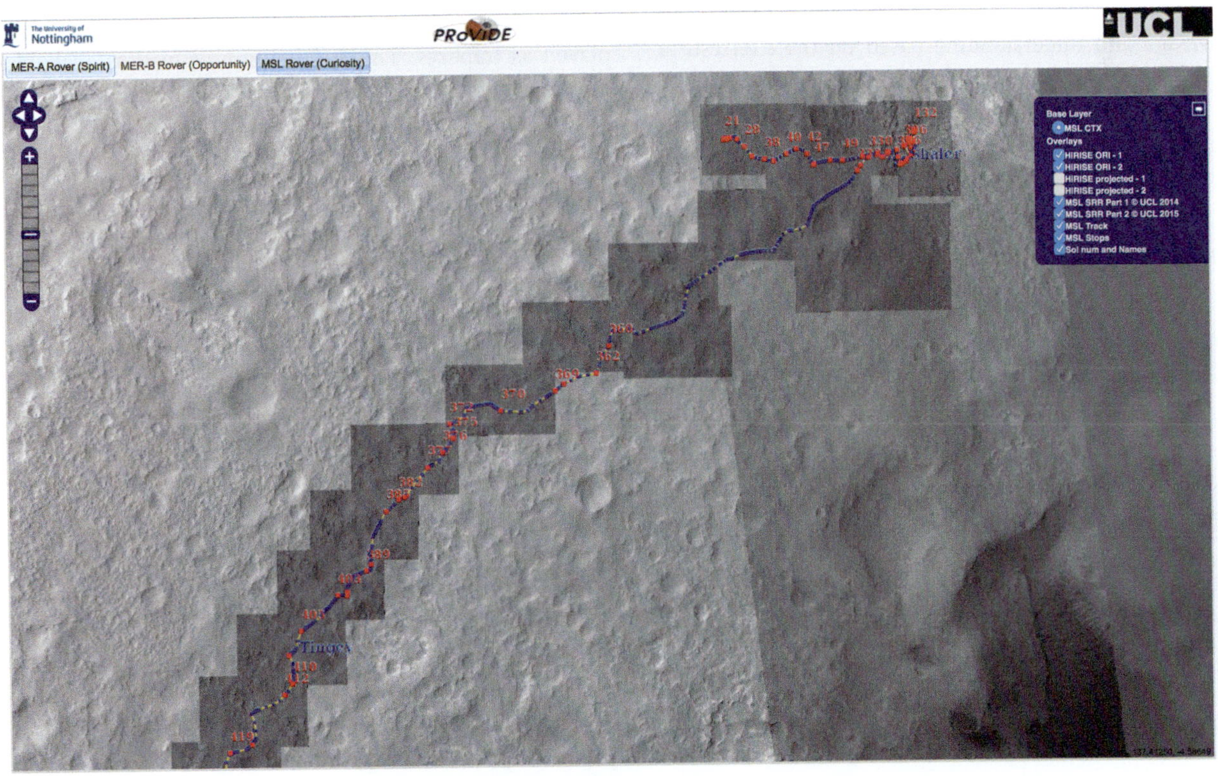

Figure 3.8 GPT SRR products for MSL were integrated into PRoGIS 1.0 to give access to SRR datasets to the planetary science community and visualization in a multiresolution co-registered context including CTX and HRSC/MOLA.

(see Figure 3.8), which allows a virtual exploration of the Martian surface using PRoGIS at 6.25 cm SRR of areas that are hundreds of meters away from the rover track at resolutions comparable to being 5 m or more away from the MSL rover. See Figure 3.8 for scenes processed so far for MSL and inserted in the PRoGIS for wider accessibility. It should be noted that two separate SRR systems, called OpTiGAN (Tao and Muller, 2021) and MARSGAN (Tao et al., 2021), using dense optical flow and deep learning techniques, for multi-image (without viewing angle differences) SRR and single-image SRR, respectively, have been developed at UCL and implemented on a GPU system. OpTiGAN and MARSGAN have achieved significant speedups and are able to process full HiRISE scenes of 11 × 22 km in a few hours (OpTiGAN) and in minutes (MARSGAN), with slightly lower output effective resolutions of

~8 cm (from 25 cm HiRISE) in comparison to the original GPT–SRR system.

In addition, GPT-SRR has also been tested with MER Pancam sequences that were specially acquired for SRR (Bell et al., 2006), as well as MSL Navcam repeat images within the PRoViDE project. Experiments were performed on a stack of MER-B Pancam images and MSL Navcam images. However, without multiangle information, only a factor of 1.5–1.75× enhancements can be achieved with GPT-SRR. Similar to MER Pancam and MSL Navcam, MSL ChemCam SRR was also produced and examined using a stack of continuous views. The details of small stones and sand can be more clearly observed.

A planetary geologist and geo-morphologist would generally be interested in exploiting the highest possible resolution orbital imaging dataset. HiRISE SRR will assist them in formulating and testing hypotheses about planetary surface processes, as they will be able to apply their knowledge and understanding based on their terrestrial fieldwork. Geologists can achieve more reliable classification and inference from super-resolution restored features such as rocks, sedimentary layers, and cliff crosscutting profiles. These SRR maps are used to enhance the visualization experience in PRo3D by fusing SRR and rover-based imagery to allow smooth exploration/transition from multiresolution datasets and products provided by the described PRoViDE framework.

3.2.5 PRoGIS–Geographic Information System for Planetary Scientists

PRoGIS (Planetary Robotic GIS) is a web-based prototype tool to present planetary surface data in a geospatial context of remotely sensed images. Users are presented with orthoimage layers and/or digital elevation models and other GIS data, overlaid with information of where to find specific surface data (e.g. rover images, derived 3D surface models, rover tracks, ground penetrating radar scans, etc.) and can call up other specific data presentation tools such as image viewers, 3D rendering engines or spectral analysis tools. Figure 3.9 shows a screenshot of PRoGIS, where a rover trajectory along the rim of Victoria crater is displayed. Image capturing locations are shown by yellow dots while the retrieved image footprints (fulcra) are displayed as trapezes. This information as well as available data products and reconstructions is queried from PRoDB.

PRoGIS exists only as a prototype demonstrated on a limited use case. Nevertheless, several design considerations were taken into account. Most importantly, it is compliant with the Open Geospatial Consortium (OGC) standards, i.e. WMS and WFS to ensure interoperability between

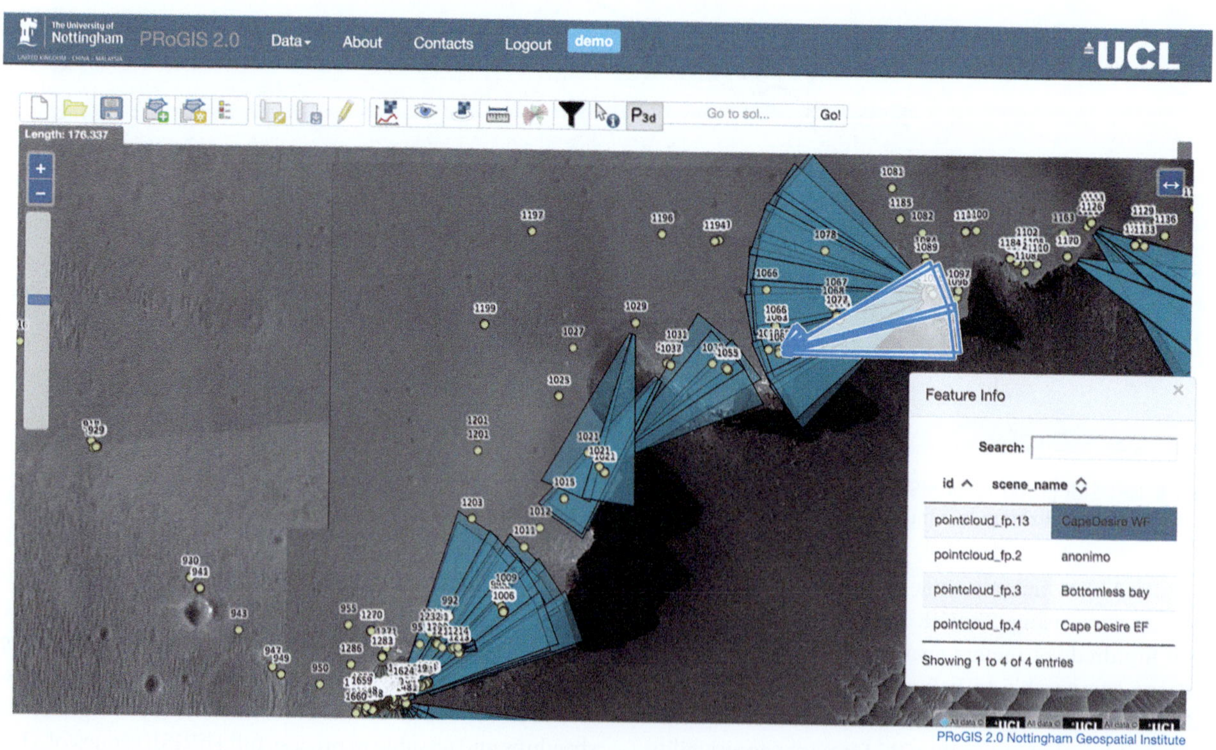

Figure 3.9 PRoGIS screenshot showing the path of the Opportunity rover along the rim of Victoria crater and along with locations where images were captured and their corresponding fulcra (dark cyan polygons).

different data sources. It supports the IAU planetary datum to achieve the integration of different resources and to make tools to measure distances and area with high accuracy.

Both the raw and derived data products can be searched by drawing a bounding box over the map layer or be searched by the data attributes. PRoGIS provides graphical annotation tools and organizes them according to groups, maps, and missions. Distances can be measured between points by drawing a line. In a similar way, the elevation profile can be calculated along a line and displayed as a graph in a separate window. Previews for point clouds and panoramas are also included as internal functions. 3D products (surface reconstructions) from PRoViP are shown as icons and can be opened up with PRo3D via its PRoGIS interface.

Figure 3.10 shows a schema of the PRoGIS 2.0 system and its building blocks. It consists of three tiers, namely, the web, application, and database tiers. Different services and interactions are provided by each of the three tiers. They are implemented using standard services for the ease of future integration with other resources. In particular, the web tier is the user front end and is accessible through the web browser. The implementation of the web tier is based on the OpenLayers and Zoomify. This tier uses a web server and a reverse proxy to provide maps and protect data from massive download requests.

The application tier is the core of the OGC data services, i.e. WMS and WFS for raster/grid data and feature data respectively. It is developed using the open source QGIS-MAPSERVER integrated with the latest Proj4 library to ensure International Astronomical Union (IAU) planetary datum support.

The database tier hosts all the data, including geographic data and image data, within standard RAID disks. A PostGIS-based geo-database and a file resource management system for raw rover images have been implemented. The file resource management system is used to store the PRoVIP processed point clouds and panorama products. It should be noted that the PostGIS database not only stores geographical data but also stores associated metadata that are used for database queries and indexing the raw and processed data.

Figure 3.11 shows the high-level architecture of the PRoGIS system, indicating how the different components are interconnected. Standard webGIS frameworks are used to ensure the best interoperability over different platforms and browsers throughout the PRoGIS implementation. These include:

1. A web client using the standard JavaScript library OpenLayers to represent maps.
2. WebGL to interact with advanced data structures that need efficient performance of big data, like very densely reconstructed point cloud products.
3. GeoCMS is the middleware that manages data access to map/missions for users/groups and handles annotations and administration tasks. It is implemented

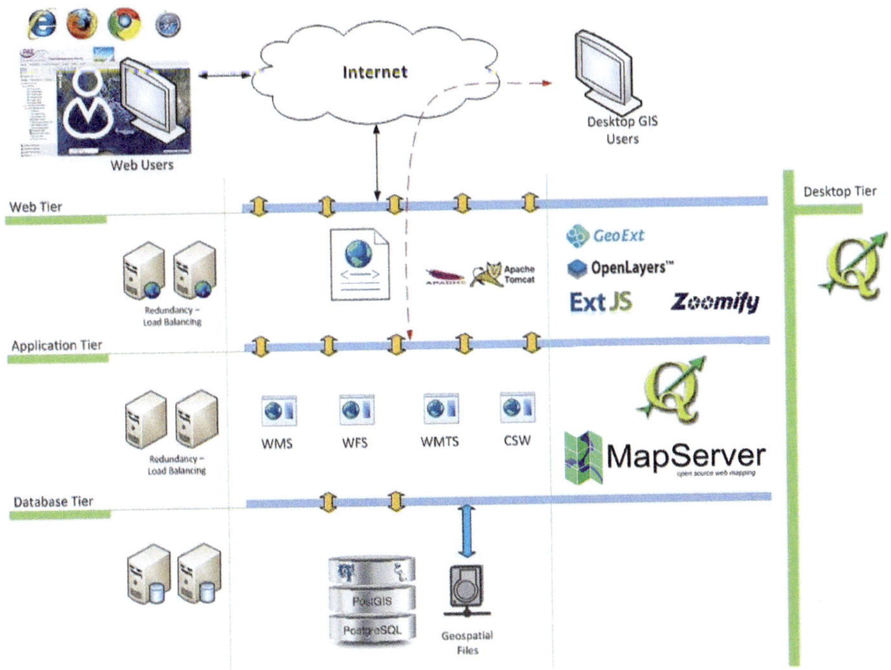

Figure 3.10 PRoGIS framework layout.

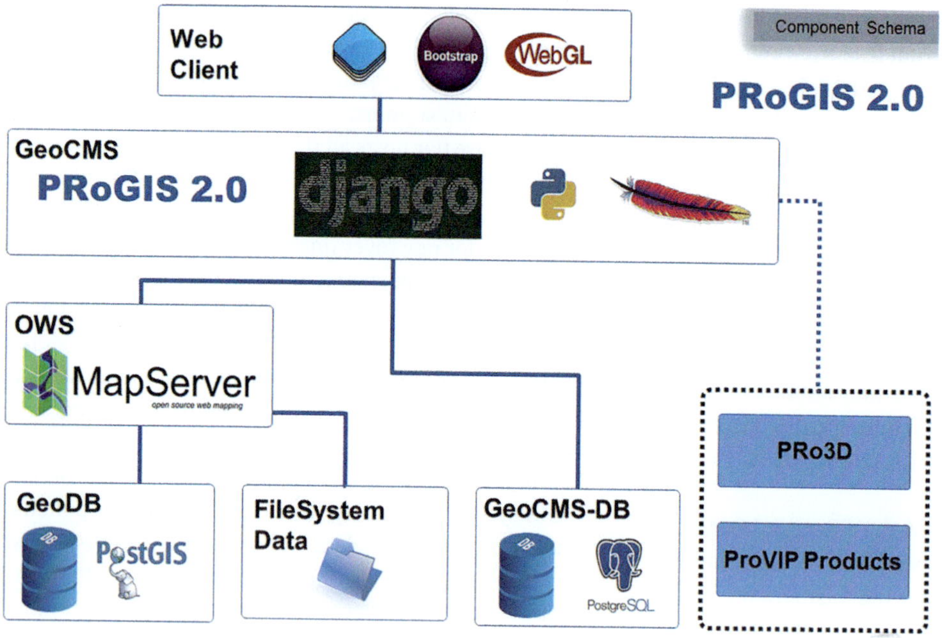

Figure 3.11 PRoGIS high-level system architecture.

with Django, a Python framework, and APACHE with a WSGI application object.

4. QGIS-MAPSERVER serves the data for the webGIS clients using the GeoDB data sharing system (as the data source) and a local (or remote) file system.
5. Data resources are managed using PostGIS as standard spatial database extension and PostgreSQL to store meta-data of both the raw and processed data.

3.2.6 PRo3D–Virtual Exploration and Visual Analysis of 3D Products

3.2.6.1 Virtual Exploration

PRo3D is a real-time rendering application, based on the rendering platform *Aardvark* (Aardvark, 2021a), and is optimized for displaying large geospatial scenes. It allows users to interactively survey whole landscapes of reconstructed planetary surfaces. The surface data is represented by textured triangle meshes, which are available at multiple resolutions through the OPC data structure (see Section 3.2.3.2). For the geological interpretation of an area on Mars, geologists typically need to investigate multiple outcrops derived from rover imagery (DOMs) within their larger spatial context and provided by the orbiter imagery and the underlying 3D information (DTMs). Such a scene of 3D surfaces typically results in a scene of around two-digit gigabytes in data size. Consequently, interactive exploration of such enormous scene products has the following challenges:

- The full scene potentially does not fit into the computer's main memory.
- The actual visible data does not fit into the graphics card's memory.
- Outcrop reconstructions and orbiter reconstructions overlap in 3D space.
- Due to inaccuracies in the reference frames outcrop reconstructions and orbiter reconstructions are typically not accurately aligned.
- Putting surface reconstructions all over Mars into one coordinate system produces very large coordinates, which result in numerical issues on the graphics card only capable of processing single precision floating point numbers efficiently.

To process data that does not fit into the main memory, typically out-of-core algorithms and data structures are used as described in general by Vitter (2001). The basic concept is to divide the data into chunks and only load data from the network or the hard disk into the main memory that is currently needed, while unnecessary data is discarded.

PRo3D roughly follows the concepts described by Varadhan and Manocha (2002), who applied out-of-core techniques to large geometric environments. So far, we have used the terms OPC and surface synonymously, which is not entirely correct. One surface can consist of multiple OPCs, which is typically true for spatially large orbiter reconstructions, whereas outcrops can usually be represented by a single OPC. Thus, our out-core-strategy operates on the OPC level, streaming currently visible OPCs and discarding OPCs outside the current view frustum. This visibility test is performed on an OPC's pre-computed bounding box, which does not require the data to be loaded. The data size of an OPC can be directly derived from its source image, since one pixel in the image relates to one 3D position in the reconstruction. To reconstruct larger surfaces, PRoViP simply cuts the source image into smaller images and constructs one OPC for each of those subimages at the preprocessing stage.

Besides data subdivision, the most important aspect of our out-of-core implementation is asynchronous resource management. While the user navigates through the scene, "add OPC" and "discard OPC" commands are dispatched to a scheduler. The scheduler executes each "add" command asynchronously to the rendering but sequentially to avoid simultaneous accesses of the hard disk, which would result in resource thrashing and longer loading times. The memory garbage collection initiated by "discard" commands has proven to be the fastest when executed at regular time intervals.

To achieve interactive framerates, i.e., at least 25 frames per second, we need to stay within a certain triangle budget where we can render each frame. A well-established method for controlling the number of rendering primitives sent to the graphics card is rendering the data at different levels of detail (LOD), as discussed by Luebke et al. (2002). The different LODs are generated during preprocessing through subsequent simplifications of geometry and image data and encoded in the OPC data structure, as discussed in Section 3.2.3.2. Key to this strategy is to represent areas that are closer to the viewer by a higher LOD than areas that are farther away, basically boiling down to a quality vs. speed tradeoff. LODs are adapted in real-time and change smoothly during navigation. Ideally, there is as little visual difference as possible between rendering at the full level of detail and rendering multiple levels while also discarding as many primitives as possible. At preprocessing, it is specified at which distance which level should be displayed, depending on the dataset, e.g., the highest detail patch should only be loaded at a distance of 5 meters. However, the user can modify these values to accommodate slower hardware. As illustrated in Figure 3.12, the reduced quality introduced by the LODs, indicated by the different colors, is not noticeable in the non-colored rendering of the Victoria crater.

The scene a geologist needs to explore typically consists of multiple surfaces often reconstructed from images that were taken with different instruments. The Victoria crater scene for instance, as depicted in **Fi**gure 3.13, consists of the following surfaces ordered from lowest to highest resolution: DTM captured by HiRISE mounted on the Mars Reconnaissance Orbiter, high resolution DTM captured by HiRISE and enhanced by SRR (see Section 3.2.4), two DOMs captured by the Pancam of the Opportunity rover, and an enhanced close-up achieved through a wide-baseline reconstruction, also from Pancam. Since these surfaces are overlapping each other, the user can assign a depth bias to each surface, essentially prioritizing higher resolution surfaces over lower resolution ones. This bias only manipulates the depth test and tells the graphics card which surface occludes which. In this way PRo3D provides data fusion at the visualization stage without the need for cumbersome multiresolution stitching at the data level.

Additionally, users have the possibility to actually transform surfaces by using translation, rotation, and scale controls, similar to those used in 3D modelling tools. Although scaling of surfaces might sound like introducing an additional error, wide-baseline reconstructions have an inherent inaccuracy in scale (depending on the length of the baseline), which needs to be corrected by aligning and

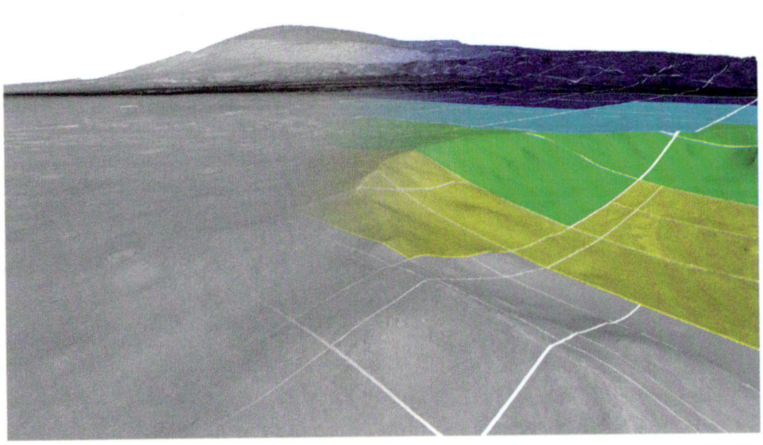

Figure 3.12 This depiction of the Victoria crater illustrates the use of different levels of detail (LODs). The left side shows the seamless transition between LODs while the right side uses color coding to emphasize the borders between different levels, where red represents the finest and blue the coarsest resolution. The cutout on the right shows the difference in geometric resolutions between the red and the yellow level through wireframe rendering, i.e. only rendering triangle edges.

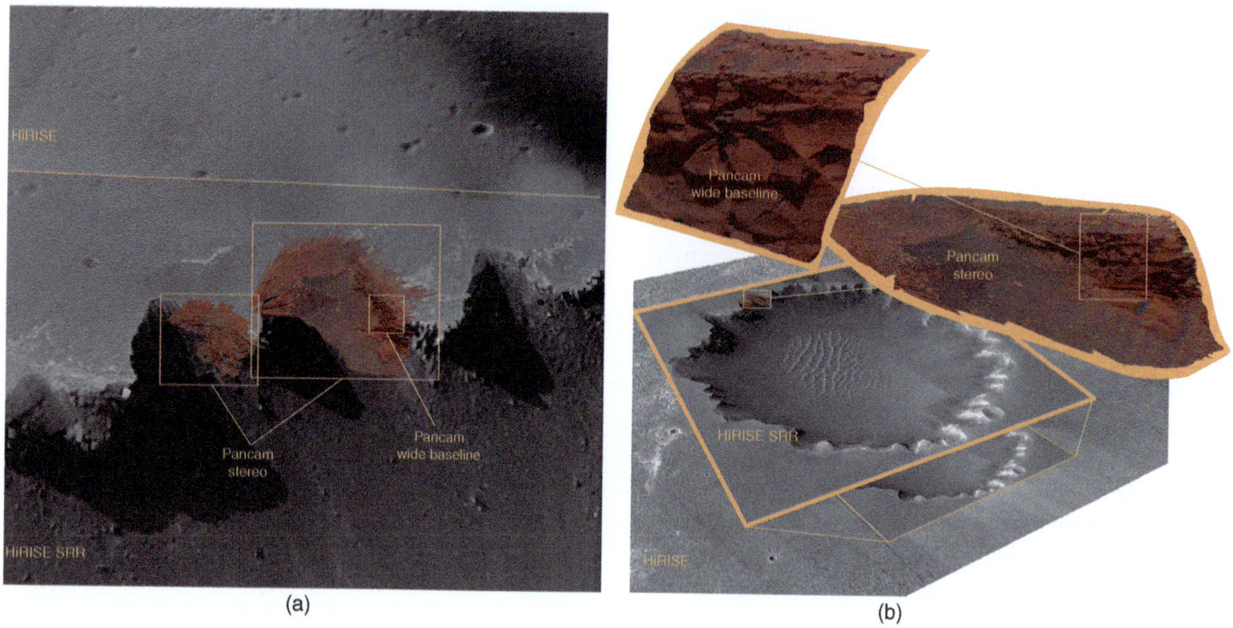

Figure 3.13 HiRISE, HiRISE SRR (super-resolution restoration), Pancam stereo, and Pancam wide baseline datasets in the Victoria crater area. (a) Top-down view of fused datasets. (b) Individual datasets in detail.,

scaling the data with respect to their spatial context, for instance an HiRISE DTM.

Showing all data of Mars in a common global coordinate system (IAU spherical coordinate system; see Duxbury et al., 2002) leads to extremely large 3D coordinates, for instance the Cape Desire DOM contains a point at $x = 3{,}376{,}372.058677169$ meters. Floating point numbers are represented by an integer part and a fraction part. This is not an issue as long as we process this data in the CPU, where floating point numbers are represented in double precision. Despite the fact that graphics cards meanwhile are able to perform double precision, floating point operations can be performed 24 times faster on consumer hardware (see ArrayFire, 2021). As described by Cozzi and Ring (2011), the conversion of large coordinates to single precision floating point values leads to a truncation of the fractional parts, which results in jitter, i.e. jumping of vertices when moving the viewport.

As mentioned in Section 3.2.3.2, each OPC patch is saved in its local coordinate system, allowing the machine to utilize the full fractional part of the floating-point representation. With each patch, there is a local to global transformation attached. When the vertex position in the local coordinate system, the camera transformation (Viewprojection Trafo), and the local to global transformation (Model Trafo) are loaded up to the graphics card separately, the large coordinates in the Viewprojection Trafo and the Model Trafo cancel each other out and the data near the user can be rendered at maximum precision, avoiding jitter even for very small scales. Here we also benefit from the LOD approach, which reduces the necessary precision by switching to a lower LOD for data farther away. This also avoids frequency artefacts similar to texture mipmaps (Möller et al., 2008). For overlapping surfaces and rendering annotation lines, as will be discussed in the next chapter, we also have to cope with z-fighting, i.e. the graphics card does not know which object is closer to the viewer for objects that are near each other. To compensate for that we employ dynamic near and far planes depending on the closest distance to the surface and a logarithmic depth buffer (Kemen, 2009).

Mapping image textures from the original imagery onto the meshes has the advantage that material properties and shading effects are preserved. Besides that, the image resolution is much higher than the geometric resolution, revealing minute features of the surfaces. On the downside, the shading is baked into these image textures, showing illumination conditions at a certain time of day, including direct and indirect lighting as well as shadows. PRo3D allows users to switch off image textures and render the geometry with simple material properties and an artificial light source. By interactively changing the direction of this light source, structures can be emphasized by illuminating them from an oblique angle as depicted in Figure 3.14a.

To support, for instance, multispectral imagery, the OPC format can hold multiple textures for each patch (essentially multiple image pyramids for each OPC). If this data is available, the user can switch between these textures interactively. Attribute maps, on the other hand, specify an additional value for each vertex. The OPC format allows multiple attribute maps. For instance, Figure 3.14b shows

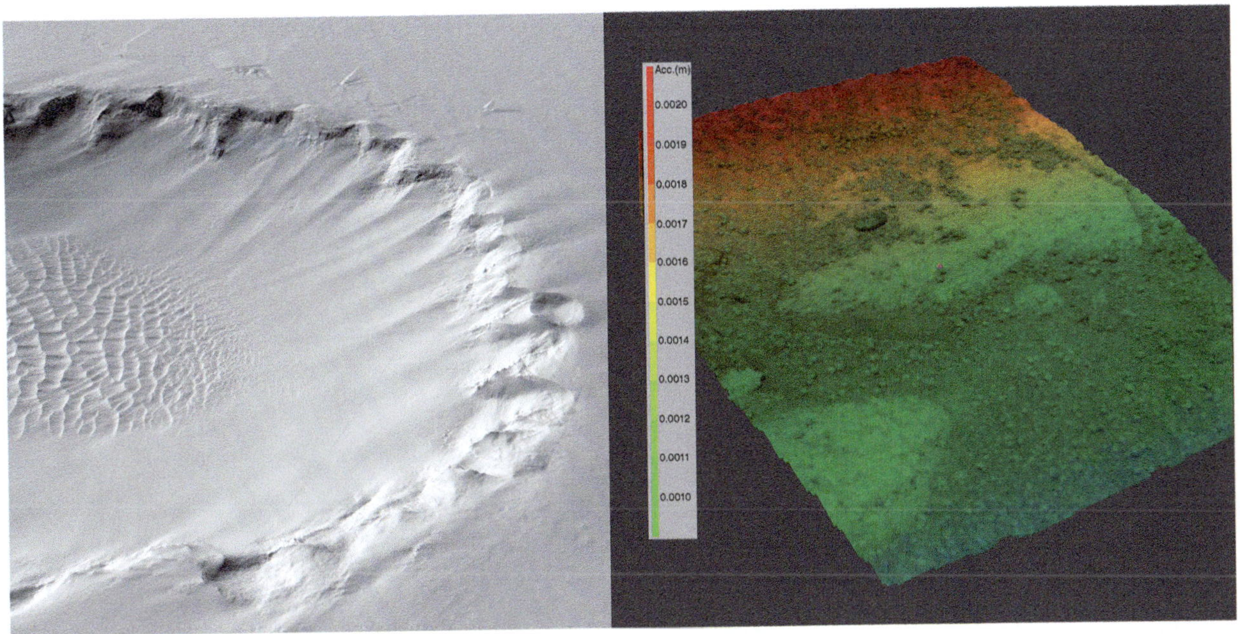

(a) (b)

Figure 3.14 (a) Rendering of the rim of the Victoria crater without texture. (b) Illumination from an oblique angle enhanced the geological features captured in the reconstructed surface. An accuracy map is rendered on top of a surface, which contains an accuracy value for each vertex. The reconstruction accuracy values range from 0.001 m (green) in the front to 0.002 m (red) in the back.

the reconstruction accuracy for each position illustrated by false color coding blended with a black and white texture.

3.2.6.2 Tools for Measurements and Geological Annotations

PRo3D offers planetary scientists a variety of tools for taking measurements and for the geological annotation of digital surface models represented through OPC data. We developed these tools in iterative design cycles under constant feedback from geologists specialized in remote structural geology. In the first iteration, the measurement tools were based on specific interactions common in GIS applications, such as distance and line-of-sight measurements. We soon realized that tools for geological annotation require a higher flexibility than that. To cover a wide range of use cases, we enable users to create their own annotations by specifying the following parameters: geometric primitive, projection, color, thickness, and an optional screen aligned text (Figure 3.15a).

Annotation is based on one of the following geometric primitives: point, line, polyline, and closed polyline. After selecting the primitive type users specify its form and location by selecting one, two, or multiple points on the 3D surface, which we will refer to as picking. The projection defines the behavior of line segments, i.e. how the picked points are connected. PRo3D features three projection modes: linear, which connects two picked vertices by a straight line, orthographic, which projects the line orthogonally from the sky onto the surface, and viewpoint, projecting the line onto the surface from the viewer's position (Figure 3.15b). Further, users are able to change the color and thickness of lines, allowing them to visually group features together and encode a hierarchical order or other semantics. Finally, every measurement can be annotated with text displayed as a screen-aligned billboard. This construction kit approach is extremely versatile and allows geologists to express a multitude of annotations by varying the described parameters.

For instance, to measure the horizontal or vertical extent of an outcrop a geologist may select the line primitive with linear projection. After picking two points they can attach a billboard showing the calculated linear distance.

In addition to the linear length of a measurement we derive a multitude of values as there are: way length, i.e. the projected length along the surface equivalent to topographic or geodetic distances, the maximum height value and the difference in height, and a measurement of the azimuth and slope in relation to the north vector and horizontal plane respectively.

Based on the polyline primitive, PRo3D supports geologists in the estimation of paleocurrent flow directions through taking dip and strike measurements. Users pick points to trace a geological feature by polyline. On completing this polyline, a plane is fitted using linear regression, which exhibits the least squared distance to each point picked. The plane is represented by a colored disc and its

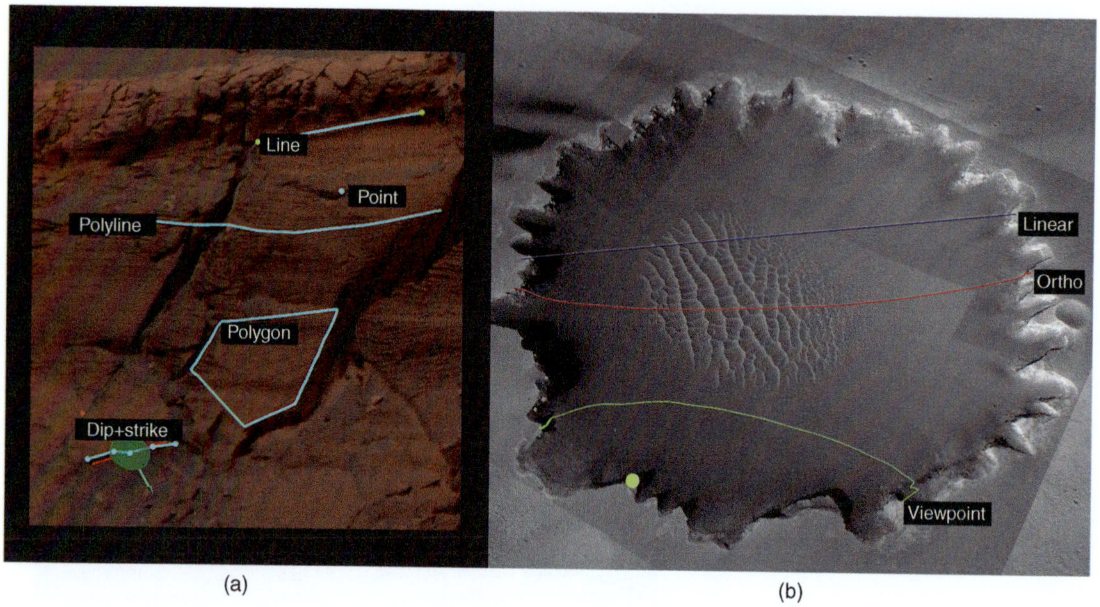

Figure 3.15 (a) The individual geometric measurement primitives range from a single point, over a line and polyline to the more complex dip and strike tool. In the latter case, a plane is fitted to a user drawn polyline. This plane is represented by a disc and the dipping direction is indicated by an arrow, while their color encodes the amount of dipping. (b) Illustration of the three different projection modes based on the two-point line primitive. The blue line is a linear connection between the two points, the red line is projected orthogonally onto the surface, while the green line is projected from a viewpoint, indicated by the green dot.

dip angle is encoded by its color, ranging from blue for 0° dip to red for 90° dip. The dip and strike measurement contains an additional set of derived values, as there are azimuths for dip and strike directions, and the dip angle itself, which is shown in Figure 3.16.

Since even analyses of small-scale outcrops may contain thousands of annotations, PRo3D allows users to group measurements hierarchically, where each group can be named and may contain an arbitrary number of subgroups. This enables geologists to group measurements according to their location in an outcrop, their type, or their stage in the analysis process. Geologists also used measurement grouping for scientific storytelling by successively showing and hiding groups and thus leading an audience through the entire analysis process by transitioning from bookmark to bookmark.

To integrate PRo3D with the general workflow of building large-scale geological models, users are able to export all or certain groups of annotations as .csv files. This is especially relevant to use information visualizations, such as stereo nets or rose diagrams, for characterizing for instance the distribution of dip angles or orientations. The export feature also allows geologists to verify and compare their findings with existing publications.

Taking a measurement or drawing an annotation requires users to pick points on the surface. Clicking into the scene just selects a pixel on the screen. To determine the correct 3D position on the surface corresponding to the clicked pixel, a geometric projection from 2D to 3D space is necessary. Therefore, we construct a ray originating from the current eye point and pointing towards the clicked pixel on the image plane, and intersect it with the geometry in the scene. Since it is necessary to capture a geological feature in the highest available accuracy, such an intersection would involve testing against millions of triangles, resulting in long delays for each picking operation. Consequently, we employ a kd-tree as a spatial hierarchical data structure subdividing 3D space into subspaces. Testing against subspace-bounding boxes is extremely fast and reduces the number of necessary triangle intersections to a minimum. The construction of the kd-tree is performed by PRoViP and is readily available when loading a surface reconstruction.

3.2.6.3 Implementation Decisions and Technological Choices

PRo3D, an open-source 3D viewer (PRo3D, 2021a, 2021b), is built on top of the open-source visualization platform Aardvark (Aardvark, 2021a, 2021b). This platform has been developed in-house at VRVis since 2005, has undergone many iterations, and is grounded on a strong research foundation being the backbone of 10+ research projects, with PRo3D being one of them. At its core we were able to efficiently separate semantic concerns from rendering details

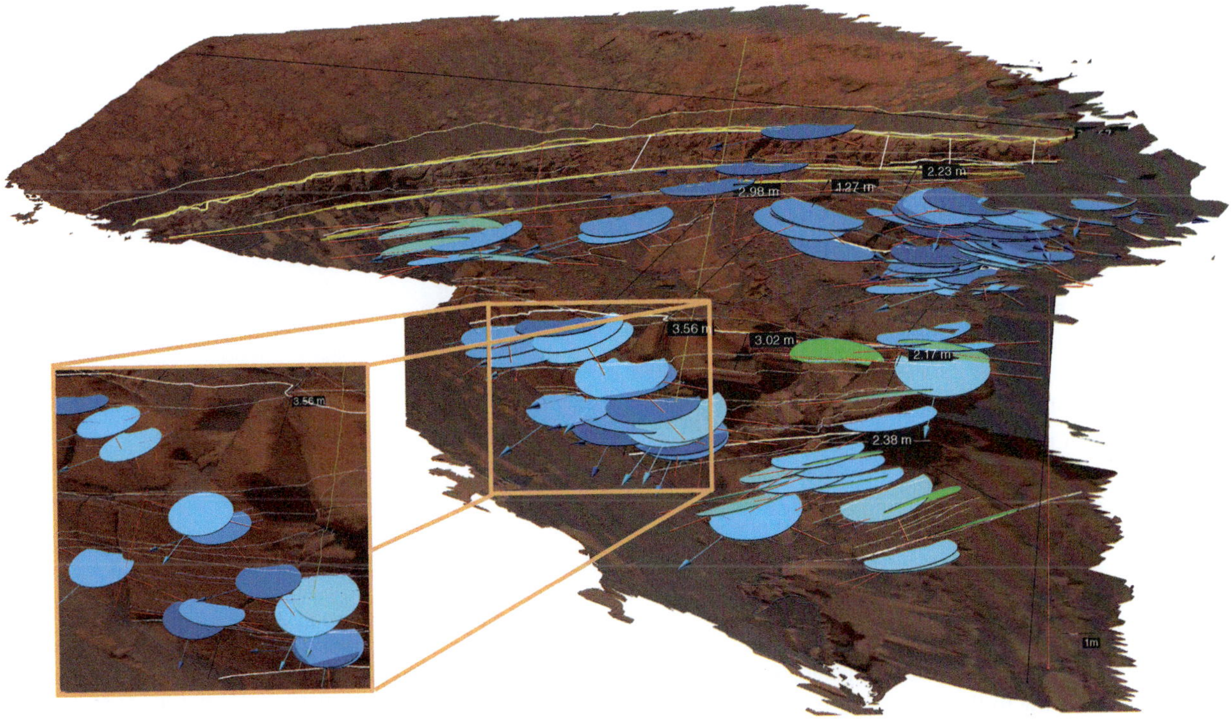

Figure 3.16 Full-scale geological analysis of the Cape Desire outcrop. This analysis contains hierarchical layer delineations indicated by yellow and white thicker and thinner lines. At the top, there are several distance measurements with annotations to determine the layer thickness. Further, there are many dip and strike measurements exhibiting similar dipping angles in their local neighborhood. The cutout on the left illustrates the screen-space scaling of discs, arrows, and line thicknesses.

in the form of a semantic scene graph (Tobler, 2011). As the traversal of increasingly complex scene graphs became too expensive, we employed incremental computation and lazy evaluation of render caches to only re-evaluate parts that have actually changed (Wörister et al., 2013), for instance, the selection state of an annotation and the respective graphical highlighting. An additional abstraction from standard graphics APIs such as DirectX or OpenGl, as presented in Haaser et al. (2015), offered an additional performance increase also for reasonably dynamic scenes, while in Steinlechner et al. (2019) we could introduce domain-specific languages into our system that handle change effectively and are declarative. Most of the Aardvark Platform and PRo3D are written in F#, a functional programming language. Immutable data and side-effect free functions are core paradigms of functional programming and come with several key benefits: Conciseness, Convenience, Correctness, Concurrency, and Completeness (Wlashin, 2021). We also built a purely functional front-end following the Elm Architecture (Elm, 2021), which combines an html-based GUI, high-performance 3D graphics, and a functional backend. Using Microsoft's .NET Core runtime and a platform-agnostic GUI allows us to deploy and run PRo3D on Windows, OSX, and Linux.

3.2.7 Typical Workflow

The PRoViDE framework supports the entire workflow of an image-based investigation of planetary surfaces, which consists of the following steps:

1. Gather all relevant orbital imagery and (if needed) co-register them so that they are transformed into a common coordinate system (Tao et al., 2016). This also contains a Super-Resolution Restoration (SRR) process (Tao and Muller, 2016) to produce a higher resolution context base on repeat views of the finest available orbital imagery. These images are stored in the geo-database PRoDB.
2. Start by investigating small-scale maps of the region of interest, which are derived from orbiter imagery. This is done using PRoGIS. It can also display planned rover paths and locations where images were captured and with which instruments, along with their fields-of-view (fulcra).
3. Find suitable images taken by rover cameras of the region of interest. Panoramas and image pairs from stereo cameras can be directly identified from their metadata and automatically inserted into PRo3D, while for wide baseline stereo reconstruction (observations from two or more rover locations) matching image pairs

must be identified. Image selection is also supported by PRoGIS.

4. Run the selected images through a computer vision processing chain to generate a 3D reconstruction. This is done here by means of PRoViP, which subsequently harvests all imagery as available in PRoDB and allows their efficient batch processing into 3D vision products such as DOMs by parallel computing on a large Linux cluster.
5. Survey available 3D products in PRoGIS, where they can be shown as icons coinciding with corresponding rover locations. Scientists can then select a 3D product to load it into PRo3D.
6. Interactively explore the chosen 3D reconstruction in PRo3D. Perform measurements on the 3D surface and make annotations to complete an extensive geological analysis and interpretation.

3.3 Geological Interpretations of DOMs

The 3D environment in PRo3D allows a geologist to follow a comparable workflow to that employed in field geology. The DOM can be rotated, panned, and zoomed to allow for multiple perspectives, the scale of topographic or geological features can be directly measured, and interpretations can be digitized straight onto the OPC. Geometrical analysis is also possible using the dip and strike tool. Measurements can be exported for analysis outside PRo3D. The following section presents two case studies that were used to test the application of PRo3D to geological analysis of Martian rock outcrops. The workflows applied to geological analysis of 3D data in Pro3D are described in detail in Barnes et al. (2015), and a summary of those results are presented here.

Interpretation of the stratigraphy, as well as measurement of the layer thicknesses, evaluation of the textural features, such as grain size and morphology, and collection of dip and strike data, was completed in PRo3D. Sedimentary structures present were interpreted, and the thickness of each package of cross-bedding (bedset) was measured, together with the dip and strike of laminations within each bedset. These represent preserved dune systems, and geometrical analysis, together with the textural observations, are key observations used to determine the environment in which the rocks were deposited.

3.3.1 Victoria Crater

The Victoria crater (Figure 3.17) is a ~ 750 m wide, moderately degraded, simple crater (Grant et al., 2008) located at 2.05°S, 5.50°W in the equatorial Meridiani Planum region of Mars. It was visited by the MER-B Opportunity Rover, between Sols 952 and 1634 of operation.

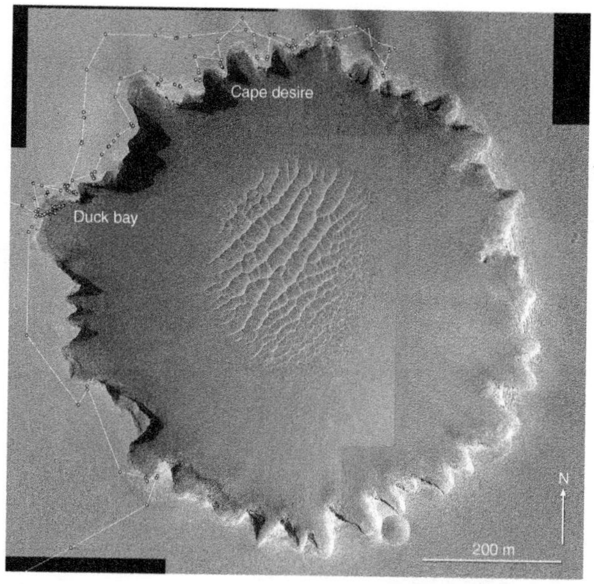

Figure 3.17 Super-resolution HiRISE image (processed by UCL-MSSL) of the Victoria crater, which was visited by MER-B Opportunity between September 2006 and August 2008. Erosional widening has resulted in a ~ 750 m wide crater with a scalloped morphology, consisting of numerous capes and bays. Sols 952 to 1634 of the mission were spent traversing the rim of the crater, imaging up to 12 m high rock outcrops of the Upper Burns Formation as well as searching for a safe ingress path to enter the crater. Investigations within the crater commenced at Duck Bay on Sol 1293 until Sol 1634.

Erosional widening has created well-preserved outcrops of sulphate-cemented sandstone (Edgar et al., 2012) in the crater wall, which provided the opportunity to look at significant sections of the stratigraphy in the area.

Duck Bay and Cape Desire (locations in Figure 3.17) were chosen as sites to test the application of PRo3D. Stereo panoramas taken by the Pancam instrument on Sols 1385 and 1423 were merged to create the OPC of Duck Bay where the rover ingressed into the crater. The rock outcrops within the scalloped bays and capes around the Victoria crater were typically imaged from a distance of 50–70 m, precluding the use of fixed-baseline stereo processing due to heavy degradation of the fixed (= 30 cm structural available) baseline stereo results at such a large distance. Cape Desire was imaged on Sols 1060 and 1061 of the MER-B campaign. These locations were spaced approximately 1 m apart, thereby improving the distance-to-baseline ratio of the imaging.

The PRo3D interpretation of Duck Bay is shown in Figure 3.18a. Three individual stratigraphic members were identified following Edgar et al. (2012), the Lyell, Smith, and Steno members. The boundaries were digitized based on variations in relative color, texture, presence, and character of laminations, and the weathering profile. Edgar

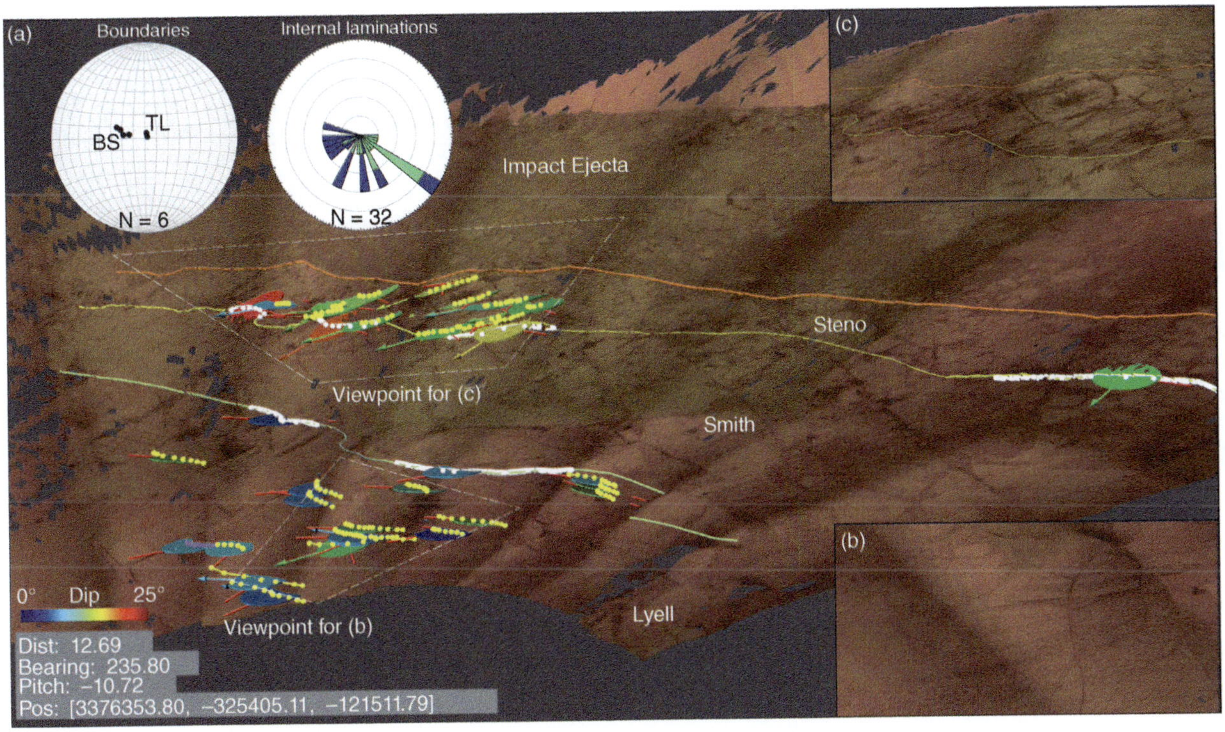

Figure 3.18 (a) 3D rendering of an interpreted scene of Duck Bay in PRo3D, highlighting a lower, relatively dark, pinstripe cross-laminated sandstone (Lyell unit), overlain by a diagenetically altered unit of the same material (Smith unit). There is an erosional boundary at the top of this diagenetic unit, indicated by the irregular topographic expression of the boundary, and the changes in measured dip in the finely laminated sandstone (Steno Unit), above that boundary. The boundary dip and strikes were measured along the thick white lines. Internal laminations in the Lyell and Steno are the thin yellow dotted lines. The best-fit dip planes are color coded by dip. Inset are the stereonet, indicating the attitude of the main boundaries (BS = Base Steno, TL = Top Lyell) and the laminations in the Lyell and Steno members measured at Duck Bay. (b) Blowup image of pinstripe cross-laminations in the Lyell member. The location of the image is shown in (a). (c) Blowup image of trough cross-laminations in the Steno member. The location of this image is shown in (a). Source: NASA/JPL-Caltech/Cornell U./ASU/PRoViDE.

et al. (2012) made detailed sedimentological descriptions of these units from analysis of Microscopic Imager (MI) images of fresh surfaces exposed by the Rock Abrasion Tool (RAT). The Lyell member (Figure 3.18b) is characterized by pinstripe cross laminations and a darker appearance in the outcrop. The laminations are truncated in some locations, but the low topographic slope and incomplete preservation in the outcrop model precludes mapping of truncations to locate set boundaries. The rough texture of the Lyell Member is caused by the numerous hematite concretions present in that unit, identified in MI and Pancam imagery (Edgar et al., 2012). The Smith member is lighter in appearance and structureless. The Steno member is also light colored, but darker than the Smith member, and contains trough cross laminations (Figure 3.18c), with a smoother texture in the outcrop, inferred to be the result of a lesser concentration of haematite concretions than the Lyell member. Topographic slopes of 12° and 25° in the Smith and Lyell members respectively, and up to 29° in the Steno Member, were measured, indicating that the Lyell and Steno members may be more resistant to erosion and weathering than the Smith member. True thicknesses of 0.9 m in the Lyell, 0.5 m in the Smith, and 0.6 m in the Steno were calculated as the vertical distances between the lower and upper bounding surfaces of each unit. The Lyell thickness value can be considered a minimum value, as the base of the unit is not exposed in the PRo3D scene. This shows 0.1–0.3 m deviation from the vertical values presented in Edgar et al. (2012).

The upper boundary of the Lyell member was calculated in PRo3D to be 2° to the SW (~246°) based on two measurements (TL "Boundaries" stereonet, inset in Figure 3.18a). Visualization of the dip and strike planes allowed for assessment of their intersection with the OPC surface. They showed a consistent fit with the topographic expression of the upper Lyell boundary. Dips calculated in PRo3D on the Smith-Steno contact were averaged to ~24° to the ESE (~110°), but showed some variation (BS "Boundaries" stereonet, inset in Figure 3.18a) along the boundary, steepening to the west. This is higher than the 10° to the SSE presented in Edgar et al. (2012). Dip and strike measurements were also made on laminations

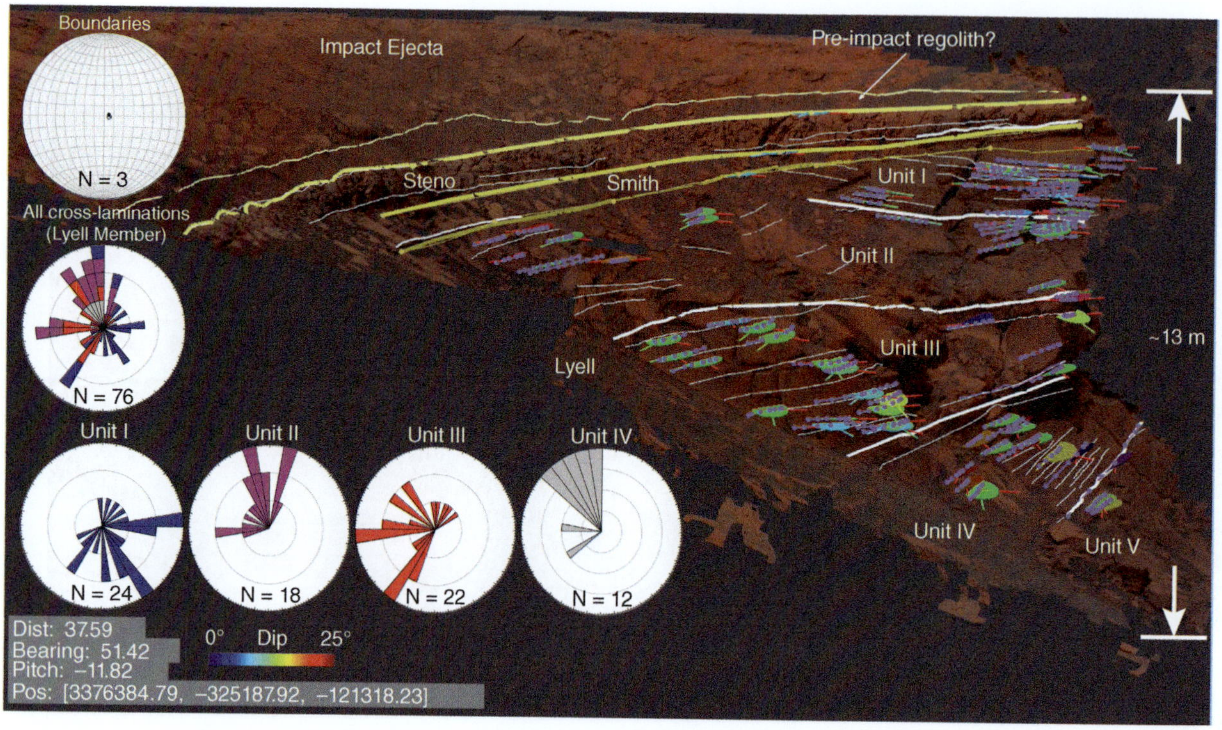

Figure 3.19 Interpreted 3D scene of Cape Desire in PRo3D. The stratigraphy has been correlated with the Duck Bay reference section. Cross-lamination patterns were mapped (thin white lines) in order to locate the bedset boundaries (thick white lines). These form 1.7–3.5 m thick preserved bedsets. The dip and strike of the boundaries and laminations have been calculated and show a steepening down the section. At the base of the outcrop, dip values reach up to 43°, exceeding the angle of repose, inferring some kind of steepening as a result of rotation or faulting (Hayes et al., 2011). They are visualized as circular best-fit planes. The red line on each dip measurement shows the strike direction. Inset are the stereonet of the boundary dip and strike values, showing them to be roughly conformable, in contrast to Duck Bay, as well as rose diagrams of the dip directions measured at Cape Desire, showing all values, as well as the individual values for Units I–IV. Dip and strike values were not measured in Unit V due to a lack of 3D exposure of laminations there. *Source*: NASA/JPL-Caltech/Cornell U./ASU/PRoViDE.

in the Lyell and Steno members and show a dominant dip direction to the SE (green petals in the "internal laminations" rose diagram, inset in Figure 3.18a) in the Steno, and to both the SE and WSW in the Lyell (blue petals in the "internal laminations" rose diagram, inset in Figure 3.18a).

3.3.1.1 Analysis at Cape Desire

Measurement of the dimensions of the Cape Desire outcrop in PRo3D (Figure 3.19) shows that a maximum of 13.75 m vertical section is exposed, and the lateral extent at the top of the outcrop is 30 m, down to 15 m mid-section, and 10 m at the base. The stratigraphic succession mapped at Duck Bay was also identified here, but the thicknesses are different. The Lyell Member is up to 12.8 m thick vertically and is characterized by abundant pinstripe laminations in cross-strata. The 0.4–0.7 m thick Smith member is identified by its light color and weathering profile, which is more apparent than in the 2D images. The Steno member is thinly laminated and pervasively fractured, varying in thickness from 0.8 to 1 m. The dip and strike of the stratigraphic boundaries showed a more conformable geometry than at Duck Bay, with all layers dipping 5° to the WNW (boundaries stereonet, inset in Figure 3.19).

The Lyell member (Figure 3.19) consists of five sets of cross-bedding (bedsets), which were identified using lamination truncations and observations of the weathering profile in 3D. This functionality is currently unique to PRo3D for martian 3D rock outcrop datasets. Bedset thicknesses ranged from 1.7 to 3.6 m thickness.

Three maxima were observed in the dip directions of laminations in the Lyell unit ("All cross-laminations" rose diagram, inset in Figure 3.19), trending towards 270°, 320°–360°, and 216°, with lesser maxima observed towards 10°, 80°, and 140°, comparable to what was observed by Hayes et al. (2011). The paleotransport directions in bedsets I–IV (Figure 3.19) are variable down the section, with Unit I showing peaks at 140° and 90°. This swings to 340° and 10° in Unit II and values changed again in Unit III, showing maxima at 270° and 220°. Unit IV shows a dominantly N-NW paleotransport direction, trending between 310° and 360°.

3.3.1.2 Discussion

The difference in dip value of the base of the Steno member between Duck Bay and Cape Desire indicates that the boundary is undulating at the decimeter scale, but is continuous around the crater, as viewed at Cape Desire. The light-toned Smith member can be picked out in various other locations not visited by the rover, as seen in the SRR HiRISE images in Figure 3.20. The upper boundary is abrupt, and as described in Edgar et al. (2012), it does not show the effects of diagenetic alteration as was described in the Smith member. This evidence is consistent with the interpretation of an erosion surface at the base of the Steno member. The thickness of the bedsets in the Lyell member at Cape Desire, combined with the pinstripe laminations at Duck Bay (Figure 3.18c), fine grain size and good sorting (visible in Microscopic Imager data, not shown here) are consistent with an eolian paleoenvironment as interpreted by Edgar et al. (2012). It is inferred that the Smith member was deposited in the same environment as the Lyell, but was subject to a diagenetic episode that affected a 0.4–0.7 m thick preserved section prior to deposition of the Steno member. Sufficient data was not present to determine the depositional environment of the Steno member at both Duck Bay and Cape Desire (see Edgar et al., 2012, for a detailed description of the Steno member sedimentology). The variations in dominant transport directions shown in the rose diagram in Figure 3.19 in units I–IV of the Lyell Member at Cape Desire could be the result of changing wind directions through time, or a result of the deposition of sinuous crested sand dunes, forming trough cross-stratified sets with variable dip directions.

3.3.2 Yellowknife Bay

Yellowknife Bay (Figure 3.20) was visited by the MSL Curiosity rover shortly after it landed in Gale crater, between Sols 54 and 330 of operation (Vasavada et al., 2014). The textures and sedimentary structures within the rocks found there have been interpreted by previous authors (Grotzinger et al., 2014; Edgar et al., 2018) to be indicative of both flowing water in a river (Shaler, Figure 3.20), distal fluvial fan deposits (Gillespie Lake sandstone; Grotzinger et al., 2014), and standing water in a lacustrine environment (Grotzinger et al., 2014) (Figure 3.20). These were determined to be paleoenvironments, which may have been habitable in the past, and therefore an ideal location to develop the tools in PRo3D for analysis of those features. In general, the MSL Curiosity rover was able to image targets from closer range than the MER rovers, with a more advanced camera system. This has resulted in a higher resolution of the processed OPCs.

An erosional window in the Yellowknife Bay area (Figure 3.20) was chosen as a case study for investigation, as it shows an excellent example of an irregular, erosional contact between distal fluvial fan sandstones (Gillespie Lake member) and an underlying lacustrine mudstone member (Sheepbed) (Grotzinger et al., 2014). A merged dataset was used in this paper compiled from Mastcam stereo-data taken between Sols 94 and 301.

3.3.2.1 Analysis at Yellowknife Bay

Geologic evaluation of the rock outcrops at Yellowknife Bay was carried out in the same manner as at Victoria Crater. Four separate stratigraphic intervals were identified (Figure 3.21). At the base of the succession, mudstones were identified, which were drilled at the Cumberland and John Klein targets, to reveal a grey material underneath the rusty colored patina on the surface of the rock and tailings for chemical analysis (Vaniman et al., 2014; Bristow et al., 2015; Ming et al., 2014; Farley et al., 2014). The OPCs of these drill sites were fused into the DOM, providing a spatial context for them. This mudstone—the

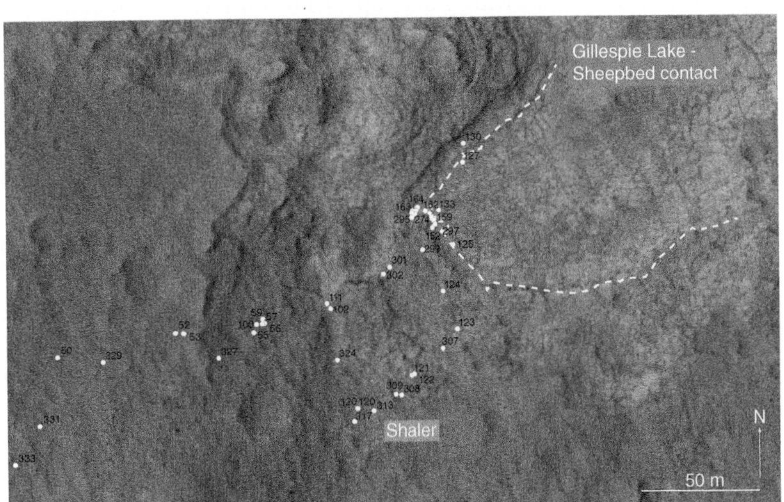

Figure 3.20 HiRISE image of Yellowknife Bay in the Gale crater, where MSL Curiosity investigated the orbitally mapped stratigraphic units there. Grotzinger et al. (2014) interpreted the sedimentary layers present to be indicative of ancient distal fluvial fan and lake sediments.

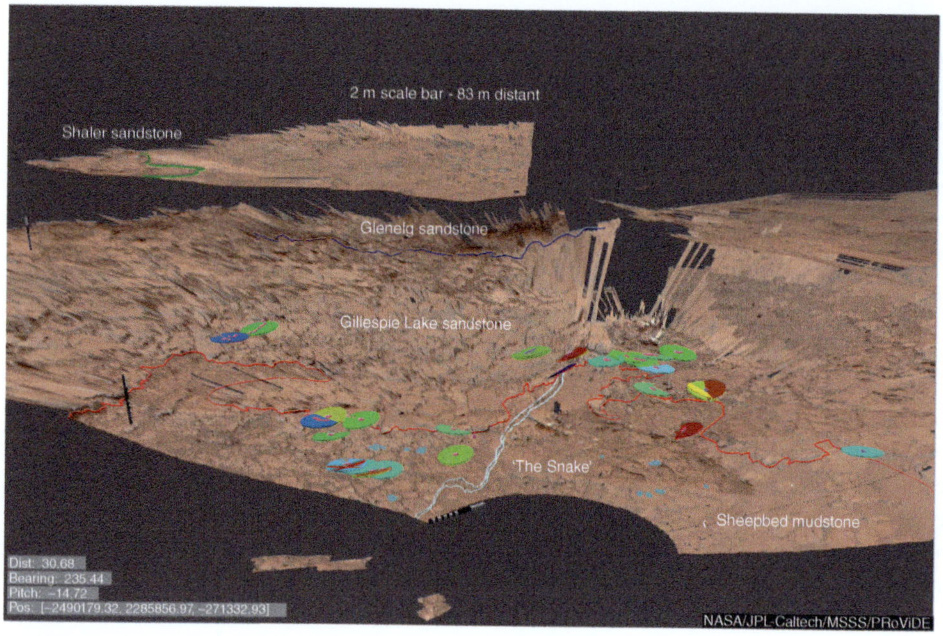

Figure 3.21 Full interpretation of a merged Mastcam dataset of the exposed stratigraphy at Yellowknife Bay carried out in PRo3D. The red boundary indicates the location of the base of the Gillespie member and the light blue boundary shows the location of the linear feature known as 'The Snake' (Grotzinger et al., 2014). The dark blue boundary is the base of the Glenelg sandstone and the dark green line shows the position of the base of the Shaler outcrop. The scene was compiled from Mastcam imagery acquired between sols 94 and 301. *Source:* NASA/JPL-Caltech/MSSS/PRoViDE.

Sheepbed mudstone (Grotzinger et al., 2014; Vaniman et al., 2014; Schieber et al., 2016)—was also characterized by abundant diagenetic features, namely nodules up to 2.8 mm in diameter, and synaeresis cracks (Léveillé et al., 2014; Stack et al., 2014; Siebach et al., 2014; Schieber et al., 2016). The boundary between the Sheepbed mudstones and overlying Gillespie sandstone is an irregular, abrupt one, which shows evidence that the Gillespie sandstones overlie an erosional surface, with scours up to 6 cm deep, cut into the mudstones. The Gillespie member is a poorly sorted coarse to very coarse sandstone (average 0.92 mm from 28 measurements) with grains up to 1.6 mm measured from Mastcam OPCs in PRo3D. The lower boundary of the Sheepbed mudstone is not visible (Grotzinger et al., 2014), but exposed thicknesses show an average of 0.3 m, with a minimum of 0.2 m and a maximum of 0.4 m from four measurements. The Gillespie member has an average thickness of 0.7 m, ranging from 0.6 m to 1 m from four measurements. The Glenelg member shows more variability and was measured between the boundary at the Point Lake outcrop (Figure 3.21) and its uppermost boundary. Values measured were between 1.3 m and 2.1 m in three measurements.

3.3.2.2 Discussion

Grain size variations between the Sheepbed and Gillespie Lake members, as well as the geometry of the boundary and diagenetic features, are consistent with previously proposed models stating that the stratigraphy at Yellowknife Bay was deposited in a lake bed (Grotzinger et al., 2014), which was subsequently buried in this location by deposition from an alluvial fan, with deposits showing signs of deposition in more proximal environments towards the top of the stratigraphy (fluvial deposits at Shaler) (Edgar et al., 2017). The thicknesses of each unit measured are consistent with previously published work. As at the Victoria crater, these observations are greatly enhanced in PRo3D by the ability to easily measure grain size, bed and unit thickness, and make dip and strike measurements. PRo3D observations at Yellowknife Bay are covered in more depth in Barnes et al. (2018).

3.4 Conclusions

Using the PRoViDE framework, planetary scientists can explore and analyze planetary surface reconstructions with unprecedented accuracy. High-resolution digital outcrop models based on rover imagery make it possible to investigate geological features at a range of scales from sedimentary bedding architecture at tens of meters scale to small-scale features such as sedimentary structures, grain size distributions, and fractures. Data fusion with larger terrain models based on orbiter imagery allows relating these small features to the larger geological context. Moreover, PRo3D provides a variety of interactive measurement and annotation tools to obtain real dimensions of features, inclination of layers, delineate structures, and mark regions of interest. 3D models are essential to fully characterize geological structures. The interactive visual analysis makes the interpretation much more efficient and reliable than pure 2D image analysis.

The results show a good agreement with previously published observations, but further validation is required. This is ongoing within a United Kingdom Space Agency funded project in which an emulator for the ExoMars 2022 Rover PanCam is taken to field locations to collect 3D outcrop data. Field measurements of the outcrops will be compared with measurements taken from PRo3D. It is important to know whether the measurements taken from DOMs of the Martian surface are reliable, and ground-truthing on Mars is presently not possible. Values are consistent within PRo3D, and can be manually checked, but further work is needed on data collection and processing.

The PRoViDE framework is currently in use for the Mars 2020 mission (Mastcam-Z Instrument) and is planned to be used for imaging instruments on the forthcoming ExoMars 2022 (PanCam Instrument) mission.

In addition to planetary science applications, the PRoViDE framework can also be used for terrestrial geological investigations supplementing the traditional field work. Measurements taken in the field can be transferred to the digital representation to continue the interpretation in virtual space. This yields additional insights and allows an efficient comparison with interpretations from peers.

Acknowledgments

The research leading to these results has received funding from the European Community's Seventh Framework Programme (FP7/2007-2013) under grant agreement no. 312377 PRoViDE.

We would like to thank our partners in the PRoViDE project who brought in their research results and valuable expertise to realize the framework presented in this chapter, among them: Michele Giordano and Jeremy Morley then at the University of Nottingham, Laurence Tyler and the late Dave Barnes from Aberystwyth University, Tomas Pajdla from the Czech Technical University, Ender Tasdelen from Technical University Berlin, Irina Petrovna Karachevtseva from MIIGAiK, and Extraterrestrial Laboratory Research (MExLab), together with their team members and academic PRoViDE collaborators.

We would also like to thank all members from the scientific advisory board of the PRoViDE project, who are: Jim Bell from Arizona State University, Bob Deen from NASA's Jet Propulsion Laboratory, Ron Li previously at the Ohio State University, and Derek Pullan previously at the University of Leicester.

References

Aardvark (2021a). The Aardvark Platform, website: https://aardvarkians.com, last accessed August 2021.

Aardvark (2021b). GitHub repository, website: https://github.com/aardvark-platform, last accessed August 2021.

ArrayFire (2021). Explaining FP64 performance on GPUs, website: https://arrayfire.com/, last accessed August 2021.

Barnes, R., Gupta, S., Giordano, M. et al. (2015). Geological interpretation and analysis of surface based, spatially referenced planetary image data using PRoGIS 2.0 and Pro3D. In: *European Planetary Science Congress 2015, held 27 September–2 October, 2015, in Nantes*. France: Online at http://meetingorganizer.copernicus.org/EPSC2015, id.EPSC2015-375.

Barnes, R., Gupta, S., Traxler, C. et al. (2018). Geological analysis of Martian rover-derived Digital Outcrop Models using the 3-D visualization tool, Planetary Robotics 3-D Viewer—Pro3D. *Earth and Space Science* 5: 285–307. https://doi.org/10.1002/2018EA000374.

Bell, J.F., Joseph, J., Sohl-Dickstein, J.N. et al. (2006). In-flight calibration and performance of the Mars Exploration Rover Panoramic Camera (Pancam) instruments. *Journal of Geophysical Research* 111, no. E, p. E02S03,: January 2006.

Bell, J. III, Maki, J., Mehall, G. et al. (2021). The Mars 2020 perseverance rover mast camera zoom (Mastcam-Z) multispectral, stereoscopic imaging investigation. *Space Science Reviews* 217 (1): 1–40.

Bistacchi, A., Mittempergher, S., and Martinelli, M. (2021). Digital Outcrop Model reconstruction and interpretation. In: *3D Digital Geological Models: From Terrestrial Outcrops to Planetary Surfaces*. Wiley, this volume.

Bristow, T.F., Bish, D.L., Vaniman, D.T. et al. (2015). The origin and implications of clay minerals from Yellowknife Bay, Gale crater, Mars. *American Mineralogist* 100: 824–836. doi: 10.2138/am-2015-5077CCBYNCND.

Coates, A.J., Jaumann, R., Schmitz, N. et al. (2015). *PanCam on the ExoMars 2018 Rover: A Stereo, Multispectral and High-Resolution Camera System to Investigate the Surface of Mars*, 1812. 46th Lunar and Planetary Science Conference, March 16–20, 2015, The Woodlands, Texas. LPI Contribution No. 1832.

Cozzi, P. and Ring, K. (2011). *3D Engine Design for Virtual Globes*. CRC Press.

Duxbury, T.C., Kirk, R.L., Archinal, B.A., and Neumann, G.A. (2002). Mars Geodesy/Cartography Working Group recommendations on Mars cartographic constants and coordinate systems. *International Archives of Photogrammetry Remote Sensing and Spatial Information Sciences* 34 (4): 743–748.

Edgar, L.A., Grotzinger, J.P., Hayes, A.G. et al. (2012). Stratigraphic Architecture of Bedrock Reference Section, Victoria Crater, Meridiani Planum, Mars, SEPM Special Publication 102. *Sedimentary Geology of Mars* 195–209.

Edgar, L.A., Gupta, S., Rubin, D.M. et al. (2018). Shaler: in situ analysis of a fluvial sedimentary deposit on Mars. *Sedimentology.* doi: 10.1111/sed.12370.

Elm (2021). The Elm Architecture, online article: https://guide.elm-lang.org, last accessed August 2021.

Farley, K.A., Malespin, C., Mahaffy, P. et al. (2014). In situ radiometric and exposure age dating of the Martian surface. *Science* 343 (6169): 124–166.

Giordano, M.G., Morley, J.M., Muller, J.P. et al. (2015). Sharing knowledge of Planetary Datasets through the web-based PRoGIS. In: *European Planetary Science Congress*, 2015, held on 27 September–2 October, 2015, in Nantes, France. Online at: http://meetingorganizer.copernicus.org/EPSC2015, id.EPSC2015-235.

Grant, J.A., Wilson, S.A., Cohen, B.A. et al. (2008). Degradation of Victoria crater, Mars. *Journal of Geophysical Research E: Planets* 113: E11010. doi: 10.1029/2008JE003155.

Grotzinger, J.P., Sumner, D.Y., Kah, L.C. et al. (2014). A habitable Fluvio-Lacustrine environment at Yellowknife Bay, Gale Crater. *Mars. Science* 343 (6169): 1–15.

Haaser, G., Steinlechner, H., Maierhofer, S., and Tobler, R.F. (2015). An incremental rendering VM. In: *Proceedings of the 7th Conference on High-Performance Graphics*, 51–60. Los Angeles, USA, August 7–9, 2015.

Hayes, A.G., Grotzinger, J.P., Edgar, L.a. et al. (2011). Reconstruction of Eolian bed forms and paleocurrents from cross-bedded strata at Victoria Crater, Meridiani Planum, Mars. *Journal of Geophysical Research E: Planets* 116 (4): 1–17. doi: 10.1029/2010JE003688.

Kemen, B. (2009). *Logarithmic Depth Buffer. Outerra Blogspot website*: http://outerra.blogspot.co.at/2009/08/logarithmic-z-buffer.html, last accessed: June 27,2017.

Léveillé, R.J., Bridges, J., Wiens, R.C. et al. (2014). Chemistry of fracture-filling raised ridges in Yellowknife Bay, Gale Crater: Window into past aqueous activity and habitability on Mars. *Journal of Geophysical Research: Planets* 119: 2014JE004620. doi: https://doi.org/10.1002/2014JE004620.

Luebke, D., Watson, B., Cohen, J.D. et al. (2002). *Level of Detail for 3D Graphics*. New York: Elsevier Science Inc.

Ming, D.W., Archer, P.D., Glavin, D.P. et al. (2014). Volatile and organic compositions of sedimentary rocks in Yellowknife Bay, Gale Crater. *Mars. Science* 343: 1245267. doi: https://doi.org/10.1126/science.1245267.

Möller, T.A., Haines, E., and Hoffman, N. (2008). *Real-Time-Rendering*, 3rde. A.K: Peters/CRC Press ISBN-13: 978-1568814247.

Ortner, T., Traxler, C., Piringer, H. et al. (2019). MINERVA: A 3D GIS and visual analysis framework. In: *Proceedings of 15th Symposium on Advanced Space Technologies in Robotics and Automation*. the Netherlands: (ASTRA 2019), ESA-ESTEC, Noordwijk, The Netherlands May 27–28, 2019.

Paar, G., Muller, J.P., Tao, Y. et al. (2013). PRoViDE: Planetary probes' mass vision data processing. In: *European Planetary Science Congress 2013*. held 8–13 September in London, UK. Online at: http://meetings.copernicus.org/epsc2013, id.EPSC2013-289.

Paar, G., Muller, J.-P., Tao, Y. et al. (2015). PRoViDE: Planetary robotics vision data processing and fusion. In: *European Planetary Science Congress 2015*. held 27 September–2 October in Nantes, France. Online at: http://meetingorganizer.copernicus.org/EPSC2015, id.EPSC2015-345.

Paar, G., Koeberl, C., Hesina, G. et al. (2016a). 3D Vision for Mars 2020 Mastcam-Z: Pre-assessment of processing techniques and geologic use cases. In: *47th Lunar and Planetary Science Conference*. held March 21–25, 2016 at The Woodlands, Texas. LPI Contribution No. 1903, p. 2810.

Paar, G. and the PRoViDE Consortium (2016b). PRoViDE Final Report, http://www.provide-space.eu/wp-content/uploads/2015/01/PRVD-REP-JR-Final-Report_2016-04-14-Public.pdf.

PRo3D–Planetary Robotics 3D Viewer (2021a) Website https://pro3d.space, last accessed August 2021.

PRo3D–GitHub Repository (2021b) URL: https://github.com/pro3d-space/PRo3D, last accessed August 2021.

Schieber, J., Bish, D., Coleman, M. et al. (2016). Encounters with an unearthly mudstone: Understanding the first mudstone found on Mars. *Sedimentology, The Journal of the International Association of Sedimentologists* 64 (2): February 2017, doi: 10.1111/sed.12318.

Siebach, K.L., Grotzinger, J.P., Kah, L.C. et al. (2014). Subaqueous shrinkage cracks in the Sheepbed mudstone: Implications for early fluid diagenesis, Gale crater, Mars. *Journal of Geophysical Research: Planets* 119: 2014JE004623. doi: 10.1002/2014JE004623.

Stack, K.M., Grotzinger, J.P., Kah, L.C. et al. (2014). Diagenetic origin of nodules in the Sheepbed member, Yellowknife Bay formation, Gale crater, Mars. *Journal of Geophysical Research: Planets* 119: 1637–1664. doi: https://doi.org/10.1002/2014JE004617.

Steinlechner, H., Haaser, G., Maierhofer, S., and Tobler, R.F. (2019). Attribute grammars for incremental scene graph rendering. In: *Proceedings of VISIGRAPP, 14th International Joint Conference on Computer Vision, Imaging and Computer Graphics Theory and Applications*, 77–88. LDA., Prague, Czech Republic: Scitepress Digital Library, Science and Technology Publications February 25–27, 2019. doi: 10.5220/0007372800770088.

Tao, Y. and Muller, J.P. (2016, February). A novel method for surface exploration: Super-resolution restoration of Mars repeat-pass orbital imagery. *Planetary and Space Science* 121: 103–114, Elsevier.

Tao, Y. and Muller, J.-P. (2021). *Super-resolution restoration of spaceborne ultra-high-resolution images using the UCL OpTiGAN System Remote Sensing* 13 (12): 2269.

Tao, Y., Muller, J.-P., and Poole, W. (2016, June). Automated localisation of Mars rovers using co-registered HiRISE-CTX-HRSC orthorectified images and wide baseline Navcam orthorectified mosaics. *Icarus* 280: 139–157.

Tao, Y., Conway, S.J., Muller, J.-P. et al. (2021). Single image super-resolution restoration of TGO CaSSIS colour images: Demonstration with perseverance Rover landing site and Mars science targets. *Remote Sensing* 13 (9): 1777.

Tobler, R.F. (2011). Separating semantics from rendering: A scene graph based architecture for graphics applications. *Visual Computer* 27 (6): 687.

Traxler, C., Hesina, G., Barnes, R. et al. (2016). The PRoViDE framework for the quantitative geologic analysis of reconstructed Martian terrain and outcrops. In: *Geophysical Research Abstracts*. Vol. 18, EGU2016-7196, EGU General Assembly 2016.

Vaniman, D.T., Bish, D.L., Ming, D.W. et al. (2014). Mineralogy of a Mudstone at Yellowknife Bay, Gale Crater. *Mars. Science* 343: 1243480. doi: https://doi.org/10.1126/science.1243480.

Varadhan, G. and Manocha, D. (2002). Out-of-core rendering of massive geometric environments. In: *Proceedings of the Conference on Visualization 02 (VIS 02)*, 69–76. Washington, DC, USA: IEEE Computer Society.

Vasavada, A.R., Grotzinger, J.P., Arvidson, R.E. et al. (2014). Overview of the Mars Science Laboratory Mission: Bradbury Landing to Yellowknife Bay and Beyond. *Journal of Geophysical Research E: Planets*. doi: https://doi.org/10.1002/2014JE004622.

Vitter, J.S. (2001). External memory algorithms and data structures: Dealing with MASSIVE DATA. *ACM Computing Surveys* 33 (2): 209–271.

Willner, K. and Tasdelen, E. (2015). A spatial planetary image database in the context of processing. In: *European Planetary Science Congress 2015*, id.EPSC2015-291, Nantes, France.

Wlashin, S. (2021). F# for Fun and Profit, online article: https://fsharpforfunandprofit.com/why-use-fsharp/, last accessed August 2021.

Wörister, M., Steinlechner, H., Maierhofer, S., and Tobler, R.F. (2013, July). Lazy incremental computation for efficient scene graph rendering. In: *Proceedings of the 5th High-Performance Graphics Conference*, 53–62.

4

Vombat: An Open Source Tool for Creating Stratigraphic Logs from Virtual Outcrops

L. Penasa[1,2], M. Franceschi[2], and N. Preto[2]

[1]Center of Studies and Activities for Space, CISAS, "G. Colombo", University of Padova, Via Venezia 15, 35131 Padova, Italy
[2]Dipartimento di Geoscienze, Via G. Gradenigo, 6, 35131 Padova, Italy

Abstract

An open source tool, *Vombat*, is presented that is designed to operate on Virtual Outcrop Models of sedimentary rocks, with the specific aim of assisting the stratigraphic analysis and interpretation. Vombat makes it possible to estimate the average attitude of the bedding and to create one or more attitude-aligned stratigraphic reference frames. This allows Vombat to extract continuous stratigraphic logs of any property associated with the point clouds (e.g. the lidar intensity or RGB color). Stratigraphic logs produced by Vombat can be compared and correlated to typical outcrop logs and petrophysical logs obtained from boreholes (e.g. gamma ray logs) and can provide information about the lithological variations in a stratigraphic succession. Furthermore, Vombat stratigraphic reference frames can be used to associate a stratigraphic position (a depth in the stratigraphic column) to any observation made on the outcrop, allowing visualization in 3D (on the virtual outcrop model) and 1D (on a stratigraphic column) for any collected data. All the geological objects created in the virtual environment can then be saved. The tool has been developed to be user-friendly and is constituted by a dynamically loaded plugin for the open source software CloudCompare.

4.1 Introduction

Three-dimensional imaging techniques have increasingly been applied to the realization of Virtual Outcrop Models (VOMs), especially in the oil and gas industry, to be used as surface analogs for the investigation of buried reservoirs (Hartzell et al., 2014; Martinsen et al., 2011; Kurz et al., 2008; Rarity et al., 2014; Bellian et al., 2005; Marques et al., 2020). Terrestrial Laser Scanners (TLSs) and photogrammetry (see Wilkinson et al., 2016; Corradetti et al., 2021; Buckley et al., 2010) are often used to produce VOMs or DOMs: these two methods allow three-dimensional representations of an outcrop's surface to be obtained in the form of *point clouds* or *meshes*, thus providing a *virtual duplicate* of the outcrop, which can be visualized, navigated, and interpreted with appropriate tools (see Chapter 2 by Bistacchi et al., 2021; Tavani et al., 2011; Buckley et al., 2017; Caravaca et al., 2020).

The potential of laser scanner- and photogrammetry-generated VOMs was also explored in the perspective of quantifying lithological properties in a fast, semiautomated and geometrically accurate way. Franceschi et al. (2009) and Burton et al. (2011) investigated the relationship between litologies and the intensity recorded by laserscanners, Penasa et al. (2014) demonstrated a method for the automated detection of chert from TLS intensity, multi- and hyper-spectral data, fused with three-dimensional models, which gave promising results (Kurz et al., 2012a; Krupnik et al., 2016; Thiele et al., 2021), and hyperspectral lidars are currently being developed (Suomalainen et al., 2011; Malkamäki et al., 2019). A sound review of geologic applications of TLSs, which in some measure can be extended to photogrammetry, can be found in Telling et al. (2017).

Scalar properties associated to the VOM (e.g. lidar intensity, RGB) can be studied in the form of logs that display the variations of the property along a stratigraphic succession (Franceschi et al., 2011, 2015; Penasa et al., 2019). These logs can be considered a remote-sensing equivalent to stratigraphic logs obtained by sampling in the field and then measuring some properties (e.g. chemical or mineralogical composition) and reporting them on a stratigraphic column, or to petrophysical logs realized in boreholes (Ellis and Singer, 2007). These latter logs refer to a range of properties (e.g. gamma ray, resistivity, spontaneous potential, etc.) that are used to obtain information about the rocks crossed during core drilling, basically

3D Digital Geological Models: From Terrestrial Outcrops to Planetary Surfaces, First Edition.
Edited by Andrea Bistacchi, Matteo Massironi, and Sophie Viseur.

describing how some measurable rock-property varies with the stratigraphic depth (Miall, 2016).

The logs obtained from VOMs could be used in a similar way and also provide means of correlating surface outcrop analogs and their buried counterparts. Despite this potential, however, there are no specific tools to construct property logs directly from 3D models, and their realization is complicated and limited by the fact that most software packages are not designed for this purpose. The development of a tool able to quickly and accurately provide such kind of logs from VOMs could be relevant both for stratigraphic investigations and for the industry.

In this contribution, we present Vombat, and open source software whose primary goal is to process lidar intensity, RGB data, or any scalar field associated with three-dimensional outcrop models, to create high-resolution and accurate stratigraphic logs of these properties. To reach this goal, we developed a simplified 3D modelling strategy, based on the identification of outcrop regions with constant attitude.

Vombat was developed as a dedicated toolbar within CloudCompare (Girardeau-Montaut, 2014), an open source software that makes it possible to visualize and modify point clouds. Far from being a complete toolbox, Vombat may set the stage for the development of a more complete data analysis and storage platform that could be used for exploiting the full potential of VOMs, whose applications range from the study of inaccessible outcrops to the mentioned correlation to boreholes, to the preservation of key outcrops in a virtual form, and may further expand in the broad field of geosciences.

4.2 Vombat

To provide a virtual counterpart to the operations commonly carried out by stratigraphers in the field, we first identified a suitable rendering engine for the visualization of point clouds. Point clouds are in fact the raw output of both lidar acquisitions and photogrammetric surveys. Many commercial software produced nowadays allow highly optimized point cloud visualization to be obtained, and are normally coupled with advanced editing tools. These tools are mainly closed-source proprietary solutions that leave little room for customization. We thus chose the open source software CloudCompare as a hosting application, which has been extended by means of a new plugin, Vombat.

CloudCompare (Girardeau-Montaut, 2014) is a software written in C++ and Qt libraries (e.g. Summerfield, 2011), dedicated to the editing, visualization, and comparison of point clouds. Being open source, its code can be freely extended or modified by the user, thus making this tool the ideal candidate to host third-party plugins. Vombat (Virtual Outcrop Basic Analysis Tool) has been developed to provide a tool for operating stratigraphic tasks on Virtual Outcrop Models. CloudCompare also features several third-party plugins of geologic interests: the qCompass plugin (Thiele et al., 2017) allows for automatically tracing linear features on VOMs and takes fast and easy attitude measurements. The qFacet plugin (Dewez et al., 2016) allows the user to perform the automatic extraction of planar facets from point clouds that can be used for structural analysis. The qCanupo (Brodu and Lague, 2012) suite is a simple yet efficient way to automatically classify a point cloud on the basis of its geometric characteristics, through the use of Support Vector Machine classifiers (Mountrakis et al., 2011), in order to separate vegetation from exposed rock. Finally, qM3C2 makes it possible to estimate differences between point clouds in a robust way, with the aim of identifying changes in the outcrop through time (Lague et al., 2013). Vombat has been designed as a two-components tool:

- An underlying C++ library taking care of most of the computations, which can be also accessed by the user with programming skills by creating an executable linked to the library itself.
- A plugin for CloudCompare, which provides a toolbar for creating and editing the objects defined in the C++ library and allows any user to interact in a graphical way with the software through a Graphical User Interface (GUI).

Vombat is displayed as a new toolbar in CloudCompare (Figure 4.1), where each tool can be activated by clicking on the correspondent icon. Each tool in the toolbar (Figure 4.2) takes as input the objects that are selected within the database tree of CloudCompare and processes them, generating or modifying the entities. Vombat also provides a specific tool for plotting curves that can be used to obtain a visual feedback of the generated continuous stratigraphic logs. We will not discuss all the implementation details of the Vombat toolkit, but we will rather give a brief idea of the underlying computations; however, the code is open source and can be freely accessed and modified.

4.2.1 Example of Workflow

Vombat has been designed to be as generic as possible and the specific workflow is mostly dependent on the user needs. In Section 4.3 some examples with real data will be demonstrated, but a typical workflow including the use of our plugin can be depicted as follows:

1. The stratigrapher uses a TLS (or photogrammetric methods) to produce a virtual outcrop in the form of one or

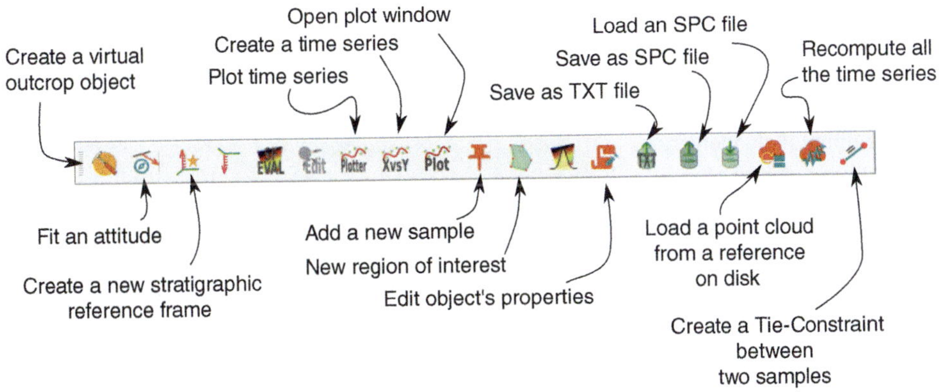

Figure 4.1 The Vombat toolbar as it appears in CloudCompare. Each icon corresponds to a specific operation that can be performed on the currently selected object.

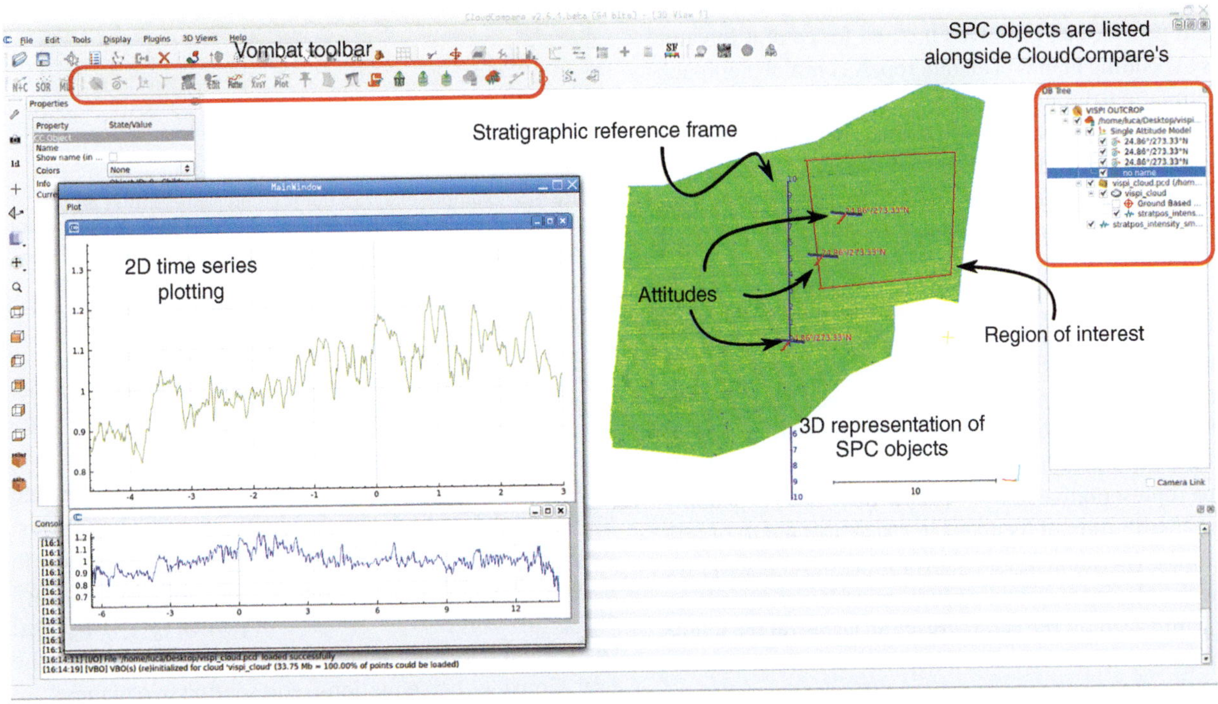

Figure 4.2 A screenshot of CloudCompare with the Vombat plugin. Basic tools for plotting logs are provided. Each Vombat object is represented as a hierarchical object, alongside those of CloudCompare in the DB Tree window (red square at right).

more point clouds. This data might be complemented by other investigations done in the field (e.g. rock samples that will undergo laboratory analysis and a stratigraphic column).

2. The point clouds are imported in CloudCompare, where geological linear or punctual features (important horizons and the location of collected samples) can be easily traced with the tools provided by CloudCompare itself.
3. The user then sets up one or more *Stratigraphic Reference Frames (SRF)*, which allows a *Stratigraphic Position (SP)* to be assigned to any linear or punctual feature that was identified. The setup of SRFs is very flexible: they can be forced to respect various external constraints (e.g. placing the stratigraphic zero in correspondence with a particular horizon).
4. Thanks to the SRFs the user is now able to directly compare any new observation made on the virtual outcrop with the observations collected in the field. For example, he can measure the stratigraphic position of some cherty levels that were unreachable in the field, and place them in the stratigraphic column previously drawn.
5. The user then uses Vombat to produce a stratigraphic log of any remotely sensed scalar property, e.g. lidar intensity, which can be used as a proxy for clay content

(Franceschi et al., 2009). This log can then be compared with any other log obtained within the outcrop or nearby.

6. The virtual outcrop, together with the collected geological data, can then be shared to unambiguously identify key stratigraphic features on the outcrop, e.g. to plan a field campaign on the same outcrop.

In this work, we present the basic issues that must be faced when creating stratigraphic logs from Virtual Outcrops, exposing some of the functions that have been implemented in Vombat to deal with these specific problems. Later sections are dedicated to the discussion and conclusion of the presented toolkit.

4.2.2 Estimation of the Average Bedding Attitude

Vombat provides an improved method for computing attitudes of bedding planes of an outcrop. A common approach for estimating the normal of a stratigraphic feature is to obtain a subset of points which *lay* on the plane of interest and then compute the best fitting plane for these points (see, for example, Fernández, 2005; Woodcock, 1977). In a similar way, a trace of the bedding plane can be digitized and then used in the best-fitting procedure to obtain an attitude.

When considering point clouds representing outcrops exposed on vertical cliffs, the approach to selecting points on a bedding plane does not perform well, because the bedding planes can be poorly exposed. Instead, using the digitized trace of a bedding plane, the estimated attitude will be representative of that specific trace and, depending on the distribution of its nodes in space, can led to unreliable results (Fernández, 2005). In order to be representative of the whole outcrop the attitude must be computed by taking into account the geometry of the layers on the whole outcrop.

Vombat tries to solve this issue by providing a tool for normal estimation that takes into account as many digitized bedding traces as possible. This may prove particularly useful in defining a more representative average attitude computed by taking into account many layers. Figure 4.3 shows an example of a point cloud for which an attitude should be estimated. The user identifies the bedding by using polylines, providing sets of points, each representing a specific horizon. Vombat then computes a best-fitting attitude using the information provided by all these point sets: in this way the estimated attitude will be representative of a larger portion of the outcrop.

The problem of estimating a unique attitude by using several bedding traces at once (instead than just one) is solved in Vombat as a minimization problem. The user defines two or more sets of points, each set representing a

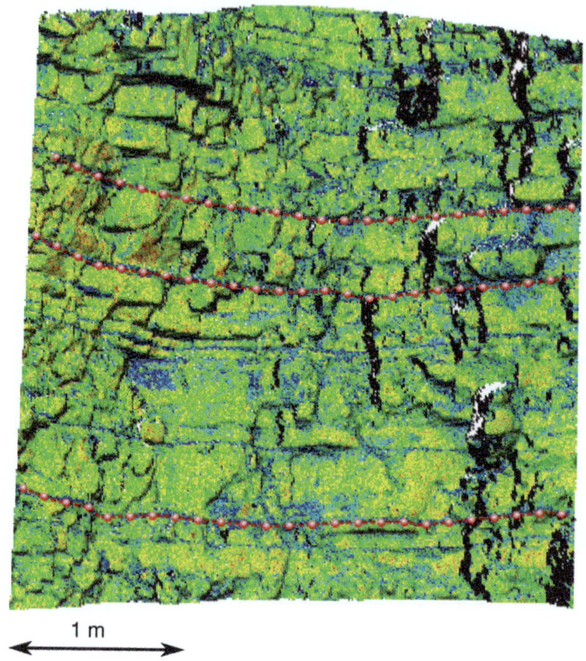

Figure 4.3 Example of a TLS point cloud representing an undeformed stratigraphic sequence for which an attitude should be estimated. Colours depend on intensity values retrieved by a TLS. The user creates one or more sets of hand-picked points, each representing a bedding plane, and then a dedicated algorithm computes the best fitting attitude for the outcrop. Attitudes can be computed for each bedding separately or a unique normal can be fitted to ensure representativeness of the attitude for the whole outcrop or outcrop region.

different bedding plane, parallel to the others, as depicted in Figures 4.3 and 4.4. For simplicity, we will consider just two sets of points in the following, but the extension to more sets is straightforward and can be found in Shakarji and Srinivasan (2013). Two sets of points, $\{\boldsymbol{p}_1, \dots, \boldsymbol{p}_n\}$ and $\{\boldsymbol{p}_{n+1}, \dots, \boldsymbol{p}_{m+n}\}$, represent two linear features traced by the user. Consider two equations representing two planes:

$$\hat{\boldsymbol{n}} \cdot \boldsymbol{p}_A - d_1 = 0$$

$$\hat{\boldsymbol{n}} \cdot \boldsymbol{p}_B - d_2 = 0$$

where $\boldsymbol{p}_A$ and $\boldsymbol{p}_B$ are points of their respective planes, $\hat{\boldsymbol{n}}$ is the unit normal defining the orientation of the planes and d_1 and d_2 are the distances of these planes from the origin. The problem is thus to find $\hat{\boldsymbol{n}}$, d_1, and d_2 minimizing the distance of the points from their respective plane by means of a least squares solution:

$$\sum_{i=1}^{n+m} r_i^2 = \min \tag{4.1}$$

where r_i is the distance of the point to the first plane for $1 \le i \le n$ or to the second plane for $n+1 \le i \le n+m$.

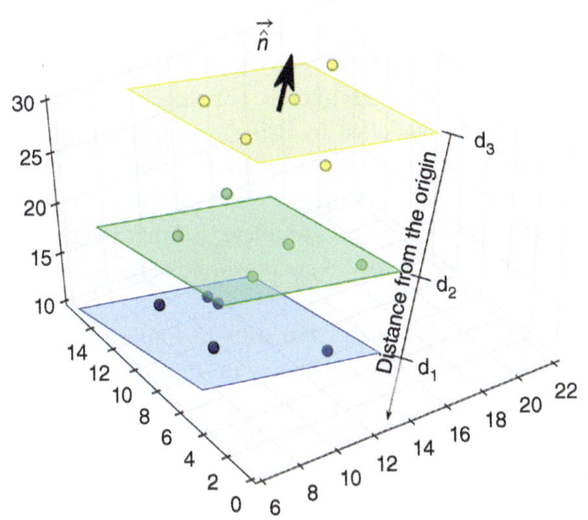

Figure 4.4 Vombat can compute the attitude of a layered sequence by using a multiple set of points lying on parallel planes. The normal $\hat{n}$ and multiple d parameters, one for each set of points to be fitted, are determined by means of a least squares solution.

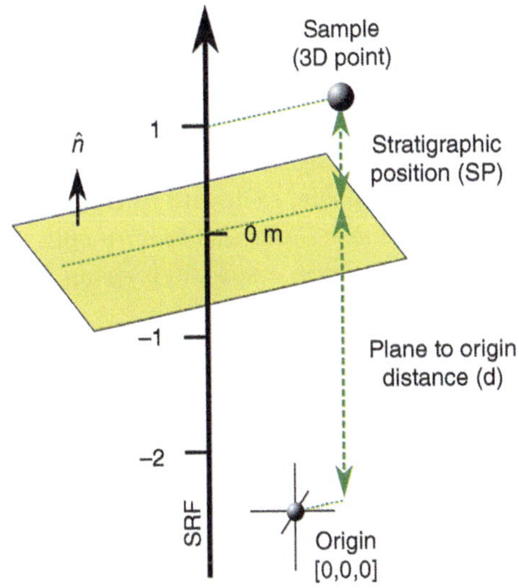

Figure 4.5 An SRF can be viewed as a plane in space: its orientation is determined by a unit normal $\hat{n}$ and the parameter d corresponds to the distance of the plane from the origin of the three-dimensional reference system. Any 3D point (e.g. the location of a sample) can be projected in the stratigraphic domain by computing its SP as the distance of the point from the plane.

This problem has a simple linear solution (see, for example, Shakarji and Srinivasan, 2013), which can be extended to any number of parallel planes to be fitted at once.

Vombat can solve this problem for any number of point sets, providing a reliable estimate of the average attitude of an outcrop. Estimated attitudes are then shown in the 3D window of CloudCompare as a dip vector and a direction line. Figure 4.2 shows three attitudes that were estimated on the point cloud. These are treated as generic CloudCompare's objects and listed in the DB Tree window.

4.2.3 Stratigraphic Reference Frames

We implemented in Vombat a specific object, called a *Stratigraphic Reference Frame* (SRF), which can be used to project any 3D point in the one-dimensional stratigraphic domain. Figure 4.2 shows how an SRF is represented in the three-dimensional window of CloudCompare. An SRF is basically a ruler that can be used to project any observation in a direction orthogonal to the bedding. In its simpler formulation, an SRF in Vombat is fully defined by a plane in space: when the plane is oriented as the bedding, the distance from the plane itself is a scalar value that can be used as a metric in the stratigraphic domain. The stratigraphic position of a point is thus computed as:

$$SP(\boldsymbol{p}) = \alpha \cdot \hat{\boldsymbol{n}} \cdot \boldsymbol{p} - d \tag{4.2}$$

where $\hat{\boldsymbol{n}}$ is the (unit) normal to the bedding plane, $\boldsymbol{p}$ is any 3D point and SP is the estimated Stratigraphic Position for that point. The parameter d is needed to define a plane in space and corresponds to the distance of the plane from the origin. Adding or subtracting any value from it corresponds to shifting the whole reference frame, while preserving the overall orientation. This simplified model is illustrated in Figure 4.5. The parameter α of Equation (4.2) is a scale factor that is normally equal to 1, but can also be modified, with the effect of *stretching* (>1) or *compressing* (<1) the reference frame, to accommodate specific needs.

Being defined by a plane, an SRF can be created in Vombat by using any `Attitude` object previously created by the user. The user selects the `Attitude` of interest in the DB Tree of CloudCompare, then activates the corresponding tool, which will initialize a new SRF accordingly oriented. Any newly created SRF is represented as a sort of ruler in the 3D window and is listed in the DB Tree of CloudCompare.

4.2.4 Vombat Objects and Their Stratigraphic Positions

The Stratigraphic Reference Frame introduced in the previous section allows a specific Stratigraphic Position (SP) to be associated with any point in space through Equation (4.2): for example, a stratigrapher might be interested in determining where a specific point on the three-dimensional model is positioned on the stratigraphic column. For this task, Vombat provides two different kinds of objects for which a Stratigraphic Position can be evaluated. The first one is the *Sample* object, as a generalization of a *rock sample*, which materializes any 3D point that might be of interest. An example of Samples on the outcrop can be seen later in Figure 4.9: each Sample is rendered in 3D as a small sphere, each with its SP value. The second one is the *Linear Geological Feature* object, which can be used to materialize a specific layer on the outcrop, for which the SP will be computed.

Each Sample object can also be used to represent a user-provided stratigraphic position, for example, because it was previously used as stratigraphic zero in former campaigns. In such a case the user can use the properties dialog provided by Vombat to force the specific Sample to have a fixed SP, which Vombat will maintain constant.

4.2.5 Stratigraphic Constraints to Build Composite Reference Frames

When dealing with large outcrops, it is desirable to deal with deformations and faulting in a consistent way. This can be done in many high-end 3D geological modelling software packages through the interpretation and the modelling of the depositional and tectonic geometries. Although this approach would produce the most exhaustive representation of the underlying geometries, it is often a highly time-consuming and specialized task, not within everyone's grasp.

Vombat implements a simpler way to deal with faulted or deformed sequences by permitting the creation of as many SRF as needed, each representing a locally undeformed portion of a larger outcrop. For example, a faulted sequence can be modelled by using two different SRFs, one valid in the footwall block and another one to be used in the hanging wall block. These two SRFs must then be aligned by changing their *d* parameter, so that the corresponding layers in the two regions will be at the same SP (Figure 4.6).

To do so, Vombat allows the user to impose specific *Stratigraphic Constraints*, which can thus be used to produce a composite stratigraphic reference frame. This concept is illustrated in Figure 4.6. Two types of constraints can be defined:

- A *Tie Constraint* can be created between two Sample objects, and correspond to providing the software with the information that those two points lie on the same stratigraphic horizon. This kind of constraint is used to establish a direct and known stratigraphic correlation between samples on separated parts of an outcrop.
- The known stratigraphic position of a Sample object can also be used as a constraint. This operation corresponds to fixing the position of that Sample in the stratigraphic column to a specific stratigraphic position. This feature is useful, for example, to define on which horizon the stratigraphic zero must be located.

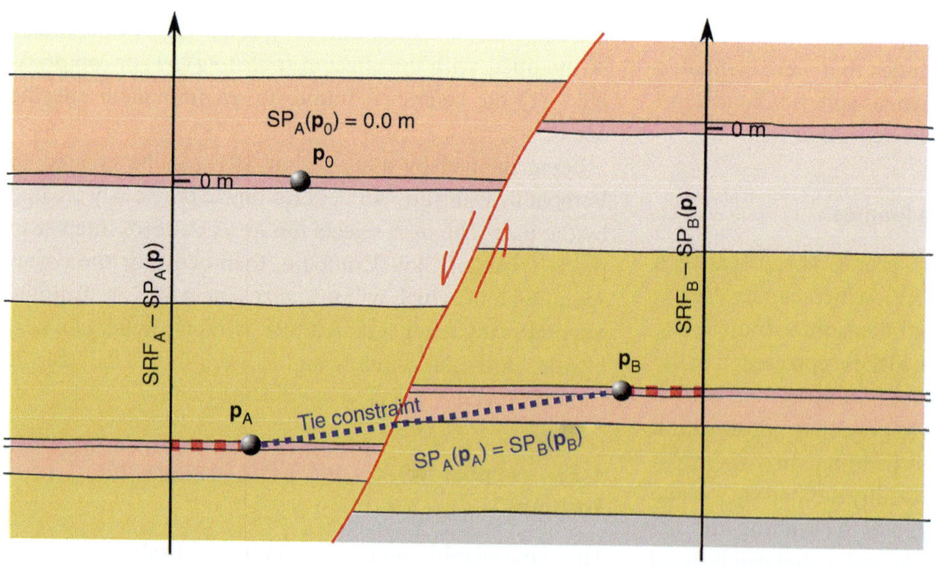

Figure 4.6 Representation of a faulted sequence for which two different SRFs are considered. To constrain the two SRFs, the user imposes the stratigraphic position of the Sample at $\mathbf{p}_0$ to be equal to 0.0 m. Such a constraint will suffice to fully determine the *d* parameter of SRF_A. To align SRF_B, the user instead establishes a *Tie Constraint* between two Samples, each associated with a different SRF ($\mathbf{p}_A$ and $\mathbf{p}_B$). Vombat will thus take care of aligning all the SRFs to match the constrained Stratigraphic Positions. Any number of constraints can be added: in such a case, the solution is provided by means of non-linear least squares. The two SRF, here represented as vertical lines, can be freely oriented in space in order to accommodate any geometry that can be locally approximated with a plane-parallel bedding.

Once the user has defined the Stratigraphic Constraints, Vombat recomputes the *d* parameters of the two SRFs, in order to accommodate the required shift between the two. The software takes care of any number of stratigraphic constraints that the operator might want to impose and will try to honor them in a least squares sense.

4.2.6 Creation of Continuous Stratigraphic Logs

The three-dimensional models obtained from TLS or photogrammetry often come associated with additional properties. TLSs normally provide intensity data that represent the recorded power of the back-scattered laser beam, while in the case of photogrammetry the models are often equipped with per-vertex Red–Green–Blue (RGB) color. Furthermore, the output of hyperspectral cameras can be mapped on the three-dimensional representation of the outcrop (see Sterzai et al., 2010; Kurz et al., 2012a, 2012b). These additional properties are normally in the form of scalar values associated with the vertices of the 3D model and can be used to create pleasant visualizations of the model by mapping the values to a color scale, but they can also provide insights about the lithologies.

Provided a point cloud with an associated scalar field, for example the intensity, was derived from the TLS, the user first estimates the average attitude of the outcrop with the method described in Section 4.2.2. This attitude will then be used to define a Stratigraphic Reference Frame, orthogonal to the bedding plane, that will be used to estimate the stratigraphic positions for each point of the cloud. From these two fields (the stratigraphic position and the intensity), it is then possible to estimate how the lidar intensity varies at varying stratigraphic positions: an example is shown in Figure 4.7.

To create continuous stratigraphic logs of a scalar field, Vombat uses a *Kernel Smoother* (Hastie et al., 2009), which makes it possible to estimate a continuous scalar function (the intensity log) from scattered observations of its value (the points of the cloud). For simplicity, we will consider the creation of a TLS intensity log, but the same method can be applied to any other scalar field. Let X : $\{x_i, \ldots, x_N\}$ be the set of computed stratigraphic positions for each vertex of the model and Y : $\{y_i, \ldots, y_N\}$ be the corresponding intensity value at the same location. Kernel smoothing can be seen as a weighted running average of the two scalar fields: the function $\hat{Y}(x_0)$ predicts the value of the lidar intensity at the

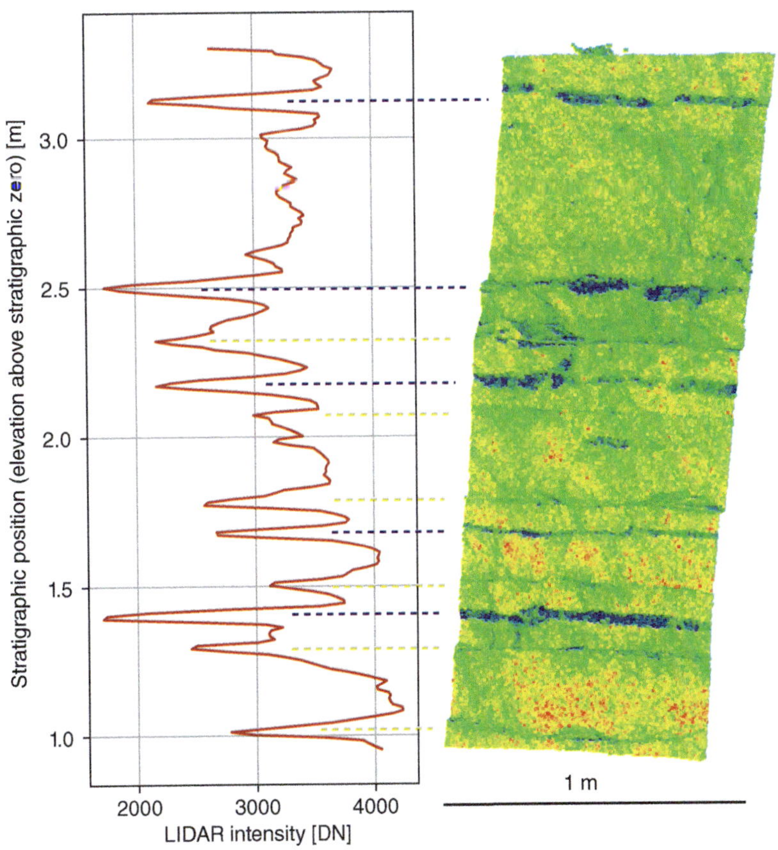

Figure 4.7 At right, a small extract of a TLS point cloud, colored by the intensity values obtained from the scanner, representing an outcrop composed of limestone (green and red colors) with some horizons and nodules of chert (blue color). The complete point cloud is also shown in Figure 4.9. Chert is characterized by a typical low-intensity response at the wavelength of this TLS. At left, the intensity has been extracted to a stratigraphic log. The log was computed along a direction orthogonal to the bedding by using Vombat. Chert nodules are clearly visible as local minimums in the log (blue dotted lines). Also notice that bedding planes, which are not easily identifiable in the point cloud, are characterized by a low-intensity response, probably due to a higher clay content (yellow dotted lines).

stratigraphic positions x_0:

$$\hat{Y}(x_0) = \frac{\sum_{i=1}^{N} K_\lambda(x_0, x_i) y_i}{\sum_{i=1}^{N} K_\lambda(x_0, x_i)} \tag{4.3}$$

where N is the total number of observations and y_i is the observation at x_i. The weights are provided by the kernel function $K_\lambda(x_0, x)$, centred in x_0:

$$K_\lambda(x_0, x) = D\left(\frac{|x_0 - x|}{\lambda}\right) \tag{4.4}$$

The size of the kernel (the averaging window) is controlled by the λ parameter (kernel radius), which thus determines the smoothness of the kernel smoother. $D(t)$ is a real-valued function, usually decreasing with increasing distance between x and x_0, which determines the shape of the weighting function. Different kernels $D(t)$ can be used, for example the Epanechnnikov kernel:

$$D(t) = \frac{3}{4}(1 - t^2)\mathbf{1}_{\{|t| \leq 1\}} \tag{4.5}$$

Note that the notation $\mathbf{1}_A$ is a short form for the indicator function (1 for $t \leq 1$ and 0 otherwise).

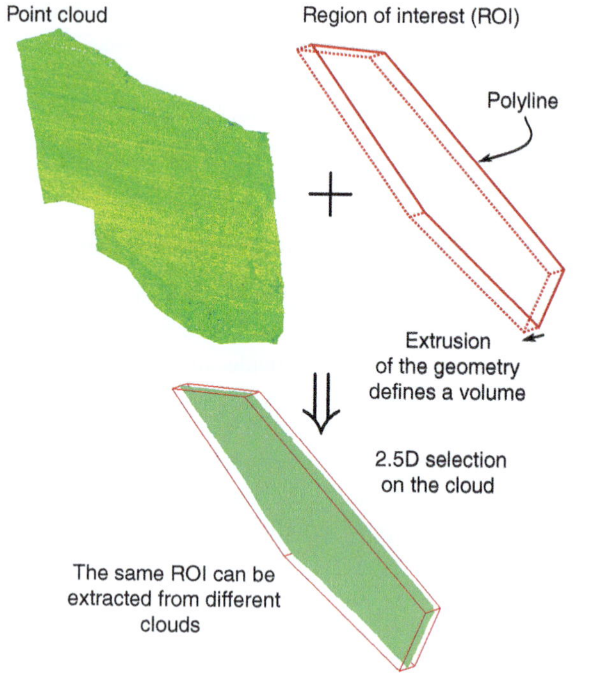

Figure 4.8 An ROI in the case of point clouds can be defined as a 2D polyline, extruded to define a volume. This kind of ROI, being halfway between a real *volume of interest* and an *area of interest* may be regarded as a 2.5D ROI. This approach allows replication in a consistent way of any operation of data selection on different models.

4.2.7 Regions of Interest

A *Region of Interest* (ROI) is a Vombat object that provides repeatable selections of a portion of a point cloud. It is created by the user as a closed polyline, which is then extruded at a specific depth to create a volume. Figure 4.8 illustrates this concept. This object allows regions of the outcrop to be unequivocally identified that might be of interest, e.g. undeformed regions bounded by faults on which stratigraphic computations will be performed. Being independent of the specific cloud, an ROI can be used to extract the same portions of the 3D model from different point clouds. This feature is especially useful when a VOM is updated, for example when a model at higher resolution becomes available: a ROI will make it possible to replicate on the new dataset the same operations of data selection that were previously done on the lower-resolution model.

4.2.8 Input/Output and Log Plotting

Vombat objects can be serialized (i.e. saved to disk files) as JSON, XML, and binary files. Some of the objects (e.g. the continuous stratigraphic logs) can also beexported as an ASCII text file for facilitating the use of the data in other software packages. A specific object, called Virtual Outcrop, represents a container for all data and objects related to a single outcrop. It is a hierarchical object that can be used to group together all the elements the user creates (Samples, ROIs, SRFs, stratigraphic logs, etc.). This object can be saved to disk and then reloaded on a different machine granting cross-platform exchange of the dataset. Basic log plotting functions have also been implemented in Vombat to provide immediate visual feedback of the obtained stratigraphic logs, as illustrated in Figure 4.2.

4.3 Examples

Vombat has been successfully used to prepare and to process several datasets that have been at the basis of cyclostratigraphical analyses, for example in Franceschi et al. (2015), while CloudCompare was also employed in the preparation and the analysis of the dataset presented in Penasa et al. (2014). In that work, use was made of intensity information to automatically classify a point cloud, with the goal of extracting chert layers in a limestone series in Central Italy. We show here how the same datasets can be further interpreted with Vombat.

In the next sections we will first show how Vombat can be used to extract stratigraphic logs of the intensity, in a case in which there is a demonstrated correlation between the TLS intensity and the lithology. In a second case we will show how a continuous log of TLS intensities can be easily correlated with a dataset that was collected in the field. As an example, we will use the total carbonate content curve of a pelagic succession exposed in the Smirra section (northeastern Apennines of Italy), previously published in Franceschi et al. (2015).

4.3.1 Locating Samples on a TLS Intensity Log

Figure 4.9a shows a TLS point cloud representing a sedimentary sequence at outcropping near Gubbio in Central Italy (see Penasa et al., 2014). The colors represent the measured TLS intensities: higher values of intensity (green to yellow) correspond to limestone while cherty and marly layers are characterized by lower values (blue). We used CloudCompare and Vombat to provide a simple interpretation of the geometries directly on the VOM. We started by tracing some bedding planes by using polylines that were then employed to estimate the average attitude of the outcrop, which is needed to initialize a Stratigraphic Reference Frame aligned with the direction perpendicular to the layering.

In the upper part of the point cloud, an iron protection net was also captured during the scanning operations. To exclude the net, we created a specific ROI that has been used to extract a subset of the point cloud. We then used this portion of the virtual outcrop to reconstruct a continuous log, with a sampling step of 1 cm, representing how the intensity varies at varying stratigraphic positions. The result of this operation is shown in Figure 4.9c, where the continuous stratigraphic log of the intensity is plotted together with the collected Samples at the right stratigraphic positions. No additional stratigraphic constraints were used in this example.

By using Vombat it is possible to identify specific locations on the outcrop by using Sample objects (purple spheres in Figure 4.9a), which are automatically projected in the stratigraphic column (dotted lines in Figure 4.9c), providing a geometrically accurate way of describing stratigraphic features both in the 3D and in the corresponding stratigraphic domain (1D). In this example, we placed a Sample in correspondence with some cherty horizons, which can be easily recognized on the colored TLS point cloud. Vombat automatically computes a stratigraphic position for each, using the created SRF. In a more general case, Sample objects could represent any punctual information that needs to be reported on a stratigraphic

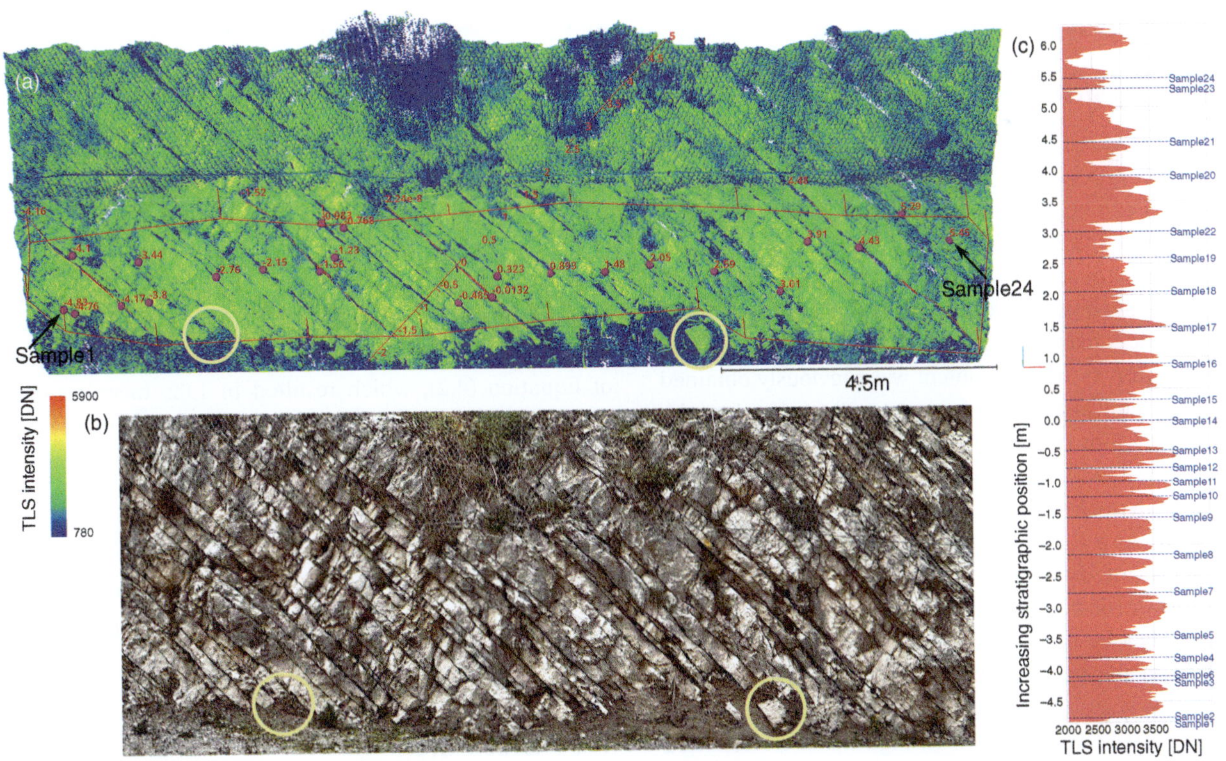

Figure 4.9 (a) TLS scan of an outcrop of cherty limestone (Lower Cretaceous Maiolica Formation) in Central Italy, nearby Gubbio, showing some of the objects that can be defined in Vombat. This same dataset was used in Penasa et al. (2014). Chert is characterized by lower intensity values in the infrared. (b) The same outcrop in the color domain; chert is mostly undetectable. As a reference, the yellow circles highlight corresponding points in the two views. (c) The continuous stratigraphic log of TLS intensities for the region defined with an ROI object.

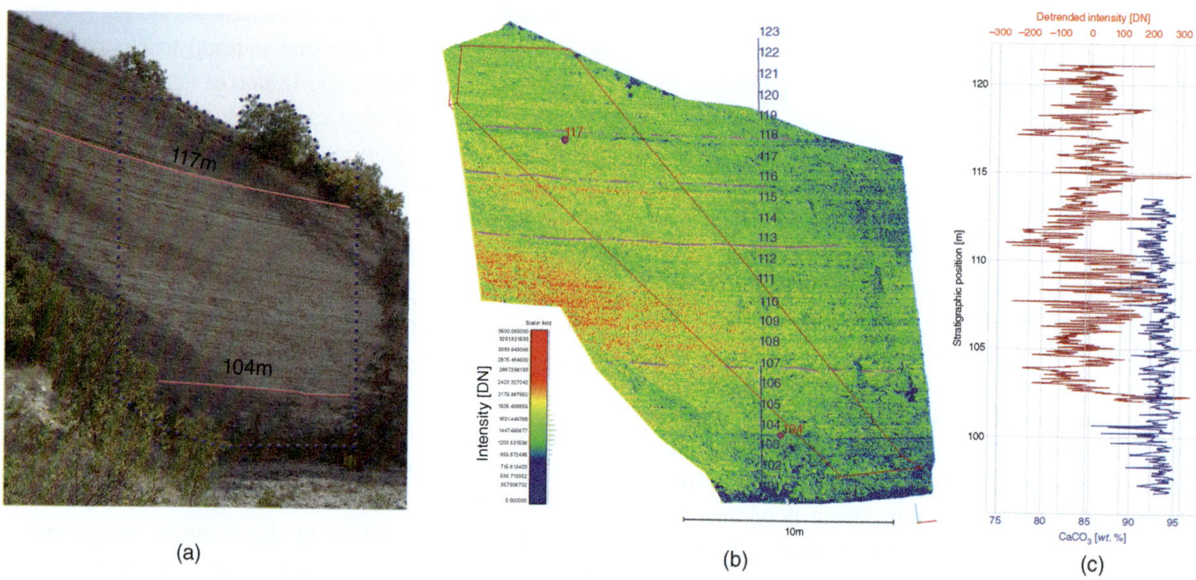

Figure 4.10 (a) The pelagic succession at the Smirra quarry, which comprises the Ypresian–Lutetian boundary (see also Franceschi et al., 2015, for additional details). The region highlighted by the dashed lines was acquired with TLS. The red lines are two horizons where the stratigraphic position was measured in the field during the collection of $CaCO_3$ rock samples. (b) A screenshot from CloudCompare with the Vombat objects that were created for processing the data. (c) The resulting intensity log, in alignment with the in-field measured $CaCO_3$ log.

scale (e.g. an annotation about a notable punctual feature observed in the field).

4.3.2 Using Stratigraphic Constraints to Match Field Data

With a second example, we illustrate how stratigraphic constraints can be used to align a stratigraphic reference frame to data that were previously collected in the field. We will use the dataset described in Franceschi et al. (2015) to show how an SRF can be forced to obey external stratigraphic constraints. In this specific case, a detailed log of the calcium carbonate content was previously obtained from the chemical analysis of samples collected in the field. At the same time, it was known that in a limestone-marl alternation, TLS intensity values should correlate to carbonate content (Franceschi et al., 2009). We thus wanted to compare how the calcium carbonate log correlates with a log of TLS intensities.

The data collected in the field are constituted by a log of the carbonate content in wt.%. Each sample that was collected in the field was positioned on the stratigraphic column by measuring its stratigraphic distance from a specific horizon with a ruler. In Figure 10a we reported the position of two stratigraphic horizons, as measured in the field, that were identified both on the field log and on the virtual outcrop.

We used this information to create a "virtual counterpart" to the metric framework that was used in the field to measure sample positions. This can be easily done in Vombat, as depicted in Figure 4.10b. A Stratigraphic Reference Frame was set up using some traces of the bedding planes to obtain the average attitude. Then the same horizons of Figure 4.10a were recognized on the scan and marked using two Sample objects. We then imposed these Sample objects to have a fixed Stratigraphic Position of 104 and 117 meters, respectively. Vombat then automatically aligned the Stratigraphic Reference Frame to these constraints. Additionally, to compensate for the measurement errors that cumulate when taking thickness measures in the field, we also computed the α parameter of Equation (4.2), which resulted in 1.02, meaning that the error introduced by the operator in the field must have been in the order of a couple of meters for 100 m of section measured.

Finally, we reconstructed the continuous log of the intensities, as shown in Figure 4.10c, which can now be plotted in the same stratigraphic frame of the $CaCO_3$ samples, thus providing a real bridge between the intensity log from the 3D model and the samples collected in the field.

4.4 Discussion

In stratigraphy, outcrops often represent the primary source of information for geoscientists. Classical methods of investigation based on the realization of stratigraphic columns require the compression of the information collected on an outcrop, which is obviously three-dimensional, into a one-dimensional log where each observation (lithology, lateral bedding geometry, presence of fossils, etc.) is annotated by assigning it a stratigraphic position.

This procedure makes it difficult to maintain the original relationship between observations and the real outcrop. Furthermore, each research team or scientist produces its own log, so that the integration of stratigraphic logs of the same outcrop made by different researchers can be difficult. This problem has been normally solved by producing annotated photographic documentation of the outcrop: in this sense the use of three-dimensional imaging could be viewed as the natural evolution of this established practice in sedimentary geology. One of the key features of VOMs is that they provide a permanent and accurate geometric representation of the outcrop that can be preserved through time, even when an outcrop is degraded (Lozar et al., 2015; Dumitriu, 2013). This might be particularly relevant to some specific outcrops that represent the physical expression of key geologic boundaries or events and become of worldwide reference (GSSP; Salvador, 2013).

The model used in Vombat to define a Stratigraphic Reference Frame can be seen as an extremely simplified version of the modelling tools that are available in many commercial 3D geological modelling software. Although our tool does not allow complex deformed geometries to be modelled, it can be used to create fairly complex setups. For example, it is possible to produce composite sections by linking multiple reference systems. The simplicity of this tool permits stratigraphic measurements to be easily carried out, without the need to produce an exhaustive modelling of the underlying geometry, which is normally a highly specialized and time-consuming task. Furthermore, the implementation is made with the aim of being as generic as possible, thus allowing, in the future, to extend its capabilities to account for more complex outcrop geometries.

Due to the simplified geological modelling technique used in the definition of SRFs, this system is expected to perform better when applied to undeformed plane-parallel bedding. In the case of sediments that were not originally deposited as plane-parallel layers (e.g. carbonate platform slope deposits), it might still be possible to define smaller portions of the outcrop that can be approximated by a parallel bedding, by considering one or more narrow ROIs on the outcrop. The generated stratigraphic logs suffer from the same problems that might be present in petrophysical logs obtained from well data: e.g. faults and hiatuses/unconformities that might be present are not directly taken into account by the software. On the other hand, they can be more easily identified thanks to the VOM itself. Being open source, Vombat could be implemented to account for some of the listed problems.

Vombat was originally designed for obtaining long TLS intensity logs to be used for cyclostratigraphical analysis, for which long and continuous stratigraphic logs are needed to perform time-series analysis (Hinnov, 2013). In fact, three-dimensional models, possibly associated with other remotely sensed data, are also a potential source of high-resolution lithological information that can complement classical methods of investigation. When dealing with inaccessible outcrops, remote sensing methods might be the only chance to obtain information on a valuable stratigraphic record. The continuous stratigraphic logs that can be obtained using Vombat allow the variations of specific properties to be quantified at varying stratigraphic depths directly from outcrop data, mimicking the information obtained by means of petrophysical logs, which can be normally obtained only from a well log. TLS intensity logs in the infrared domain, for example, have been demonstrated to be a valid proxy for lithological variations in some instances (Franceschi et al., 2011, 2015; Burton et al., 2011).

4.5 Conclusions

The open source Vombat toolkit allows some of the most common operations to be performed in a simple way that might be needed by a stratigrapher dealing with Virtual Outcrop Models. The source code is made freely available to the community on the project page https://github.com/luca-penasa/vombat, where a precompiled version of CloudCompare with Vombat can be found. Vombat has been developed for the purpose of providing a user friendly and easily extensible 3D environment for the analysis of VOMs by making it possible to:

- Estimate attitudes by means of bestfitting planes on multiple coplanar set of points.
- Create stratigraphic reference frames, and merge multiple stratigraphic sections into a single and continuous measurement framework.
- Automatically compute stratigraphic positions for samples and for bedding traces.
- Generate continuous stratigraphic logs of the properties attached to the point clouds, as the intensity or RGB values.
- Create an exchangeable database of geological features directly on the outcrop.

Commercially available point cloud processing software is generally designed for engineering and architecture applications. Although some dedicated solutions have been recently developed, they are difficult to extend due to their closed-source nature. Vombat is open source and also provides a simple C++ programming interface that can be used to perform geological-related operations by the researcher with programming practice. Although still being a simple tool, Vombat is novel in its approach to the extraction and storage of stratigraphic logs that may display high-resolution 3D-property logs or any observation linked to a VOM.

Being able to compute a stratigraphic position for any three-dimensional stratigraphic feature that might be reported on the VOM, Vombat makes it possible to produce a highly accurate dataset of observations, with real thickness measures, while still maintaining the three-dimensional nature of the data itself. Relevant data and observations can then be preserved on the VOM as they were collected directly on the outcrop and visualized as a stratigraphic log. This aspect might be of particular relevance when dealing with outcrops of special chronostratigraphic value, as the ones selected to represent specific time boundaries (GSSP) by the International Commission of Stratigraphy. These outcrops might be made more accessible thanks to tools similar to Vombat, which allows the stratigraphic data to be merged with their original, three-dimensional context.

Acknowledgment

The authors thank Daniel Girardeau-Montaut, creator of CloudCompare, for the help provided during Vombat's development.

References

Bellian, J., Kerans, C., and Jennette, D. (2005). Digital outcrop models: applications of terrestrial scanning lidar technology in stratigraphic .odeling. *Journal of Sedimentary Research* 75 (2): 166–176. doi:https://doi.org/10.2110/jsr.2005.013.

Bistacchi, A., Mittempergher, S., and Martinelli, M. (2021). Digital outcrop model reconstruction and interpretation. In: *3D Digital Geological Models: From Terrestrial Outcrops to Planetary Surfaces*. Wiley, this volume.

Brodu, N. and Lague, D. (2012). 3D terrestrial lidar data classification of complex natural scenes using a multi-scale dimensionality criterion: Applications in geomorphology. *ISPRS Journal of Photogrammetry and Remote Sensing* 68: 121–134.

Buckley, S., Ringdal, K., Dolva, B. et al. (2017). LIME: 3D visualisation and interpretation of virtual geoscience models In *EGU General Assembly Conference Abstracts. EGU General Assembly Conference Abstracts* 19: 15952.

Buckley, S.J., Enge, H.D., Carlsson, C., and Howell, J.A. (2010). Terrestrial laser scanning for use in virtual outcrop geology. *The Photogrammetric Record* 25 (131): 225–239.

Burton, D., Dunlap, D.B., Wood, L.J., and Flaig, P.P. (2011). Lidar intensity as a remote sensor of rock properties. *Journal of Sedimentary Research* 81 (5): 339–347. doi: https://doi.org/10.2110/jsr.2011.31.

Caravaca, G., Le Mouélic, S., Mangold, N. et al. (2020). 3D digital outcrop model reconstruction of the Kimberley outcrop (Gale crater, Mars) and its integration into Virtual Reality for simulated geological analysis. *Planetary and Space Science* 182 (104): 808. doi: https://doi.org/10.1016/j.pss.2019.104808.

Corradetti, A., Seers, T., Billi, A., and Tavani, S. (2021). Virtual outcrops in a pocket: The Smartphone as a fully equipped photogrammetric data acquisition tool. *GSA Today* 31 (9): 4–9. doi: https://doi.org/10.1130/GSATG506A.1.

Dewez, T.B., Girardeau-Montaut, D., Allanic, C., and Rohmer, J. (2016). Facets: A Cloudcompare plugin to extract geological planes from unstructured 3D point clouds. *ISPRS–International Archives of the Photogrammetry, Remote Sensing and Spatial Information Sciences, XLI-B5* 799–804. doi: 10.5194/isprsarchives-XLI-B5-799-2016.

Dumitriu, T.-C. (2013). Towards digital preservation of Repedea's big quarry outcrop using "Structure-from-Motion" photogrammetry tools. *Analele Stiintifice de Universitatii AI Cuza din Iasi. Sect. 2, Geologie* 59 (2): 19.

Ellis, D.V. and Singer, J.M. (2007). *Well Logging for Earth Scientists*, 2nde. Dordrecht, Literaturangaben: Springer.

Fernández, O. (2005). Obtaining a best fitting plane through 3D georeferenced data. *Journal of Structural Geology* 27 (5): 855–858.

Franceschi, M., Teza, G., Preto, N. et al. (2009). Discrimination between marls and limestones using intensity data from terrestrial laser scanner. *ISPRS Journal of Photogrammetry and Remote Sensing* 64 (6): 522–528. doi: https://doi.org/10.1016/j.isprsjprs.2009.03.003.

Franceschi, M., Preto, N., Hinnov, L.A. et al. (2011). Terrestrial laser scanner imaging reveals astronomical forcing in the Early Cretaceous of the Tethys realm. *Earth and Planetary Science Letters* 305 (3–4): 359–370. doi: https://doi.org/10.1016/j.epsl.2011.03.017.

Franceschi, M., Penasa, L., Coccioni, R. et al. (2015). Terrestrial laser scanner imaging for the cyclostratigraphy and astronomical tuning of the Ypresian–Lutetian pelagic section of Smirra (Umbria–Marche Basin, Italy).). *Palaeogeography, Palaeoclimatology, Palaeoecology* 440: 33–46. doi: https://doi.org/10.1016/j.palaeo.2015.08.027.

Girardeau-Montaut, D. (2014). *Cloudcompare, a 3D point cloud and mesh processing free software*. Telecom Paris Tech: EDF R&D.

Hartzell, P., Glennie, C., Biber, K., and Khan, S. (2014). Application of multispectral LiDAR to automated virtual outcrop geology. *ISPRS Journal of Photogrammetry and Remote Sensing* 88: 147–155. doi: https://doi.org/10.1016/j.isprsjprs.2013.12.004.

Hastie, T., Tibshirani, R., and Friedman, J. (2009). *The Elements of Statistical Learning*, 2nde, vol. 1. New York: Springer Series in Statistics.

Hinnov, L.A. (2013). Cyclostratigraphy and its revolutionizing applications in the earth and planetary sciences. *Geological Society of America Bulletin* 125 (11–12): 1703–1734. doi: https://doi.org/10.1130/B30934.1.

Krupnik, D., Khan, S., Okyay, U. et al. (2016). Study of Upper Albian rudist buildups in the Edwards Formation using ground-based hyperspectral imaging and terrestrial laser scanning. *Sedimentary Geology* 345: 154–167. doi: https://doi.org/10.1016/j.sedgeo.2016.09.008.

Kurz, T.H., Buckley, S.J., Howell, J.A., and Schneider, D. (2008). Geological outcrop modelling and interpretation using ground based hyperspectral and laser scanning data fusion. *International Archives of Photogrammetry, Remote Sensing and Spatial Information Sciences* 37: B5.

Kurz, T.H., Dewit, J., Buckley, S.J. et al. (2012a). Hyperspectral image analysis of different carbonate lithologies (limestone, karst and hydrothermal dolomites): The Pozalagua Quarry case study (Cantabria, North-West Spain): Hyperspectral image analysis of carbonate lithologies. *Sedimentology* 59 (2): 623–645. doi: https://doi.org/10.1111/j.1365-3091.2011.01269.x.

Kurz, T.H., Buckley, S.J., and Howell, J.A. (2012b). Close range hyperspectral imaging integrated with terrestrial LiDAR scanning applied to rock characterization at centimeter scale. *International Arch. Photogramm. Remote Sens. Spat. Inf. Science* 34: 417–422.

Lague, D., Brodu, N., and Leroux, L. (2013). Accurate 3D comparison of complex topography with terrestrial laser scanner: Application to the Rangitikei canyon (N-Z). *ISPRS Journal of Photogrammetry and Remote Sensing* 82: 10–26. doi: https://doi.org/10.1016/j.isprsjprs.2013.04.009.

Lozar, F., Clari, P., Dela Pierre, F. et al. (2015). Virtual tour of past environmental and climate change: The Messinian Succession of the Tertiary Piedmont Basin (Italy). *Geoheritage* 7 (1): 47–56. doi: https://doi.org/10.1007/s12371-014-0098-8.

Malkamäki, T., Kaasalainen, S., and Ilinca, J. (2019). Portable hyperspectral lidar utilizing 5 GHz multichannel full waveform digitization. *Optics Express* 27 (8): A468–A480. doi: https://doi.org/10.1364/OE.27.00A468.

Marques, A., Horota, R.K., de Souza, E.M. et al. (2020). Virtual and digital outcrops in the petroleum industry: A systematic review. *Earth-Science Reviews* 208 (103): 260. doi: https://doi.org/10.1016/j.earscirev.2020.103260.

Martinsen, O.J., Pulham, A.J., Haughton, P.D., and Sullivan, M.D. (2011). *Outcrops Revitalized: Tools*. SEPM: *Techniques and Applications*.

Miall, A.D. (2016). *The Stratigraphic-Sedimentologic Data Base*, 33–75. Cham: Springer International Publishing doi: https://doi.org/10.1007/978-3-319-24304-7_2.

Mountrakis, G., Im, J., and Ogole, C. (2011). Support vector machines in remote sensing: A review. *ISPRS Journal of Photogrammetry and Remote Sensing* 66 (3): 247–259. doi: https://doi.org/10.1016/j.isprsjprs.2010.11.001.

Penasa, L., M. Franceschi, M., Preto, N. et al. (2014). Integration of intensity textures and local geometry descriptors from Terrestrial Laser Scanning to map chert in outcrops. *ISPRS Journal of Photogrammetry and Remote Sensing* 93: 88–97. doi: https://doi.org/10.1016/j.isprsjprs.2014.04.003.

Penasa, L., Franceschi, M., Gattolin, G. et al. (2019). Cyclostratigraphic investigations in the Calcare Massiccio (Early Jurassic, Umbria-Marche Basin) through photogrammetry. In: *250 Million Years of Earth History in Central Italy: Celebrating 25 Years of the Geological Observatory of Coldigioco, Geological Society of America Special Paper*, vol. 542, 87–103.

Rarity, F., van Lanen, X.M.T., Hodgetts, D. et al. (2014). LiDAR-based digital outcrops for sedimentological analysis: Workflows and techniques. *Geological Society, London, Special Publications* 387 (1): 153–183. doi: https://doi.org/10.1144/SP387.5.

Salvador, A. (ed.) (2013). *International Stratigraphic Guide: A Guide to Stratigraphic Classification, Terminology, and Procedure*. Boulder, Colorado: Geological Society of America doi: 10.1130/9780813774022.

Shakarji, C.M. and Srinivasan, V. (2013). Theory and algorithms for weighted total least-squares fitting of lines, planes, and parallel planes to support tolerancing standards. *Journal of Computing and Information Science in Engineering* 13 (3): 031,008.

Sterzai, P., Vellico, M., Berti, M. et al. (2010). LiDAR and hyperspectral data integration for landslide monitoring: The test case of Valoria landslide. *Italian Journal of Remote Sensing* 42 (3): 89–99.

Summerfield, M. (2011). *Advanced QT Programming: Creating Great Software with C++ and QT 4*. Upper Saddle River, NJ: Prentice-Hall.

Suomalainen, J., Hakala, T., Kaartinen, H. et al. (2011). Demonstration of a virtual active hyperspectral LiDAR in automated point cloud classification. *ISPRS Journal of Photogrammetry and Remote Sensing* 66 (5): 637–641. doi: https://doi.org/10.1016/j.isprsjprs.2011.04.002.

Tavani, S., Arbues, P., Snidero, M. et al. (2011). Open Plot Project: An open-source toolkit for 3-D structural data analysis. *Solid Earth* 2 (1): 53–63. doi: https://doi.org/10.5194/se-2-53-2011.

Telling, J., Lyda, A., Hartzell, P., and Glennie, C. (2017). Review of Earth science research using terrestrial laser scanning. *Earth-Science Reviews* 169: 35–68. doi: https://doi.org/10.1016/j.earscirev.2017.04.007.

Thiele, S.T., Grose, L., Samsu, A. et al. (2017). Rapid, semi-automatic fracture and contact mapping for point clouds, images and geophysical data. *Solid Earth* 8 (6): 1241–1253. doi: https://doi.org/10.5194/se-8-1241-2017.

Thiele, S.T., Lorenz, S., Kirsch, M. et al. (2021). Multi-scale, multi-sensor data integration for automated 3-D geological mapping. *Ore Geology Reviews* 136 (104): 252. doi: https://doi.org/10.1016/j.oregeorev.2021.104252.

Wilkinson, M., Jones, R., Woods, C. et al. (2016). A comparison of terrestrial laser scanning and structure-from-motion photogrammetry as methods for digital outcrop acquisition. *Geosphere* 12 (6): 1865–1880. doi: https://doi.org/10.1130/GES01342.1.

Woodcock, N.H. (1977). Specification of fabric shapes using an eigenvalue method. *Geological Society of America Bulletin* 88 (9): 1231–1236.

5

Interpretation and Mapping of Geological Features Using Mobile Devices in Outcrop Geology: A Case Study of the Saltwick Formation, North Yorkshire, UK*

Christian Kehl[1,2,4], *James R. Mullins*[3], *Simon J. Buckley*[1,2], *John A. Howell*[3], *and Robert L. Gawthorpe*[2]

[1] *Uni Research AS, Bergen, Norway*
[2] *Department of Earth Science, University of Bergen, Norway*
[3] *Department of Geology and Petroleum Geology, University of Aberdeen, UK*
[4] *Utrecht University, The Netherlands*

Abstract

Digital outcrops have seen widespread adoption in geoscience applications for direct interpretation and analysis tasks. Handheld devices are becoming powerful computers capable of displaying outcrop models in 3D, creating new opportunities for interpreting geology directly in the field. This chapter describes such handheld-based digital techniques and their use in field geology. A workflow from the direct outcrop interpretation to a statistical reservoir model via Multiple Point Statistics (MPS) is given in a field study on the Yorkshire coastline in the UK. The workflow allows large parts of formerly office-based tasks to be performed directly in the field. Computationally expensive operations are conducted in the basecamp via laptops, which is facilitated by export options to common outcrop and reservoir modelling packages. The result is a streamlined workflow that enriches the fieldwork. Prevalent handheld limitations demand careful quality control when transferring field results to subsequent modelling stages. The case study of Saltwick Formation in this chapter merits in a static reservoirs model, conditioned to field-documented sedimentary logs, which is analysed in terms of geobody connectivity and lithofacies distribution. The resulting outcrop-based reservoir model can be queried for reservoir quality characteristics for analogous, major Brent Group hydrocarbon fields.

5.1 Introduction

Digital tools on mainstream office workstations, such as Google Earth, ArcGIS, and other open-source cartographic services, are widely applied in geoscience domains as means of studying and analysing geological and geophysical problems, as shown by McCaffrey et al. (2005) and Buckley et al. (2019). LiDAR and photogrammetry are state-of-the-art technologies that capture dense, high-precision surface geometry of natural surfaces draped with photographic imagery (i.e. "photorealistic" 3D models in geosciences), as demonstrated by multiple research groups in digital outcrop geology (Buckley et al., 2008; Jones et al., 2009; Olariu et al., 2011; Gonzaga et al., 2018; Chapter 2 by Bistacchi et al., 2021). Several case studies in recent years, such as by Bistacchi et al. (2015), Siddiqui et al. (2019), Dujoncquoy et al. (2019), and Qu et al. (2021), use high-precision DOMs for interpreting and analysing structural and sedimentary architectures. In those studies, outcrop data are interpreted in the office, applying workflows for connecting field-geology observations and office-based reservoir analog modelling. Establishing this reservoir modelling workflow in the field with current technologies is impractical because of technical difficulties, the large size and weight of computing equipment, its limited battery power, and the required computing power for processing large data of high-precision DOMs.

The advent of handheld devices puts an arsenal of physical and locality sensors, high-resolution cameras, powerful general and graphics processors, and touch-technology interfaces into the hands of geoscientists. Handheld software, which makes use of these advanced technological components, is a potentially powerful instrument to collect, synthesize, and interpret field geology information, as conceptually demonstrated by Kehl et al. (2017a). The prospect advantages of mobile handheld devices for geology has been furthered by virtual field trips, which recently took center-stage, as shown by Rader et al.

*Corresponding authors: Christian Kehl, christian-kehl@web.de and James R. Mullins, james.mullins@abdn.ac.uk

3D Digital Geological Models: From Terrestrial Outcrops to Planetary Surfaces, First Edition.
Edited by Andrea Bistacchi, Matteo Massironi, and Sophie Viseur.

(2021). Recent studies in the geosciences for using mobile devices, such as Novakova and Pavlis (2017), Cagliari et al. (2017), Tavani et al. (2019), and Senger and Nordmo (2021), focus on measuring rock deposition orientations, annotating images, or auto-registering DSLR photographs via smartphone-captured orientation information, which is in line with earlier studies by Kehl et al. (2017b). Further applications for digital outcrop models on mobile devices within the geosciences are manifold: interpreted models are used to highlight zones of stress failure on landslide-prone outcrops in the field and track successive rock failures in shorter time spans, as shown by Prokop and Panholzer (2009). In petroleum geology, interpreted outcrop models are used as a basis for surface analog studies of subsurface hydrocarbon reservoirs, as described by Howell et al. (2014) and Dujoncquoy et al. (2019). A threshold to overcome for the wide adoption of mobile technologies is their supplementary incorporation in target workflows (e.g. in reservoir modelling) that enriches state-of-the-art outcrop and subsurface analysis. MPS, introduced by Strebelle et al. (2001), is a modern modelling technique that allows for stochastic scenario modelling of subsurface environments with highly heterogeneous lithological composition. The MPS modelling process is built upon smart pattern replication from input training images (TIs), conditioned to surface (i.e. sedimentary logs) and subsurface (e.g. wells and cores) measurements.

This chapter introduces novel digital techniques and technologies for field-based geological interpretations on handheld devices. The research is motivated by the need for improved field-based interpretation tools in geology to rapidly generate qualitative outcrop models, preferably performing all interpretation processes while working in the field on a specific locality, without the need for office-exclusive tools. Qualitative outcrop models hereby refer to facies-mapped outcrops, the delineation of sedimentary (e.g. channels, overbank deposits) or structural (e.g. fault cores, fractures) features, and the annotation of such objects with auxiliary measurements (e.g. dipping angle). In contrast to prevalent and recent literature, our approach not only uses mobile devices as measuring tools or digital cameras alone, but facilitates the more direct mapping of acquired information onto pre-captured 3D DOMs and the direct interpretation of the 3D photo-textured surface according to geological features on the mobile device.

This chapter also introduces a novel workflow, as an application of mobile device field technology, for the generation of stochastic subsurface geocellular models via DOM analogs that are interpreted directly in the field. Observed cross-sections of outcropping sedimentary elements are used to create a subsurface volumetric model. The creation process of such geocellular models as TI for MPS simulations using interpreted DOMs is described in detail. This mobile device-based reservoir modelling workflow is introduced by a field study of the Whitby coastline, being a surface analog for major Brent Group hydrocarbon reservoirs. Within the field study, it becomes apparent how the use of mobile technology closely links field- and office-based workflows in multiple stages of the modelling, which is another contribution of this chapter.

The structure of the chapter is as follows: Section 5.2 introduces the geological setting of the Saltwick formation at the Whitby coastline in Yorkshire, UK and its analogy to major Brent Group reservoirs. In Section 5.3, the main process and the involved techniques of going from an outcrop interpretation to a subsurface reservoir model is illustrated according to previously established workflows. Novel approaches of handheld outcrop interpretation technology are explained in Section 5.4, based on surface geometry and outcrop photograph correspondence. Section 5.5 presents the process in which the reservoir model is generated from the handheld outcrop interpretations using MPS modelling. Subsequently, Section 5.6 discusses the results of the Saltwick formation case study and Section 5.7 discusses the mobile device approach, future research directions, and further benefits for geoscience applications.

5.2 The Geological Setting: The Saltwick Formation

The Ravenscar Group (formerly named the "Lower Deltaic Series") accumulated during the Middle Jurassic in the UK's Cleveland Basin to the south of the Mid-North Sea High (Mjøs and Prestholm, 1993). The fluvio-deltaic Saltwick Formation is the oldest of the five formations that comprise the Ravenscar Group (Hemingway and Knox, 1973; Mjøs and Prestholm, 1993). The Saltwick Formation is conformably overlain by the Eller Beck, Cloughton, Scarborough, and Scalby Formations (Mjøs and Prestholm, 1993). The contact with the Eller Beck formation has been interpreted as a transgressive surface and as a major marine flooding event, resulting in a northward shift in sedimentation (Alexander and Gawthorpe, 1993).

The Saltwick Formation unconformably overlies the marine Dogger Formation (Lias Group). The basal surface has been interpreted as a major sequence boundary (sensu Van Wagoner, 1988), marking the initiation of a period of deltaic deposition over the underlying marine sediments (Mjøs and Prestholm, 1993; Alexander and Gawthorpe, 1993). The Saltwick Formation has a highly variable thickness and can be best observed in a series of 50-meter cliff section near the town of Whitby (see the

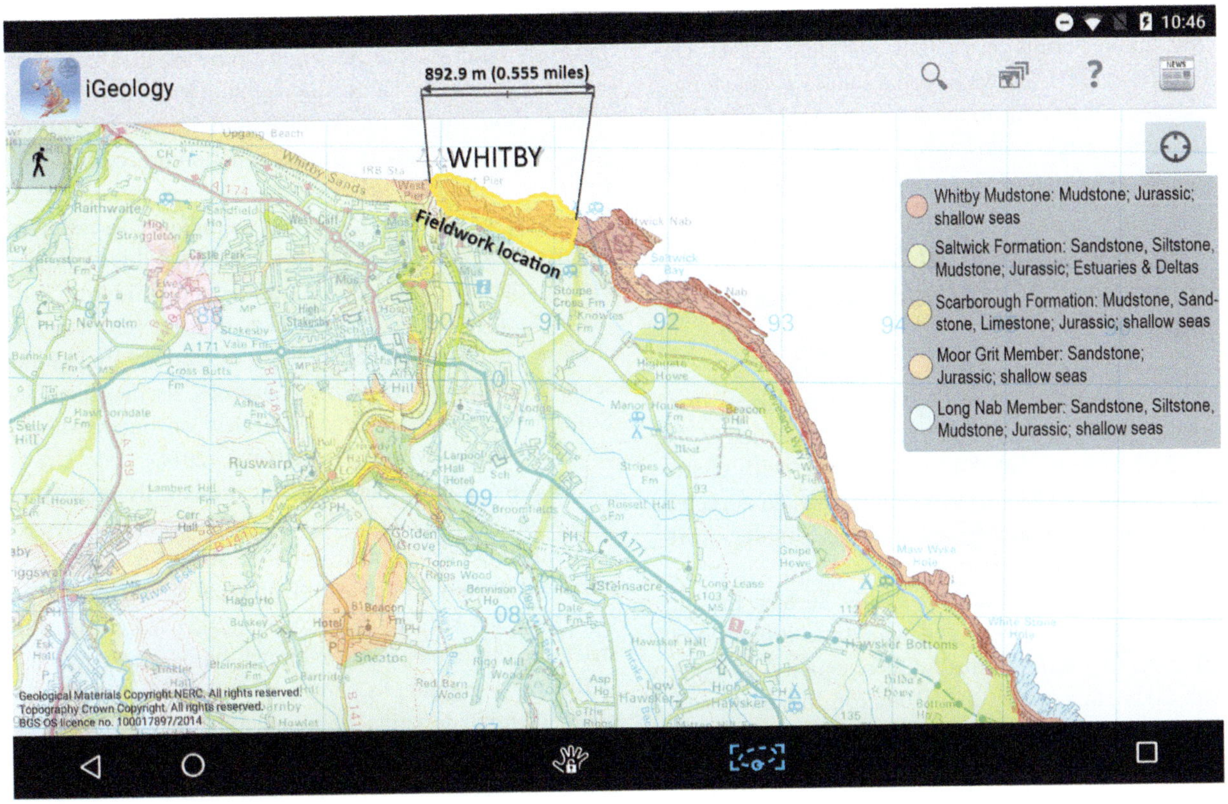

Figure 5.1 Geological map of north-east Yorkshire, UK, exported on a handheld device (Source: iGeology *BGS*, 2013).

bright yellow area in the geological map of Figure 5.1, generated with the iGeology app *BGS*, 2013). The Saltwick Formation is the delta top of a fluvio-deltaic system and is comprised of channel sandstones, crevasse splays, levee, and overbank deposits. Vertical (Knox, 1969; Livera and Leeder, 1981) and horizontal variations (Alexander and Gawthorpe, 1993) have been previously well documented. Some outcrop sections are dominated by multistory, medium-grained, cross-bedded sandstone channel bodies separated by third order minor scour surfaces (sensu Miall, 1985), whereas others consist of fine-grained sediments and isolated sandstones.

The basal channel sandstones of the Saltwick formation are generally larger composite bodies, whereas channel sandstones in the upper parts of the formation are smaller with low width/thickness and sand/mud ratios (Mjøs and Prestholm, 1993). The low width/thickness ratio of the channel sandstones towards the top of the formation has been attributed to high rates of accommodation creation and differential compaction of underlying sediments (Mjøs and Prestholm, 1993). This in turn resulted in unstable channel belts, favoring shallow incisions within the channel belt and diagonal, offset stacking, with channels preferentially favoring paleotopographical lows (Mjøs and Prestholm, 1993).

Channel sandstones in the formation tend to be encased in planar overbank deposits and crevasse splay sandstone and mudstone deposits. Crevasse splays (generally <2 m in thickness) typically display a wedge-shaped cross-sectional geometry near to the distributary channel and become more sheet-like in cross-section with increasing distance to the distributary channel (Mjøs et al., 1993). Crevasse splay sandstones are recognized in outcrops as fine to medium-grained isolated or stacked parallel laminated sandstones with an undulating erosional base, with evidence of soft sediment deformation in the underlying background sedimentation with a planar top surface. Individual crevasse splays are separated by mudstone drapes.

The Saltwick Formation is notable as being an analog for the Middle Jurassic Ness Formation (Livera and Leeder, 1981; Dreyer, 1990; Mjøs et al., 1993; Mjøs and Prestholm, 1993) and the Bathonian Sleipner Formation (Vollset and Doré, 1984). Situated close to the axis of the Viking Graben, the Ness Formation composes the largest reservoir-bearing unit contained within the Brent Field (Livera, 1989). Outcropping in the southern Viking Graben, the Sleipner Formation is believed to be broadly equivalent to the Ness Formation (Vollset and Doré, 1984). Sleipner reservoirs can be correlated into the Central Graben and

Norwegian–Danish Basin where it is contemporaneous with the Bryne Formation (Vollset and Doré, 1984). The studied section of the Saltwick Formation is located along a 980 m section of sea cliffs to the east of the town of Whitby (see Figure 5.1, bright-yellow area).

5.3 From Geological Surface Interpretation to Statistical Subsurface 3D Models

A given starting point for modern reservoir analog modelling is the presence of a surface model, embedded in 3D space, which is represented as a TIN draped with photographic images. The TIN in this chapter was constructed on the basis of a TLS pointset scan, which is combined with SLR photo texture overlays to generate a DOM. Figure 5.2 gives an impression of the outcrop's scale. Because the cliff section is vertically straight and with little dimensional variability (earlier termed "three-dimensionality" by Buckley et al., 2008), this specific digital outcrop does not benefit from the augmentation of fusing TLS and photogrammetric outcrop models, as technically introduced by Buckley et al. (2010). The resulting DOM is subsequently subject to a sedimentary interpretation based on observations from a field campaign. The field observations have been simultaneously collected with the DOM acquisition.

5.3.1 Digital Geological Interpretation Mapping

Digital geological interpretation follows the observations and measurements taken in the field. In order to synthesize and embed field data into the DOM, the model is loaded into an interpretation platform such as LIME, VRGS by Trinks et al. (2005) and Rarity et al. (2014), or GoCAD by Mallet (1992). The goal of the interpretation based on TIN geometry is the extraction of depositional elements similar to previously presented work by Enge et al. (2007), Eide et al. (2014), and Rittersbacher et al. (2014). Line segments also indicate the boundary demarcation of the bedding along the outcrop cross-section, analogous to procedures described by Labourdette and Jones (2007). Successive bedding demarcations form zones in 2D (along the outcrop) and 3D (within the resulting reservoir model).

The interpreted line segments are subsequently exported from an outcrop interpretation software to a reservoir modelling package. There, line segment interpretations are used as input for bedding surfaces and the direct extrusion into 3D facies models, including zonation procedures. The resulting zones can either be used as a coarse model or be refined using facies modelling techniques described in Ringrose and Bentley (2015). Alternatives to this workflow include the manual, fully deterministic definition of sediment architectural elements (i.e. drawing cropped-out channels in 3D) or the application of modern analogs for semi-deterministic modelling purposes, as described by Ruiu et al. (2015). A larger overview on outcrop-based geomodelling is given by Howell et al. (2014).

Currently, mobile device applications only support a subset of the necessary functionalities within referred geomodelling workflows. Mobile apps for geological purposes already make use of physical sensors, such as presented by Ferster and Coops (2013) for measuring strike-dip directions, or visual touch-based interaction for sedimentary logging, as shown by Wolniewicz (2014). In other scientific domains, handheld applications have been developed for the surface acquisition of small 3D objects using photogrammetry (as in Tanskanen et al., 2013) or presenting large 3D surface geometries on mobile devices (García et al., 2015, introduced a point-based rendering system for large surface models). Viseur et al. (2014) presented a handheld 3D geological interpretation application prototype (named "Outcrop"), which shares similarities with the approach presented in this chapter. The software requirements specific to a mobile device application for field geology can be summarized as follows, which states the boundary conditions of the technology presented here:

- **Interpretations in 3D.** A plethora of possibilities already exist for doing 2D interpretations on outcrop image content. Utilizing DOMs in 3D is important for the additional benefit of scale, consistency, interactivity, and ease of desktop integration to reservoir modelling.
- **Independency of WiFi reception.** Field work locations can be very remote. Some outcrop studies in rural areas of denser-populated regions on the globe can potentially utilize WiFi. On the other hand, field studies in remote environments can lack any WiFi reception.

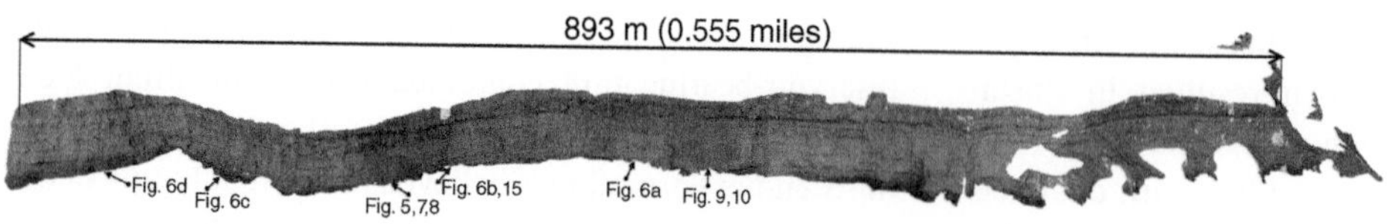

Figure 5.2 Large overview of the Whitby cliff section, spanning a distance of 650 m along the outcrop.

A mobile device app for field-based outcrop geology should operate undisturbed even in these conditions.

- **Energy-efficient operability.** A large constraint of mobile devices is the limited battery capacity. This can be extended to a given limit by external batteries, but a mobile device application for field geology should be designed in an energy-aware manner, leading to a smart, on-demand usage of energy-draining mobile components (e.g. physical sensors, computation circuits).
- **Simple interaction.** Fieldwork is a physically demanding activity, so the geologist is expected not to devote unconstrained attention constantly to the mobile device. Thus, the discussed app needs to facilitate a simple and intuitive interaction scheme to engage field experts in its utilization.
- **Desktop interoperability.** One of the largest advantages of digital procedures is to remove the need for time-consuming manual data digitization after the fieldwork. Therefore, the data of the mobile device are to be provided in an obviously interoperable fashion. If such interoperability is not provided, it is expected that the geologist returns to a paper-based workflow for reasons of convenience.

5.3.2 The MPS Facies Modelling and Simulation for Subsurface Reservoirs

Mutiple point statistics (MPS) is an alternate property modelling technique based on research by Guardiano and Srivastava (1993) exploiting a representative TI to capture lateral and vertical relationships between different facies (Strebelle, 2002). A TI is a conceptual, non-location specific numerical model of a geological (e.g. depositional, stress regime) system and should contain all of the expected heterogeneities of the reservoir under study. Data at sparse locations, as in the case of subsurface reservoirs, are generally insufficient to infer high-order or multiple statistics. Therefore, the MPS methodology derives high-order statistical information from the TI based on the concept that geometries and spatial configurations are governed by the values of neighboring nodes (sensu Guardiano and Srivastava, 1993) rather than the analytical statistics of a variogram.

Once the probability of specific facies has been established at a modelling cell, it is used to statistically simulate the occurrence of the respective facies in the simulation grid, with data from the reservoir anchoring specific patterns into place. The simulation then proceeds along a random path to assert the stochasticity of the realizations and at a cell the facies value is randomly chosen using the computed marginal local probability laws. The principal advantage of MPS is that it combines the realistic shape recreation and long-range connectivity of Boolean models with the fast, easy data conditioning of variogram-based methods. The theory underpinning MPS has been described elsewhere (Guardiano and Srivastava, 1993; Strebelle, 2000. 2002; Strebelle et al., 2001).

To date, the MPS methodology has been hampered by the lack of suitable TIs and the production of TIs on a case-by-case basis, based on the subjective criteria of the modeller. MPS simulation is inherently dependent on the conceptual training data contained within the TI. Therefore constructing representative TIs of the reservoir under study that maintain capability at both core and seismic scales is a key challenge.

Initially, conceptual models including hand drawings and field sketches were used as TIs (Strebelle, 2000), which are time consuming and difficult to reproduce. Current practice is to use non-conditional object-based modelling (Strebelle et al., 2001; Strebelle, 2012) or process-based models (Michael et al., 2010; Comunian et al., 2014), which are easy to generate. Various other methods for constructing TIs exist, including high-resolution seismic (Strebelle et al., 2002), modern analogues (Comunian et al., 2012; Lochbühler et al., 2014; Bayer et al., 2015), existing reservoir models (Hu et al., 2014) and CT scans (Comunian et al., 2012).

Pyrcz et al. (2008) proposed a library of training images for fluvial and deepwater reservoirs as a starting point for a database of realistic TIs to help standardize their development and reproducibility between studies. Reservoir modelling often leans heavily on outcrop analogs and studies of modern-day systems as aids in building 3D models. Comunian et al. (2012) proposed generating a 3D TI from a series of 2D slices. This method was successfully applied by Bayer et al. (2015) to the Herten and Descalvado aquifers in Germany and Brazil respectively. This methodology has been adapted to create TIs directly from imaging processing techniques (Ruelland et al., 2012) and DOMs (Pickel et al., 2015). All of these techniques for training image construction can be used independently or in combination with one another.

A major challenge in the MPS methodology is the selection of an appropriate TI. As TIs represent first-order uncertainty of the expected reservoir heterogeneity (Strebelle, 2012), it is essential to integrate as many sources of information as possible during TI creation. When uncertainty remains, it can be further accounted for by the provision of multiple TIs that correspond to alternative geological scenarios. As MPS is a geostatistical technique, it is bound by an underlying assumption of stationarity when inferring multiple statistical moments from the TI to the MPS model. Therefore, any pattern found in the TI can be recreated anywhere in the MPS model, provided that it matches the available conditioning data. However, as most reservoirs are not stationary, spatial variations of facies, facies proportions and the orientations and geometries of geobodies commonly occur. These variations can be inferred through the provision of secondary or soft data to modify the training image specifically to the reservoir under study. Soft data constraints can be represented by auxiliary variables (Chugunova and Hu, 2008), facies distribution maps (Harding et al., 2004; Strebelle and Levy, 2008; Strebelle, 2012) or derived from seismics (Strebelle et al., 2002; Wong and Shibli, 2003; Ronot et al., 2014).

5.4 Mobile Interpretation Using Image-to-Geometry Techniques

In this chapter, we introduce a digital approach that is designed to contribute and improve field geological procedures, in contrast to the common extension of office-based modelling toolsets, by using modern mobile handheld devices. Most recent digital studies in outcrop geology are utilizing DOMs on desktop workstations. The study of outcrops on mobile devices in this case is analogous, also utilizing DOMs in computationally simplified forms to register measurements, observations, and interpretations. The following approach allows outcrop fieldwork with as little power consumption as possible and without requiring network connectivity or power-demanding computations while facilitating a large array of geology and reservoir modelling tasks.

Using mobile handheld devices as media for digital field geology has shown potential in a previous study by Viseur et al. (2014) due to their ubiquitous availability. Although recent technological advances reduced some original limitations of mobile devices (such as primitive user interfaces, little computing capabilities, and mobile network access requirement), a few limitations (such as small data memory, the lack of dedicated graphics processor integration, and high battery consumption) still persist. Therefore, decisions are taken in the method design to balance the exploitation of novel mobile technologies (e.g. sensors and 3D graphics) with their implementation cost.

Common outcrop digitization utilizes a flexible navigation in 3D space around the scanned DOM to interpret the views, which was realized on handheld devices by Viseur et al. (2014). This interaction is already perceptually challenging for some users in an office environment, as most navigational motion constraints contradict one another. Natural limitations of mobile devices, such as reduced screen space and a fuzzy selector (i.e. mouse-, touch-, or stylus pointer), render 3D space navigation for handheld interpretations unfavorable. This approach allows field geologists to interpret images, similar to the work on outcrop photo panels, and automatically transfers the 2D photo interpretation straight to 3D space. This automatic transfer is made possible using available DOMs. The approach is not locked to a specific DOM representation (i.e. point cloud or TIN DOM).

All techniques and algorithms outlined in this section are collectively integrated into a prototypical version of the GRIT software package, which is designed to work seamlessly in combination with LIME. The software is developed purely on the basis of Open Source software libraries, such as OpenCV, Open Scene Graph, and the Google Android SDK. Figure 5.3 shows the main interface of GRIT, with a list of available outcrop photographs and essential meta-information attached to each entry.

5.4.1 Image Acquisition

The capturing of outcrop photographs is performed within GRIT, utilizing the built-in high-resolution camera that is a standard utility of modern handheld devices. Sensor measurements and EXIF information (storing physical manufacturing information of the camera as well as the GNSS info at capture time) are logged during photo acquisition. The user interface is kept simple for quick image acquisition in the field. The GNSS and the magnetic field sensor allow a globally consistent localization and orientation of the photograph, while the camera parameters are key to establishing a link between the 2D photograph and the corresponding DOM in 3D. Camera focus and zoom are fixed during the acquisition as a prerequisite for single-view geometry (see Quan, 2010, for further details).

The sensor accuracy of GNSS and magnetometer during the photo acquisition has a major influence on the later registration process. Therefore, an ample-inspired GNSS accuracy measuring tool has been integrated to give the user control over the sensor quality during image acquisition, illustrated in Figure 5.4. Subsequently, the captured photo is added to an outcrop study if the application succeeded in obtaining camera parameters, GNSS readings, and magnetic field measurements. The capturing process cannot succeed in quarry or tunnel environments (due to the lack of GNSS information) and strongly metallic or magnetically charged environments (due to the failure of stable magnetic field measurements). Blum et al. (2013) presented a study on environmental influences of handheld device sensors and their impact on the measurement accuracy of location and orientation, which can be taken as a guideline during digitally enhanced fieldwork. Furthermore, recent studies by Novakova and Pavlis (2017) and Cawood et al. (2017) assess the sensor accuracy for

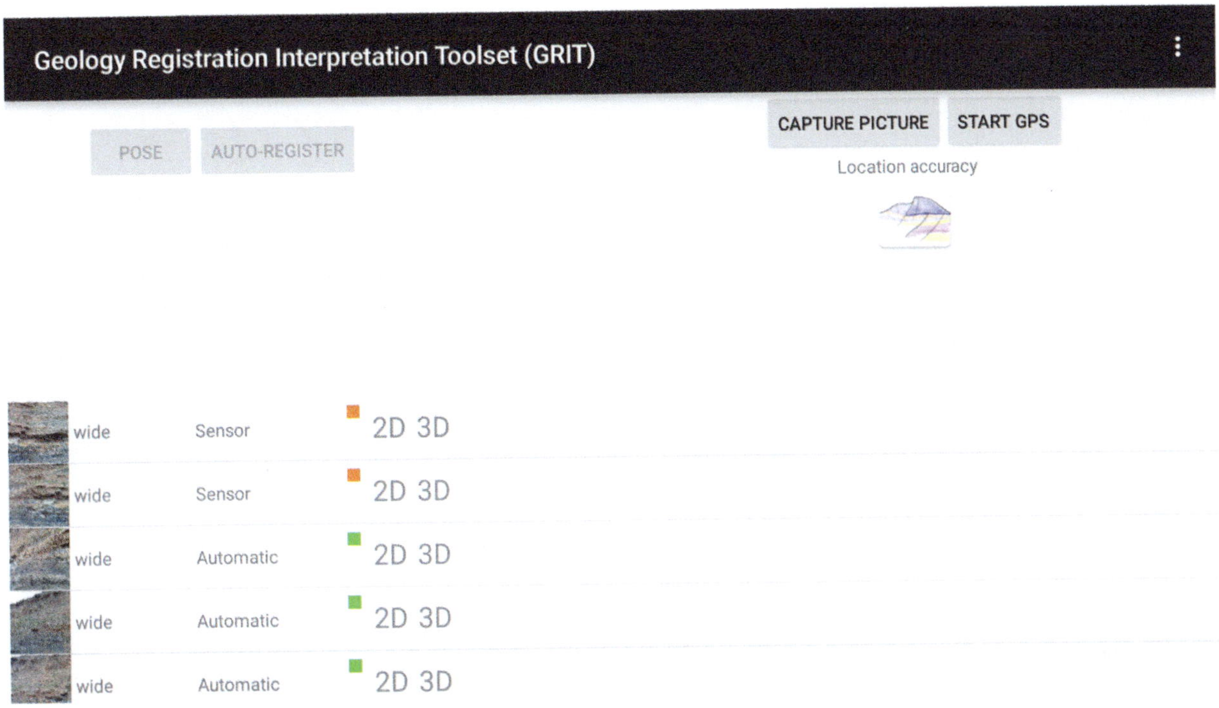

Figure 5.3 The GRIT main interface, including a preview panel (top left), a photo capture panel (top right), and a list of collected photos (bottom). Each entry contains a thumbnail, a photo orientation hint (wide/landscape or portrait), a best-pose information text, an ample-inspired indicator for the pose quality, and indicators for the interpretation availability in 2D and 3D. This compact user interface also applies to small-screen smartphones.

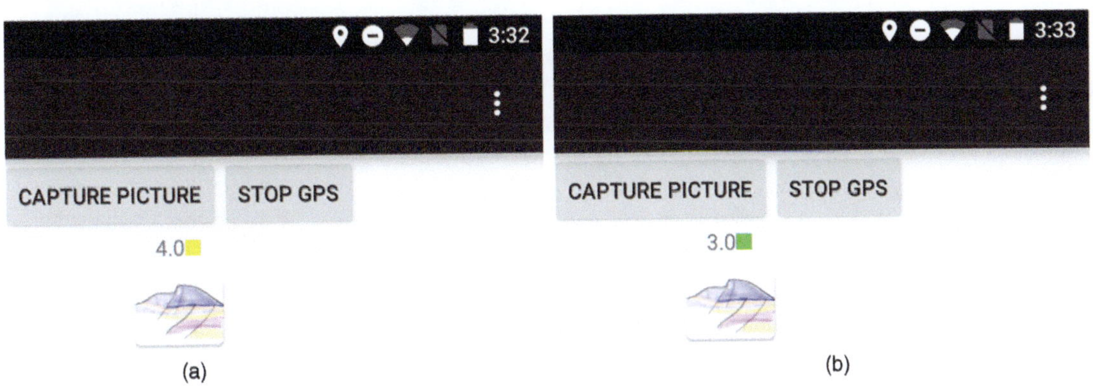

Figure 5.4 The integrated GNSS reception indicator, showing the current reception accuracy in tones of red, yellow (a) or green (b), representing a degrading scale of good (green, <3 meters), acceptable (yellow, 3–8 meters) and bad (red, >8 meters) reception accuracy variance.

measurements in structural geology on a more geoscientific level with its respective, more accessible terminology.

5.4.2 Image-to-Geometry Registration

After capturing an image, the image can be registered to surface geometry in 3D (i.e. DOMs) to establish a 2D–3D relationship. The procedure utilized by GRIT is a single-view image-to-geometry registration procedure described by Kehl et al. (2016) in detail. Its extension to run on mobile devices has been presented and performance-measured by Kehl et al. (2015). The approach is comparable to other outdoor studies by Sweeney et al. (2015) and Elias et al. (2020). Figure 5.5 shows, on one example, the fine registration process of an outcrop photo to a DOM.

The registration procedure is designed as an automatic process that establishes a coarse virtual viewpoint towards the outcrop model using device-internal sensors for location and orientation. The initial estimate of position

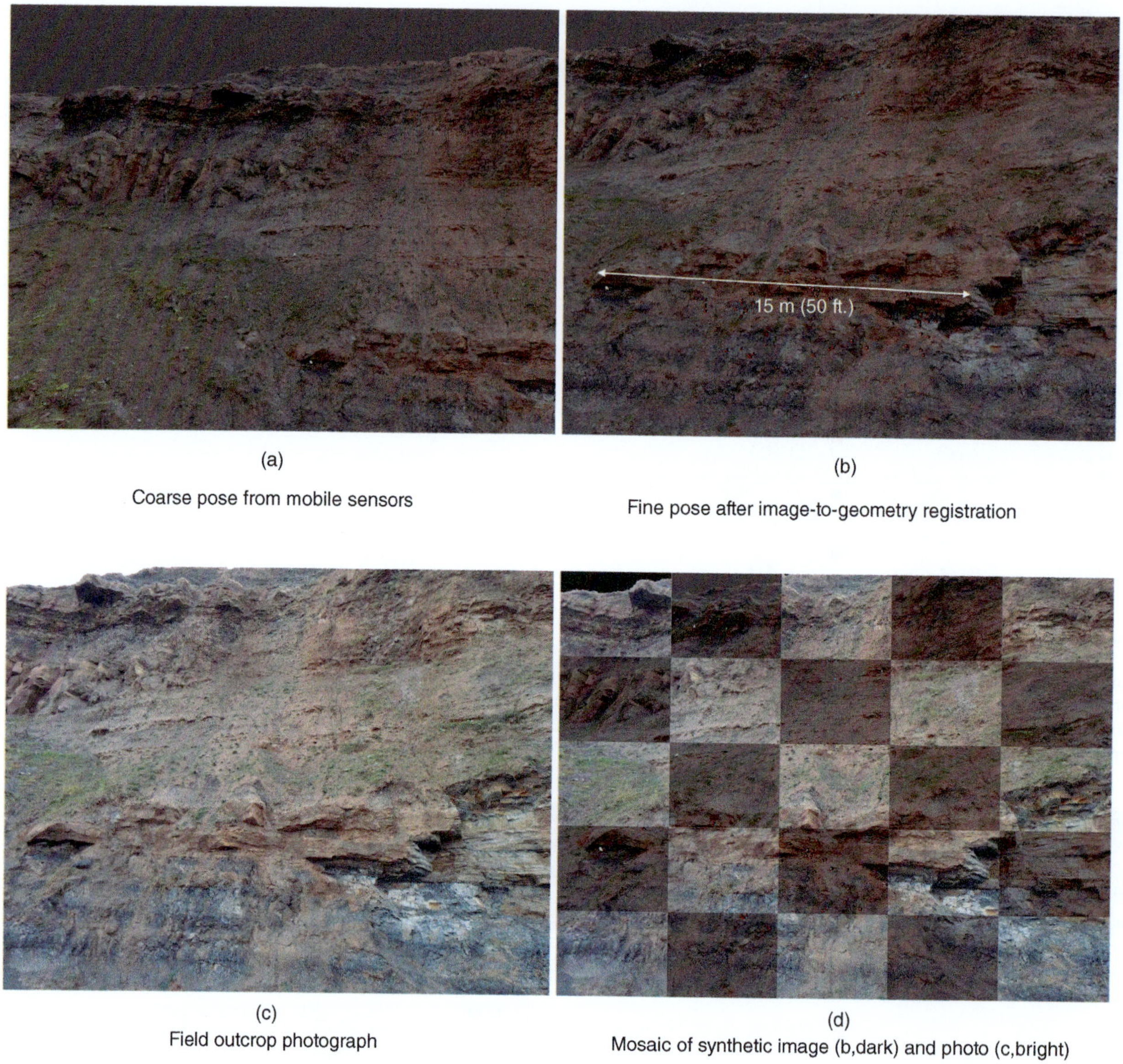

Figure 5.5 Image collection illustrating the registration procedure: first, a photograph is taken (c) while sensors record the global location and orientation. The sensor data are used to generate a synthetic image (a) of the DOM. The image-to-geometry registration establishes 2D–3D correspondences automatically and calculates a refined pose for image synthesis (b). The image mosaic (d) shows the resulting deviation between the photograph (bright squares) and synthetic DOM image (dark squares).

and orientation is further refined by means of CV, as originally presented by Kehl et al. (2015, 2016). In short, salient feature point correlations between the photograph and a rasterized image of the textured triangular surface are established using SIFT (Lowe, 2004) or alternative, illumination-robust matching procedures (as shown in Kehl et al., 2017b). Subsequently, 2D-3D correlations are constructed using the surface models. These are then used to robustly calculate a fine estimate of position and orientation for an individual image via PnP (detailed by Lepetit et al., 2009, and Li et al., 2012). This technical design allows collecting a small selection of images just where geological structures of interest are apparent, which is distinctly different to comparable alternatives on the basis of bundle adjustment (Lourakis and Argyros, 2009). The correlation of high-resolution photos previously demanded powerful equipment (laptop or computer), which is inconvenient during fieldwork and which hence restricted such automatic digital correlation to post-fieldwork procedures in the camp. With the presented technique, it is possible to compute the image-to-geometry correlation in the field on a low-power device (such as a tablet or smartphone) in a reasonable timespan.

Figure 5.6 Example field photographs of the Whitby outcrop cliff section, collected from (a) east to (d) west. The photos also vary between (c) close-up viewpoints to overviews (a, b,and d).

As discussed by Kehl et al. (2016), this automatic procedure is sensitive to external lighting conditions, which is commonly the case with photogrammetric technologies (Kehl et al., 2017b). In case of an unsuccessful automatic registration, the geologist has the option to register a given image manually by selecting corresponding tie-points between the photograph and a synthetic reference image.

For the case study of the Whitby cliffs, 87 outcrop images were taken during the field campaign. The images cover the cliff face on a 650 m (0.4 miles) parallel stretch. The photographs include detailed views near the cliffs as well as further away overview images, which can be used for tracking the general stratigraphy. A selection of photographs is shown in Figure 5.6. The presented registration method was able to extract correct automatic views for 52 of 87 photographs (a 59.77% registration success rate), which can subsequently be used for interpretation purposes. Further, a selection of 32 out of the 52 available photographs has been made to reduce the interpretation effort by excluding redundant imagery.

5.4.3 Image Interpretation

After co-registering outcrop photographs to an existing DOM, the geological interpretation on an outcrop is performed on the photographs. The interpretation follows a polyline approach, as introduced by LIME (see Rittersbacher et al., 2014, for examples in 3D), for delineation of stratigraphic bedding as well as sedimentary features. This is illustrated on the cliff section in Figure 5.7. Additionally, cross-sectionally restricted features (such as channels) can be interpreted via triangulated meshes in 2D on the photo, similarly presented by Olariu et al. (2011) in 3D. Such an interpretation is created by capturing the polygon that is selected by the user. Then, an ear clipping algorithm (see O'Rourke, 1998, and Held, 2001, for details) is used to triangulate the polygon, as shown in Figure 5.8. Ear clipping is a traditional, computationally robust technique that provides acceptable triangulations of arbitrary polygons for the majority of 2D (or 2D-projected three-dimensional) cases. In the presented application, ear clipping has the advantage to work winding-independently for user-defined polygons. Also, due to the lack of available geometry libraries on mobile platforms (i.e. smartphones, tablets), the implementation simplicity of ear clipping is a major advantage. The problematic implementation of exact mathematical, slow (i.e. $O(N \log(N))$ to $O(N^2)$complex; see Sloan, 1993) CDT in a robust manner could not be justified for the given application.

Afterwards, the line and triangle structures are converted to 3D by intersecting their 2D coordinates with the DOM via the registered pose using Raycasting. The connectivity of each delineated object persists during the 3D transformation. The resulting interpretations in 3D can subsequently be checked on the handheld device using the integrated 3D outcrop viewer (see Figure 5.9). A comparable feature is not found in available mobile application packages for field geology, such as FieldMove, that allow image capturing or sketching because of their lack in 3D DOM integration. The technology presented here allows independent geological interpretations in a way that was previously exclusive to desktop software.

The transformation does not always return coherent cross-sectional shapes due to registration problems. The impact of this error on an individual image cannot be determined prior to the photo interpretation. Furthermore, the coherent trace of bedding surfaces across photographs is not guaranteed as each photo is interpreted independently (see Figure 5.10 for an example result of multi-image interpretation merged in 3D). Such geometric errors and geometric inconsistencies need to be addressed in the post-fieldwork quality control before continuing with the geomodelling workflow.

The approach of using registered photographs for 3D interpretation differs from the "Outcrop" application, because the latter is based on the creation of 3D stratigraphic interpretations in 3D space. This demands from the user more training when navigating the model. Within GRIT, the navigation between photographs is trivial, but

Figure 5.7 Image collage of three successive images covering equal parts of the cliff section. The separate line interpretations (a, b, c) are given merit in a conclusive interpretation in 3D presented in (d). The overlay between 2D and 3D can be checked with the integrated 3D viewer on the tablet, as done in (d).

Figure 5.8 Illustration of the geometrical modelling of cross-section surfaces at outcrops on the example of a channel lens, (a) The sedimentary object is first delineated with a bounding pointset. (b) A 2D polygon is constructed. (c) By ear clipping triangulation, the polygon is converted into a mesh, which can be projected in 3D on the DOM.

the coherent fusion of separate interpretations in 3D is challenging.

After creating the collection of interpretations during the field campaign, the geometry data are made available to office-based tools succeeding the fieldwork. In this office-based environment, geological object statistics (e.g. channel width and height) are derived from the interpretation data. This statistical derivation can possibly be done directly on the mobile device in the future, but this capability is exceeding the current software objectives. Furthermore, for studies where objects intersect the outcrop multiple times, modelling parameters such as

Figure 5.9 Photo showing the use of the mobile GRIT application and its 3D digital outcrop viewer in the field. The application allows observation and interpretation checking in remote areas anywhere on the globe. The local 3D technical capabilities of modern mobile devices allow this presentation on the small device without network access and in combination with the device's motion sensors.

connectivity and sinuosity can be derived directly from the outcrop by interconnecting the related object intersections (similar to Olariu et al., 2011). All the statistics are the base for the statistical reservoir modelling with object-based modelling or their derived TIs within MPS. This process is subsequently explained in Section 5.6.

5.4.4 Office-Based Quality Control

The handheld approach enables the geologist to create line and cross-section surface interpretations in the field. A minor refinement step in the office is still necessary with the current state of technology. Geometric (MeshLab by Callieri et al., 2003) or geomatics (LIME, GoCAD) desktop software packages facilitate the geometric error correction and the connection of separate line and surface segments. Additionally, clusters of interpretations of the same geological object observed from multiple vantage points deviating by only a few centimeters can be trivially merged into one consistent outline via line averaging (see Figure 5.11, top-right inset). Technical deviations are treated in this quality control stage. The raw 3D interpretations of this case study, based on 32 outcrop photos, are presented in Figure 5.11.

Within the presented workflow, the 3D-projected interpretations are collected and exported from the tablet into LIME. Exporting the data to the desktop is facilitated via mail or cloud storage, such as Google Drive. The amount of information that can be used in further geometric analysis depends on the chosen desktop software, as not all available tools are able to handle free-form cross-section surfaces and the file formats they are distributed with.

Figure 5.11 highlights erroneous as well as correctly projected areas. Geometric interpretations errors, visible in Figure 5.11, arise from three possible error sources:

1. False control points, caused by missing ray-intersection geometry with the DOM. The issue is explained in detail in Section 5.4.3. Missing ray intersections are generally visible as holes in the LiDAR model.
2. Control points contain a defective positional offset, caused by registration inaccuracies out of missing textured LiDAR information. This is illustrated in Figure 5.12, where an incorrect camera position is retrieved as geometric information of foreground objects (i.e. boulders) is missing.
3. Control points are distorted due to dynamic camera distortions caused by the handheld device electronics. Significant radial camera distortions, as seen in Figure 5.10, occur due to optical manufacturing imprecisions in the lenses of mobile device cameras, which established camera calibration procedures cannot account for.

In the error approximation, it is to be noted that we distinguish between correctable and incorrectable geometry within erroneous segments. Correctable errors are caused by a positional offset error, which is resolved by shifting the line segments along the cliff. Radial line distortions are only correctable in a subset of interpretations. Incorrectable line segments are cropped and redrawn.

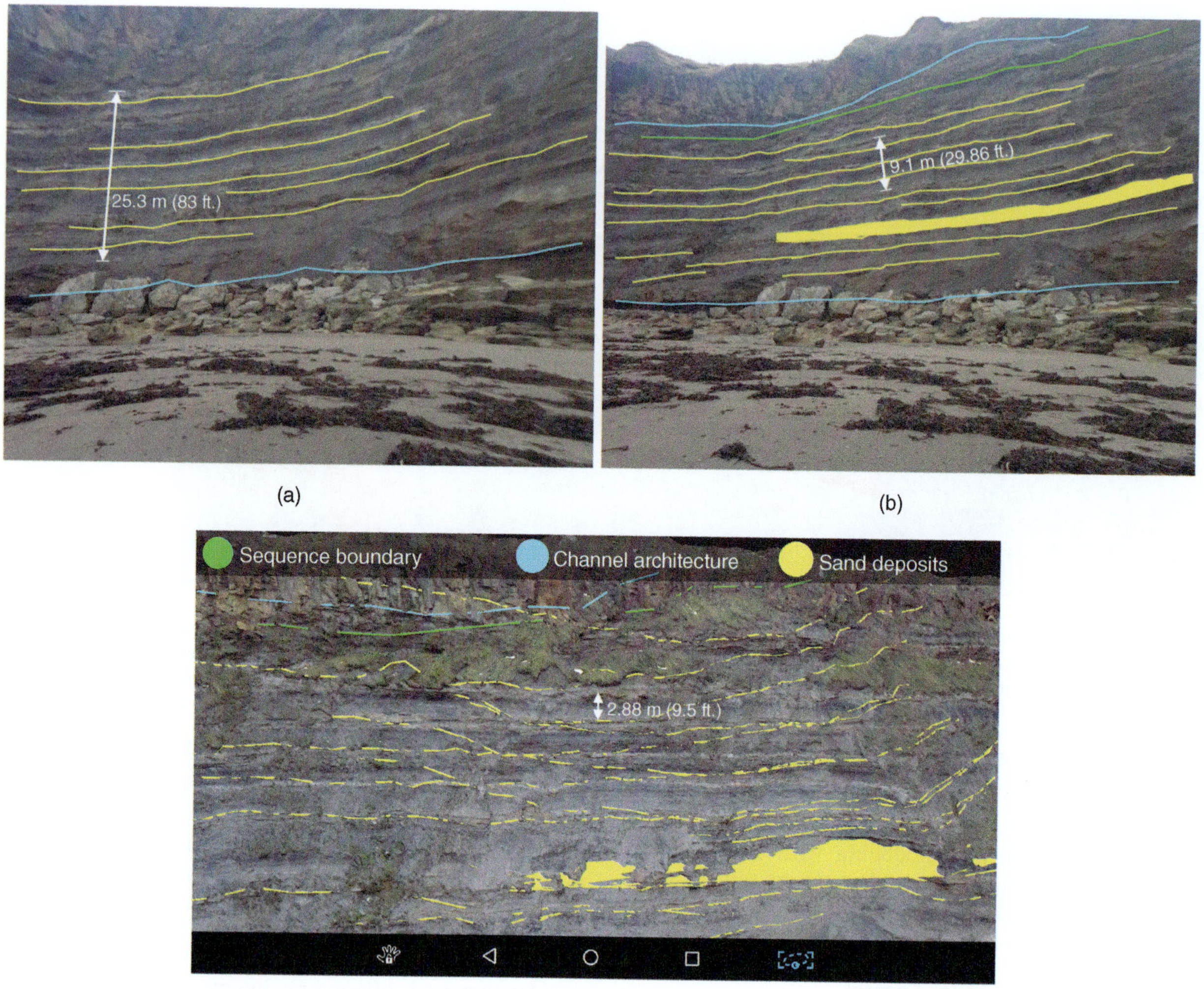

Figure 5.10 Image collage illustrating the resulting mismatch of 3D interpretations (c) caused by minor registration inaccuracies and the (partial) lack of camera internal parameter calibration for obtained outcrop images (a and b). The calculation of camera calibrations is a photogrammetric procedure that is facilitated by non-commercial open-source software.

Technical advances in the manufacturing of mobile device cameras, improved camera calibration procedures, and a more extensively covered LiDAR data acquisition will remove the need for manual QC corrections in the future. Furthermore, a manual registration process of the image (although being more time-consuming in the preparations) potentially reduces the error impact significantly (see Kehl et al., 2017b, for details). Within the Whitby cliff study area, approximately 50% of the interpretations needed quality control, of which two-thirds were position-corrected while one-third needed to be cropped and redrawn.

The resulting, correctly positioned separate line segments are subsequently merged to create a coherent and accurate line segment model, presented in Figure 5.13. The line segments can further be used for sediment object shape measurements as well as the interpolation of horizons in the subsequent geo-model.

5.5 Model Construction

Modelling of the Saltwick Formation was undertaken in Schlumberger's Petrel reservoir modelling software. The workflow for modelling of sedimentary systems from virtual outcrops was adopted from Enge et al. (2007) and shares many common elements with the modelling of subsurface systems.

Models built from outcrops often have a significantly greater number of vertical sections, in the form of sedimentary logs that can be treated as (pseudo) well datasets, than their subsurface counterparts. Surfaces defined in an outcrop can be treated with a higher degree of certainty and conceptual understanding as they can be traced between well intersections at a significantly higher resolution. The procedure for modelling this set of outcrop data is now described.

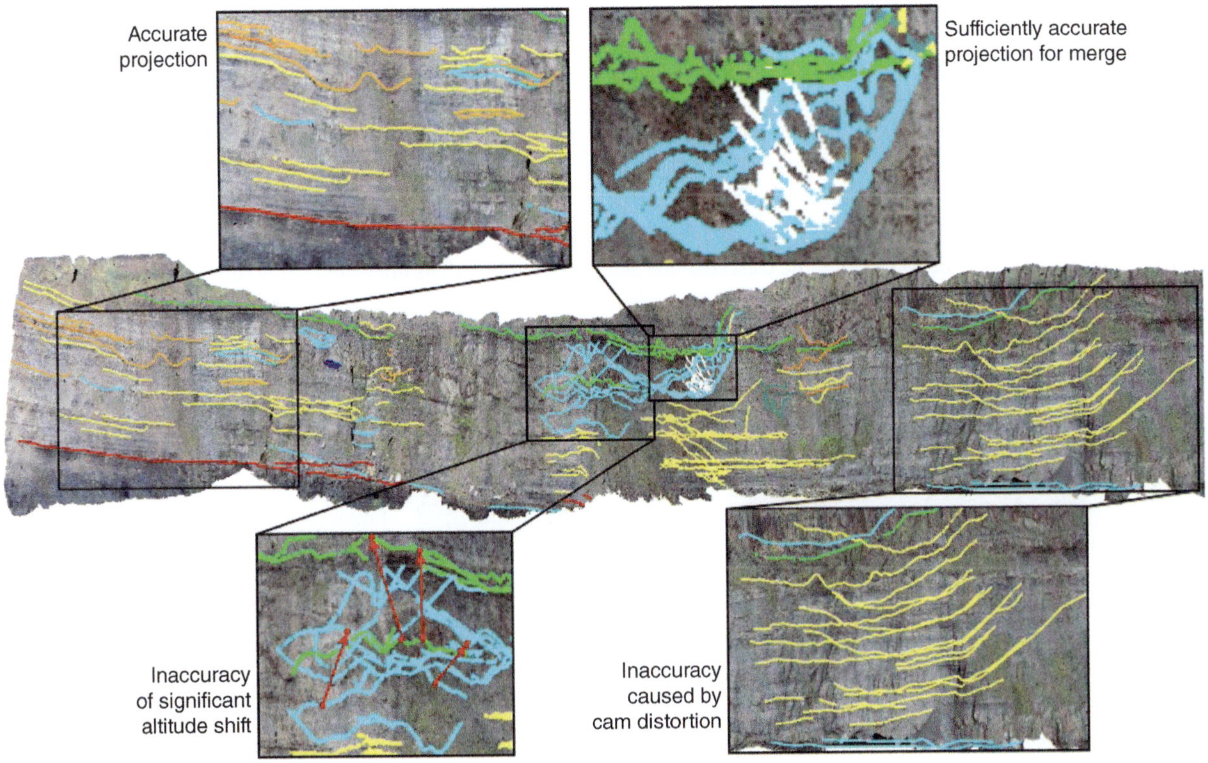

Figure 5.11 This rendering shows the imported line segments from the tablet into LIME. Further, we highlight sections of geometrically correct segments (top row), shifted segments (bottom-left), and radially distorted segments (bottom-right), which are corrected in the quality control section.

Figure 5.12 This experimental rendering shows the match between a photograph (right) and a synthetic DOM image (left), indicated by the point features and correspondences that are extracted. The resulting camera position is incorrectly estimated as parts of the foreground textured geometry (as in this case the beach and the boulders) are missing to correlate both images. This error is caused by the DOM creation process.

Polylines representing major stratigraphic horizons and key architectural elements were traced onto a series of 32 selected photographs, taken on the tablet. Once mapped, they were projected onto the 3D virtual outcrop model on the tablet, as explained in Section 5.4. Bedding surfaces were created as polygons to define the upper and lower limits of the Saltwick Formation and represent horizons for reservoir modelling (Figure 5.14). Surfaces and polylines were exported into LIME for further visual interrogation. The same dataset was then imported directly into Schlumberger's Petrel software. A textured LiDAR file was also imported to Petrel to provide a visual reference frame

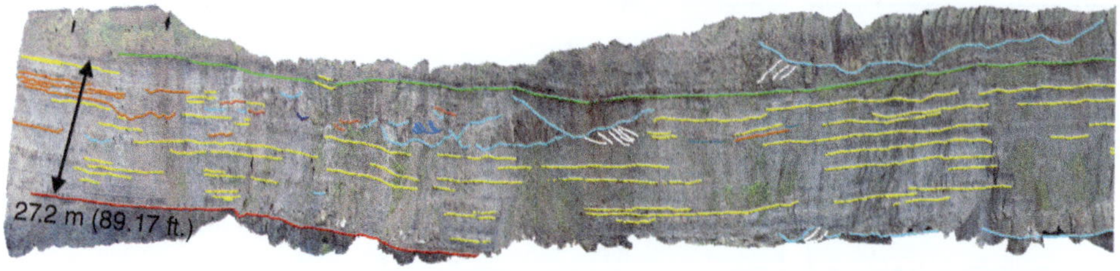

Figure 5.13 Outcrop rendering panel showing the corrected line segment interpretations after the quality control stage, in comparison to Figure 3.11.

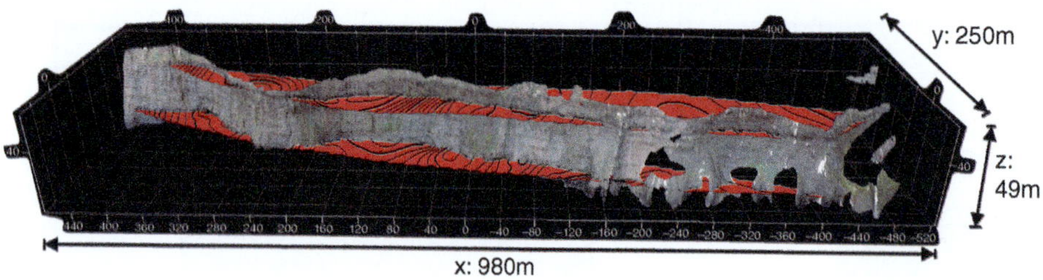

Figure 5.14 Key stratigraphic surfaces defining the reservoir zone displayed in Petrel.

during the model construction. As the LiDAR file and polylines are in project rather than global coordinates, polylines should sit in the correct stratigraphic interval in Petrel and LIME.

Surfaces in Petrel were generated to define reservoir modelling zones from the polylines from GRIT defining the contact of the Saltwick Formation with the Dogger Formation below and the Eller beck Formation above. The contact between the Dogger and Saltwick Formations is erosional and therefore relatively easy to trace. The upper contact between the Saltwick and Eller beck Formations is gradational and difficult to define. An upper surface was generated with the polyline-extrapolated surface from GRIT as an imposed datum with the aid of a series of control points. Once satisfactory surfaces were generated, they were used to define the upper and lower limits of a simple grid.

A modelling grid of 980 m × 250 m × 49 m was created. Modelling grids should ideally be parallel to geological heterogeneity; therefore the grid design used a regular thickness grid imposed at 1 m that followed the model's top surface. Cell dimensions were set at a horizontal resolution of 2 m and a vertical resolution of 1 m to capture the increasing geological heterogeneity in the vertical direction. The final grid was composed of 3,001,250 cells divided into 490 and 125 layers in the X and Y directions respectively and 49 layers in the Z direction with a total volume of 12,005,000 m^3.

Facies distributions were modelled using a Boolean or object-based approach. Three facies identified at outcrop were included in the geological model: channel sandstone, crevasse splay sandstone, and overbank deposits. The correct representation of the proportions and geometries of the outcrop facies was accomplished by conditioning the model to a plethora of width and thickness measurements taken from the DOM. Channel facies were modelled using the channel object tool that produces realistic channel geometries, while crevasse splay associations were modelled using disc-shaped ellipses. Overbank associations was kept constant as a shale background. Orientations of channels were deterministically derived from paleocurrent indicators and literature. Sinuosity of channel bodies were required as input parameters, however this cannot be ascertained itself due to the 2D nature of the outcrop. A sinuosity value was deterministically set to three, which adheres to literature documentation of low sinuosity in deltaic trunk channels. As the object model was used as a TI, a greater emphasis was placed on creating a representative model in terms of accurate geometry and architecture rather than the spatial location of channels or crevasse splays. The resulting model in this respect is fit for purpose.

A series of nine sedimentary logs at 100 m intervals was generated using Endeeper's Strataledge, a free sedimentary logging app available for Android devices, to act as conditioning data for subsequent simulation with MPS. Logs were taken systematically from a fixed datum to anchor the geometric pattern configurations in the TI to their correct spatial position. As there is no current workflow for exporting files directly out of Strataledge into Petrel, the vertical polyline created in LIME was used to define the (pseudo) well trace. To ensure accurate reproduction of

Figure 5.15 Visually interrogating sedimentary logs (created in Strataledge) within LIME.

the sedimentary logs from Strataledge, they were imported in JPEG format into LIME as a series of projected orthopanels (Figure 5.15). Well logs were subsequently exported out of LIME in ASCII formats and upscaled to the modelling grid. While in this study pseudo logs relate to known and fixed points across an outcrop, this workflow is equally valid in an exploration setting. Real-world well data locally condition geobodies, architectures, and facies relationships from conceptual outcrop analogue-based TIs into reservoir models that are representative of the subsurface, however this is outside the scope of this study.

The constructed object model was then applied as a conceptual training image for MPS (Figure 5.16). This training image is now based on a combination of outcrop-derived geometries and is a representative volume architecture of the Saltwick Formation. As the modelling focus is at outcrop scale, uncertainty between data densities is minimal.

Finally, 25 multiple-point statistical realizations were generated using the Snesim algorithm (Strebelle, 2000) in Petrel, utilizing the upscaled sedimentary logs as conditioning data. Then 20 realizations were determined to be sufficient to characterize uncertainty, while keeping ergodic fluctuations between realizations to a minimum. MPS was run as a one-step process where facies proportions, geometries, and pattern statistics are derived internally from the TI, contrary to the historic approaches MPS with patterns, where they are selected from an extracted pattern database. Target facies fractions derived from the TI are maintained via a servo-system within a tolerance that was manually set to 0.5. Due to the relative simplicity

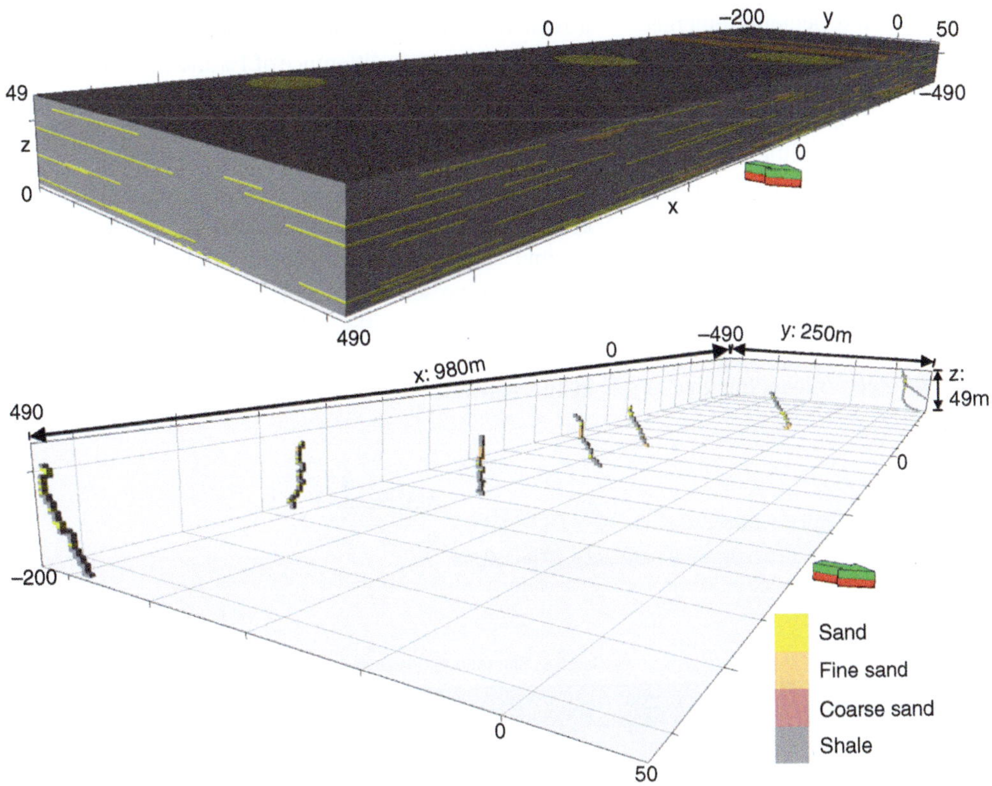

Figure 5.16 Above: conceptual training image for simulation with multiple-point statistics. Below: upscaled well logs.

of the conceptual TI, the size of the neighborhood search window was automatically selected by the algorithm.

The reservoir modelling process itself is conducted in the office environment after collecting and by using the interpretation and geometry statistics data from the field campaign. Statistical reservoir modelling and stochastic studies are computationally very involved, especially relying on large-memory workstations, which are not facilitated by mobile hardware platforms, such as tablets and smartphones. Additionally, the reservoir modelling is performed in mature, purpose-built reservoir modelling software systems (e.g. Petrel, RMS, GoCAD), which are themselves not available on mobile devices. Current research efforts are being done to develop ways of constructing reservoir models in a web-based server architecture on-demand via tablet-based requests, providing the required interpretation data with the request.

5.6 Multiple Point Statistics Simulation of the Saltwick Formation

During the field campaign, the cliff section covering 0.0613 km^2 has been interpreted with the mobile device GRIT application using the presented software. A total of 160 coherent line interpretations (representing stratigraphic horizons and sedimentary architectural elements) emerged from 209 tablet-created line segments and 43 cross-section surface interpretations on the handheld photographs. The cross-section surfaces are not used further in the geo-modelling process as they provide little additional input in relation to the line segments. Subsequently, a geo-cellular model of the depositional area has been created using the process and parameters presented in Section 5.5.

The object-based model (i.e. the TI) and the virtual outcrop are visually similar (see Figure 5.17), showing the feasibility of the stochastic object modelling that is based on the interpretations created on the handheld device. MPS realizations, of which a selection is shown in Figure 5.18, were visually quality controlled and analysed using standard static reservoir tests (e.g. connected bodies and volumes) for comparisons to the base case (i.e. the TI). These comparisons demonstrated that the average number of connected bodies is very similar (162.64 simulated; 155 for the TI), although fluctuations between realizations was high (standard deviation of 9). The average across MPS realizations of 162.64 connected bodies is within 5% of the base case. There are only a few of the 25 realizations for which the total connected volumes of channel and crevasse sandstones, 2.34–2.42 million m^3, are within a 4% margin of the TI (2.31 million m^3). The small range of potentially drainable sand equates to 80,000 m^3, illustrating the disadvantage of selecting a single realization from a suite of stochastic models for further individual analysis.

Analysis of the drainable volume was undertaken from a single, non-deviated (vertical) well, placed horizontally in the center of the model and penetrating the entire model thickness. Reservoir to rock connectivity is defined as the volume of rock intersected and drainable by a well. Over the 25 realizations the well was able to potentially drain 24% of all sandstone bodies, compared to the 17% in the base case (41% overestimation). In the TI, the well did not penetrate any of the contained channels, leading to a relatively small drainable volume. This is the potential reason for a volume difference of 7% and the resulting large volume overestimation. Of the five largest realizations in terms of drainable sand, four were in communication with at least one channel and often multiple channels. The importance of having wells intersecting channels is illustrated further by the huge range in the drainable volume across all 25 realizations, with between 71,416 and 953,22 m^3 representing 3 and 41% of total sandstone respectively. With the generally isolated channel forms, the lack of vertical stacking and low width/thickness ratios observed in the Saltwick Formation, this is perhaps unsurprising

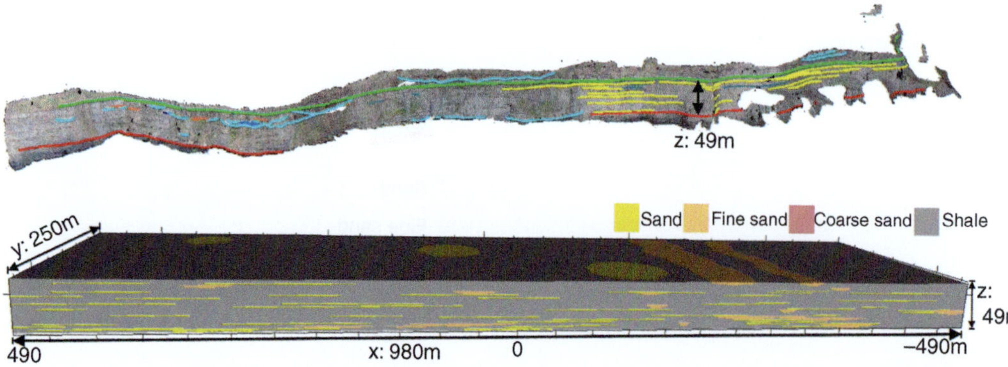

Figure 5.17 Comparison of the virtual outcrop and object model used as the training image for MPS simulation.

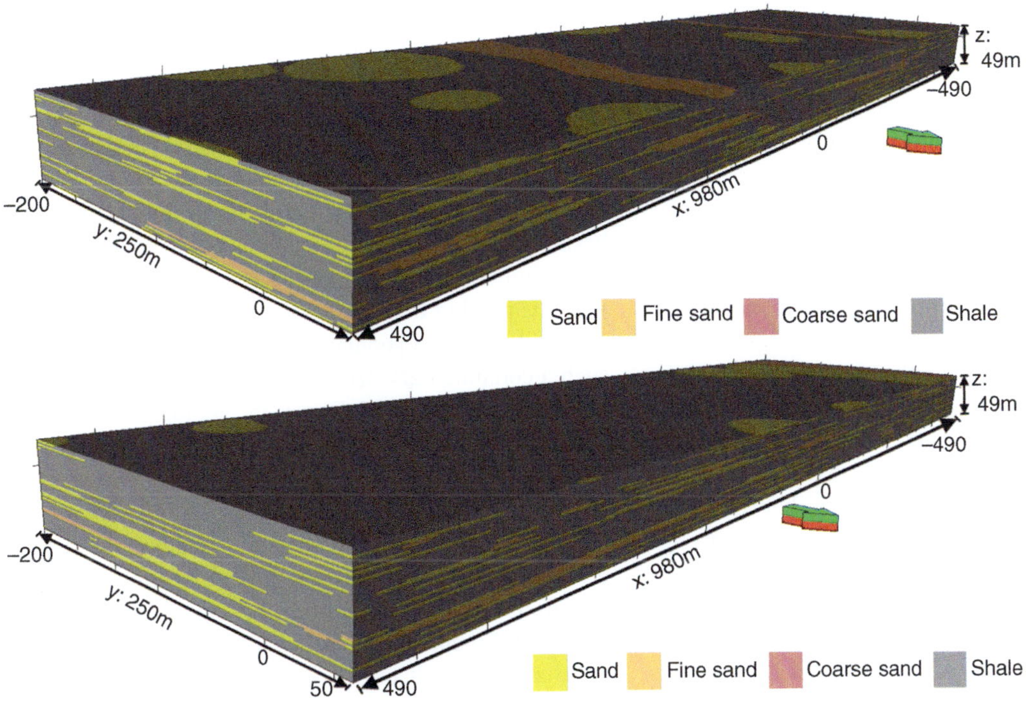

Figure 5.18 MPS realization 1 (top) and realization 2 (bottom).

as the channels are often encased in mudstone over bank deposits and therefore are not in communication with one another or to higher stratigraphic levels. Specifically, the fifth largest realization (in terms of reservoir volume in connection to the well) does not intersect a channel body, however is still in communication with 871,128 m^3 of connected sand. The well is in communication with 99% of the crevasse splay facies (Figure 5.19), which highlights the importance of accurately characterizing and modelling crevasse splay facies and including them in volumetric calculations.

5.7 Discussion

The use of virtual outcrops and their handheld-created interpretations as input for the creation of training images for reservoir modelling with MPS appears to be very successful. A stochastic object model was built using geometries and orientations extracted from the virtual outcrop that was successfully mapped on an Android tablet using the GRIT application. Digitized polylines, being projected directly onto the virtual outcrop and visually

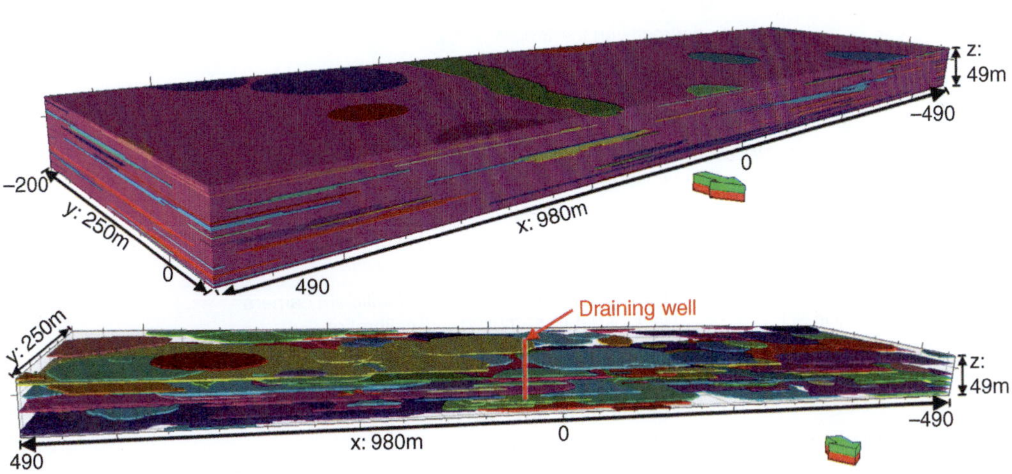

Figure 5.19 Connected bodies of realization 1 (top) and the corresponding well position for the drainable volume test (bottom), highlighting the well communication within the connected bodies of crevasse splay. Each individual connected component is attributed with a separate, random color for distinction purposes.

interrogated in the field, form the building blocks for conceptual models. Surfaces can be created directly using tablet interpretations as polygons, which are directly imported into the conventional reservoir modelling workflow. Practitioners then define zones in a reservoir by denoting the upper and lower limits of a modelling grid that is subsequently populated with facies. The MPS realizations have been shown to represent the outcrop in a realistic manner. It can be concluded that virtual outcrops can be interpreted on tablets and incorporated into the reservoir modelling workflow, which is illustrated in Figure 5.20.

From a technical point of view, the presented application (GRIT) introduces a digital, mobile, geological toolset that bridges the gap between field- and office-based DOM workflows. It seemingly integrates with an existing desktop tool (i.e. LIME) to facilitate rapid transitions between field observations and desktop-based quality control, measurements, and geo-modelling preparation. LIME provides a straightforward interface for transferring outcrop surface interpretations based on textured LiDAR DOMs to subsurface geo-modelling software, such as Schlumberger Petrel and Roxar RMS. The mobile device-based modelling workflow presented in this chapter is a significant step forward towards higher digital data utilization in the field. A distinct advantage of GRIT, compared to strictly three-dimensional DOM field apps (e.g. 'Outcrop' by Roudaut et al., 2014), is its flexibility to work on outcrop photographs directly in cases where 3D data are not initially available. In these cases, the software facilitates collecting and discussing interpretations. Available 3D data allow connecting and converting the photo interpretations in the field to true 3D geological markers. The introduced quality control step is necessary in some case studies as 2D–3D interpretations are taken on an individual image basis, which introduces geometrical ambiguities.

The automatic registration procedure, being previously presented and in this chapter assessed on a cliff-scale case study, reduces the geomatics expertise needed to realize such a combined 2D–3D workflow. The chosen approach represents the state-of-the-art in photogrammetry and the achieved results are bound by current photogrammetric and computational limitations (e.g. registration issues emerging from challenging illumination conditions). The registration procedure is subject to further practical limitations explained here. The limitations can be, in its majority, circumvented. Specifically, a more careful and mobile study-targeted DOM construction can omit issues related to 3D correspondence sampling and the lack of camera–geometry ray intersections. These adaptations include reducing the clipping of the DOM geometry in adjacent context areas that are not of actual geological relevance, as well as to improve hole filling and texturing. DEMs and dGPS localization (in fieldwork cases that are less limited by financial constraints for

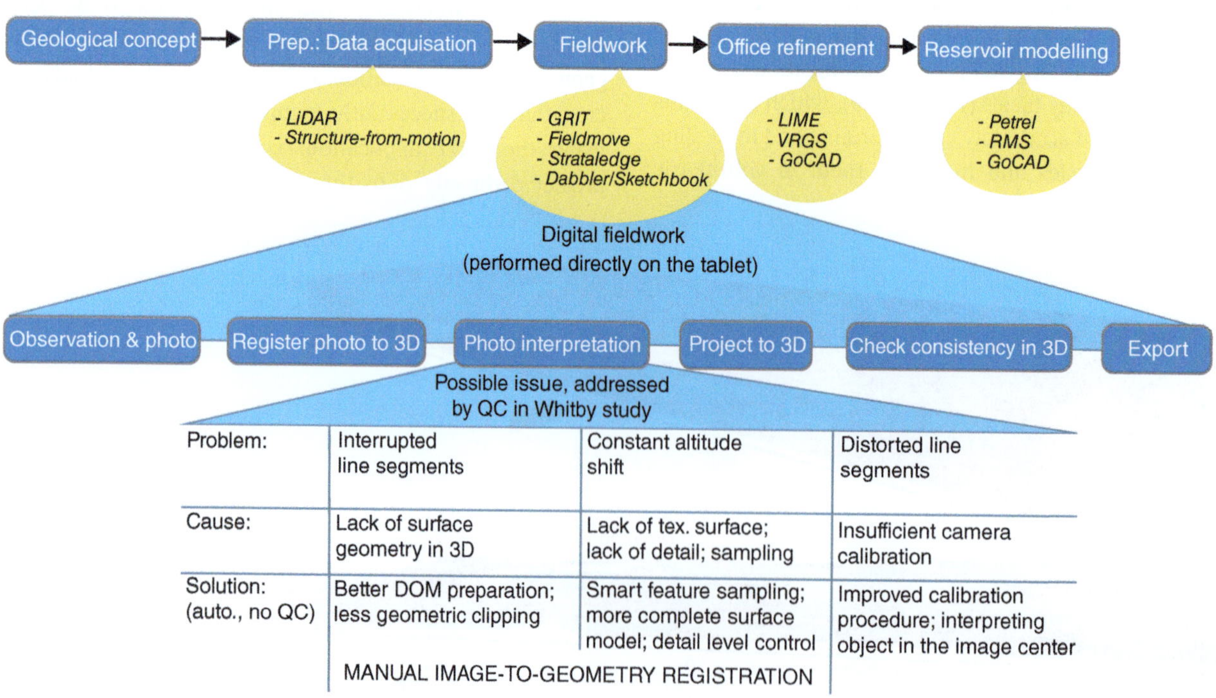

Figure 5.20 The current stage of a mobile device-integrated reservoir modelling workflow, based on field observations and interpretations.

equipment) provide sufficiently reliable initial position estimates to significantly reduce desktop-based quality control. The only issue currently remaining is the radial distortion of photo interpretations due to an inaccurate camera calibration. In this respect, our observations are in line with previous studies by Akca and Gruen (2009). The issue is subject of current research in the photogrammetric community, so that new methods in the future may remove such desktop-based corrections entirely. Additionally, upcoming technical R&D needs to account for projective occlusion, as it commonly occurs in cliff sections with significant geometric variance of the DOM surface. An improved registration mapping will allow for simple algorithmic procedures when establishing 3D segment connectivity, such as closest-point assessment.

The developed geometric toolbox allows for capturing arbitrary geological features, in so far as these can be described by line segments and (triangulated) polygonal surface boundaries. The toolbox is particularly applicable to the delineation of clearly separable geological objects, such as structural features and sedimentary elements. The delineation of sedimentary elements has been detailed here for deriving a qualitative outcrop model that is used in a reservoir modelling case study.

The body of future work and possible extensions of the technology is immense. In a first step, the object-based interpretation mapping needs to harmonize individual interpretations and connect them across images, which removes the need for large interpretation postprocessing and extensive digital quality control. The incorporation of external handheld software tools is a common request from field geologists. Existing applications facilitate geological sketching, which can be appended and co-registered as auxiliary images, and more extensive tools for geometric descriptions (via Autodesk SketchBook) that can be incorporated to the current toolbox, as well as the logging of sediment successions via Strataledge, which can be directly registered within the DOM framework. Outcrop-scale facies mapping can be realized via intuitive brushing procedures and the projection of 3D DOM vertices onto brushed segments via the camera registration. In addition to the techniques described in this chapter, original research on the topic of digital field notebooks by Cagliari et al. (2017) and Senger and Nordmo (2021) illustrate the path forward to mobile device usage in the field. Another major topic concerns extended 3D capabilities to enable geologists to connect separate interpretations along an outcrop cross-section with one another and extrude their geometry into 3D volumetric geobodies on the mobile device. Furthermore, the ability to show an upscaled preview version of a selected geocellular model (such as TIs and MPS realizations) would enable the validation of volumetric geological hypotheses and MPS modelling results during fieldwork. Some of our technical research efforts on attributed 3D surface geometry processing and large cross-scale surface algorithms, made available on mobile devices, will provide the technical basis for those extended geological studies in the future.

Analyzing the geometric heterogeneity of faults via object delineation and fault facies mapping (introduced by Braathen et al., 2009) is a logical next step to assess the applicability of the presented approach outside sedimentological cases. The usage of mobile devices as powerful, ubiquitously available tools is particularly suitable for studies in physically challenging environments, where the device's compact form factor allows new analyses to be performed in the field.

Acknowledgments

The first-authorship of this chapter is shared between Mr. Kehl and Mr. Mullins. The research was part of the VOM2MPS Project (project number 234111/E30), funded by the Research Council of Norway (RCN), FORCE consortium through Petromaks 2, and the SAFARI consortium. The Whitby dataset collection has been funded by the SAFARI project. The authors thank Sophie Viseur for insightful research discussions on the topics in this chapter. Riegl LMS GmbH is thanked for software support of the LiDAR scanning equipment, as well as Schlumberger for the academic Petrel software license and the continuing service and support. The authors collectively thank the anonymous reviewers for the constructive review feedback. Our special thanks are attributed to A. Bistacchi, S. Viseur, and M. Massironi as editors for their continued publishing support of this work.

References

Akca, D. and Gruen, A. (2009). Comparative geometric and radiometric evaluation of mobile phone and still video cameras. *The Photogrammetric Record* 24 (127): 217–245. doi: 10.1111/j.1477-9730.2009.00541.x.

Alexander, J. and Gawthorpe, R. (1993). The complex nature of a Jurassic multistorey, alluvial sandstone body, Whitby, North Yorkshire. *Geological Society, London, Special Publications* 73 (1): 123–142.

Bayer, P., Comunian, A., Höyng, D., and Mariethoz, G. (2015). High resolution multi-facies realizations of sedimentary reservoir and aquifer analogs. *Scientific Data* 2.

BGS (2013). *iGeology*. British Geological Survey (BGS): Android Application.

Bistacchi, A., Balsamo, F., Storti, F. et al. (2015). Photogrammetric digital outcrop reconstruction, visualization with textured surfaces, and three-dimensional structural analysis and modeling: Innovative methodologies applied to fault-related dolomitization (vajont limestone, Southern Alps, Italy). *Geosphere* 11 (6): 2031–2048.

Bistacchi, A., Mittempergher, S., and Martinelli, M. (2021). Digital Outcrop Model reconstruction and interpretation. In: *3D Digital Geological Models: From Terrestrial Outcrops to Planetary Surfaces*. Wiley, this volume.

Blum, J.R., Greencorn, D.G., and Cooperstock, J.R. (2013). Smartphone sensor reliability for augmented reality applications. In: *Mobile and Ubiquitous Systems: Computing, Networking, and Services*, 127–138. Springer.

Braathen, A., Tveranger, J., Fossen, H. et al. (2009). Fault facies and its application to sandstone reservoirs. *AAPG Bulletin* 93 (7): 891–917.

Buckley, S.J., Howell, J.A., Enge, H.D., and Kurz, T.H. (2008). Terrestrial laser scanning in geology: Data acquisition, processing and accuracy considerations. *Journal of the Geological Society* 165 (3): 625–638.

Buckley, S.J., Schwarz, E., Terlaky, V. et al. (2010). Combining aerial photogrammetry and terrestrial LiDAR for reservoir analog modeling. *Photogrammetric Engineering and Remote Sensing* 76 (8): 953–963.

Buckley, S., Naumann, N., Ringdal, K. et al. (2019). The digital geoscience revolution. Where are we now? In: *Conference Proceedings of APGCE*, vol. 1, 1–5. European Association of Geoscientists and Engineers.

Cagliari, J., Veronez, M.R., Heinen, F. et al. (2017). Digital field book for geosciences. In: *IEEE International Geoscience and Remote Sensing Symposium (IGARSS)*, 5674–5677. Texas, USA: Fort Worth.

Callieri, M., Cignoni, P., Ganovelli, F. et al. (2003). VCLab's tools for 3D range data processing. *VAST* 3: 5–7.

Cawood, A.J., Bond, C.E., Howell, J.A. et al. (2017). LiDAR, UAV or compass-clinometer? Accuracy, coverage and the effects on structural models. *Journal of Structural Geology* 98: 67–82.

Chugunova, T.L. and Hu, L.Y. (2008). Multiple-point simulations constrained by continuous auxiliary data. *Mathematical Geosciences* 40 (2): 133–146.

Comunian, A., Renard, P., and Straubhaar, J. (2012). 3D multiple-point statistics simulation using 2D training images. *Computers and Geosciences* 40 (0): 49–65.

Comunian, A., Jha, S.K., Giambastiani, B.M. et al. (2014). Training images from process-imitating methods. *Mathematical Geosciences* 46 (2): 241–260.

Dreyer, T. (1990). Sand body dimensions and infill sequences of stable, humid-climate delta plain channels. In: *North Sea Oil and Gas Reservoirs*, vol. II, 337–351. London, UK: Springer.

Dujoncquoy, E., Masse, P., Nicol, Y. et al. (2019). UAV-based 3d outcrop analog models for oil and gas exploration and production. In: *IEEE International Geoscience and Remote Sensing Symposium (IGARSS)*, 6791–6794.

Eide, C.H., Howell, J., and Buckley, S. (2014). Distribution of discontinuous mudstone beds within wave-dominated shallow-marine deposits: Star point sandstone and Blackhawk Formation, eastern Utah. *AAPG Bulletin* 98 (7): 1401–1429.

Elias, M., Eltner, A., Liebold, F., and Maas, H.-G. (2020). Assessing the influence of temperature changes on the geometric stability of smartphone and raspberry pi cameras. *Sensors* 20 (3): 643.

Enge, H.D., Buckley, S.J., Rotevatn, A., and Howell, J.A. (2007). From outcrop to reservoir simulation model: Workflow and procedures. *Geosphere* 3 (6): 469–490.

Ferster, C.J. and Coops, N.C. (2013). A review of Earth observation using mobile personal communication devices. *Computers and Geosciences* 51: 339–349.

García, S., Pagés, R., Berjón, D., and Morán, F. (2015). Textured splat-based point clouds for rendering in handheld devices. In: *Proceedings of the 20th International Conference on 3D Web Technology*, 227–230. New York, USA: Web3D '15, ACM.

Gonzaga, L., Veronez, M.R., Kannenberg, G.L. et al. (2018). A multioutcrop sharing and interpretation system: Exploring 3-D surface and subsurface data. *IEEE Geoscience and Remote Sensing Magazine* 6 (2): 8–16.

Guardiano, F.B. and Srivastava, R.M. (1993). Multivariate geostatistics: beyond bivariate moments. In: *Geostatistics Troia '92*, 133–144. Springer.

Harding, A., Strebelle, S., Levy, M. et al. (2004). Reservoir facies modelling: New advances in MPS. In: *Geostatistics Banff 2004, Quantitative Geology and Geostatistics*, vol. 14 (ed. O. Leuangthong and C. Deutsch), 559–568. Netherlands: Springer.

Held, M. (2001). FIST: Fast industrial-strength triangulation of polygons. *Algorithmica* 30 (4): 563–596.

Hemingway, J. and Knox, R.O. (1973). Lithostratigraphical nomenclature of the middle Jurassic strata of the Yorkshire basin of north-east England. In: *Proceedings of the Yorkshire Geological and Polytechnic Society*, vol. 39, 527–535. Geological Society of London.

Howell, J.A., Martinius, A.W., and Good, T.R. (2014). The application of outcrop analogues in geological modelling: A review, present status and future outlook. *Geological Society, London, Special Publications* 387: SP387–SP312.

Hu, L., Liu, Y., Scheepens, C. et al. (2014). Multiple-point simulation with an existing reservoir model as training image. *Mathematical Geosciences* 46 (2): 227–240.

Jones, R., McCaffrey, K., Clegg, P. et al. (2009). Integration of regional to outcrop digital data: 3d visualisation of multi-scale geological models. *Computers and Geosciences* 35 (1): 4–18.

Kehl, C., Buckley, S.J., and Howell, J.A. (2015). Image-to-geometry registration on mobile devices. n algorithmic assessment. In: *Proceedings of the 3D-NordOst Workshop*, vol. 18, 17–26. GFaI ISBN 978-3-942709-14-9.

Kehl, C., Buckley, S., Gawthorpe, R. et al. (2016). Direct image-to-geometry registration using mobile sensor data. *ISPRS Annals of Photogrammetry, Remote Sensing and Spatial Information Sciences* 3 (2): 121–128.

Kehl, C., Buckley, S.J., Viseur, S. et al. (2017a). Mapping field photographs to textured surface meshes directly on mobile devices. *The Photogrammetric Record* 32 (160): 398–423. doi: 10.1111/phor.12213.

Kehl, C., Buckley, S.J., Viseur, S. et al. (2017b). Automatic illumination-invariant image-to-geometry registration in outdoor environments. *The Photogrammetric Record* 32 (158): 93–118.

Knox, R. (1969). *Sedimentological studies of the Eller Beck bed and the lower deltaic series in north-east Yorkshire*. PhD thesis. Newcastle University.

Labourdette, R. and Jones, R.R. (2007). Characterization of fluvial architectural elements using a three-dimensional outcrop data set: Escanilla braided system, south-central Pyrenees. *Spain, Geosphere* 3 (6): 422–434.

Lepetit, V., Moreno-Noguer, F., and Fua, P. (2009). EPnP: An accurate O(n) solution to the PnP problem. *International Journal of Computer Vision* 81 (2): 155–166. doi: 10.1007/s11263-008-0152-6.

Li, S., Xu, C., and Xie, M. (2012). A robust O(n) solution to the perspective-n-point problem. *IEEE Transactions on Pattern Analysis and Machine Intelligence* 34 (7): 1444–1450.

Livera, S. (1989). Facies associations and sand-body geometries in the Ness Formation of the Brent Group, Brent Field. *Geological Society, London, Special Publications* 41 (1): 269–286.

Livera, S. and Leeder, M. (1981). The Middle Jurassic Ravenscar Group ("Deltaic Series") of Yorkshire: Recent sedimentological studies as demonstrated during a field meeting 2–3 May 1980. *Proceedings of the Geologists' Association* 92 (4): 241–250.

Lochbühler, T., Pirot, G., Straubhaar, J., and Linde, N. (2014). Conditioning of multiple-point statistics facies simulations to tomographic images. *Mathematical Geosciences* 46 (5): 625–645.

Lourakis, M.A. and Argyros, A. (2009). SBA: A software package for generic sparse bundle adjustment. *ACM Trans. Math. Software* 36 (1): 1–30.

Lowe, D.G. (2004). Distinctive image features from scale-invariant keypoints. *International Journal of Computer Vision* 60 (2): 91–110.

Mallet, J. (1992). GOCAD: A computer aided design program for geological applications. In: *Three-Dimensional Modeling with Geoscientific Information Systems*, 123–141. Springer.

McCaffrey, K., Jones, R., Holdsworth, R. et al. (2005). Unlocking the spatial dimension: Digital technologies and the future of geoscience fieldwork. *Journal of the Geological Society* 162 (6): 927–938.

Miall, A.D. (1985). Architectural-element analysis: A new method of facies analysis applied to fluvial deposits. *Earth Science Reviews* 22: 26–308.

Michael, H., Li, H., Boucher, A. et al. (2010). Combining geologic-process models and geostatistics for conditional simulation of 3-D subsurface heterogeneity. *Water Resources Research* 46 (5).

Mjøs, R. and Prestholm, E. (1993). The geometry and organization of fluviodeltaic channel sandstones in the Jurassic Saltwick Formation, Yorkshire, England. *Sedimentology* 40 (5): 919–935.

Mjøs, R., Walderhaug, O., and Prestholm, E. (1993). Crevasse splay sandstone geometries in the Middle Jurassic Ravenscar Group of Yorkshire, UK. *Alluvial Sedimentation, International Association of Sedimentologists, Special Publication* 17: 167–184.

Novakova, L. and Pavlis, T.L. (2017). Assessment of the precision of smart phones and tablets for measurement of planar orientations: A case study. *Journal of Structural Geology* 97: 93–103.

Olariu, M., Aiken, C., Bhattacharya, J., and Xu, X. (2011). Interpretation of channelized architecture using three-dimensional photo real models, Pennsylvanian deep-water deposits at big rock quarry. *Arkansas. Marine and Petroleum Geology* 28 (6): 1157–1170.

O'Rourke, J. (1998). *Computational Geometry in C*. Cambridge University Press.

Pickel, A., Frechette, J., Comunian, A., and Weissmann, G. (2015). Building a training image with Digital Outcrop Models. *Journal of Hydrology* 531, Part 1: 53–61.

Prokop, A. and Panholzer, H. (2009). Assessing the capability of terrestrial laser scanning for monitoring slow moving landslides. *Natural Hazards and Earth System Sciences* 9 (6): 1921–1928.

Pyrcz, M., Boisvert, J., and Deutsch, C. (2008). A library of training images for fluvial and deepwater reservoirs and associated code. *Computers and Geosciences* 34 (5): 542–560.

Qu, D., Frykman, P., Stemmerik, L. et al. (2021). Upscaling of outcrop information for improved reservoir modelling, exemplified by a case study on chalk. *Petroleum Geoscience* 27 (4).

Quan, L. (2010). *Image-Based Modeling*. USA: Springer.

Rader, E., Love, R., Reano, D. et al. (2021). Pandemic minecrafting: An analysis of the perceptions of and lessons learned from a gamified virtual geology field camp. *Geoscience Communication Discussions* 2021: 1–30.

Rarity, F., Van Lanen, X., Hodgetts, D. et al. (2014). LiDAR-based digital outcrops for sedimentological analysis: Workflows and techniques. *Geological Society, London, Special Publications* 387 (1): 153–183.

Ringrose, P. and Bentley, M. (2015). *Reservoir Model Design*. Cham: Springer.

Rittersbacher, A., Howell, J.A., and Buckley, S.J. (2014). Analysis of fluvial architecture in the Blackhawk Formation, Wasatch Plateau, Utah, USA, using large 3D photorealistic models. *Journal of Sedimentary Research* 84 (2): 72–87.

Ronot, C. and Ruelland, P. (2014). Using seismic attributes and multiple-point statistics to condition facies modelling of a deepwater turbidite reservoir. In: *The International Petroleum Technology Conference*. Kuala Lumpur, Malaysia Paper number IPTC-1700-MS.

Ruelland, P., Ronot, C., Wirbel, E. et al. (2012). A pre-processing workflow for multi-point statistics modelling involving training image construction and well validation. In: *Conference Proceedings, 5th EAGE St. Petersburg International Conference and Exhibition on Geosciences–Making the Most of the Earths Resources*. April 2012, cp-283-00080.

Ruiu, J., Caumon, G., and Viseur, S. (2015). Semiautomatic interpretation of 3D sedimentological structures on geologic images: An object-based approach. *Interpretation* 3 (3): SX63–SX74.

Senger, K. and Nordmo, I. (2021). Using digital field notebooks in geoscientific learning in polar environments. *Journal of Geoscience Education* 69 (2): 166–177.

Siddiqui, N.A., Ramkumar, M., Rahman, A.H.A. et al. (2019). High resolution facies architecture and digital outcrop modeling of the Sandakan Formation sandstone reservoir, Borneo: Implications for reservoir characterization and flow simulation. *Geoscience Frontiers, Special Issue: Advances in Himalayan Tectonics* 10 (3): 957–971.

Sloan, S. (1993). A fast algorithm for generating constrained delaunay triangulations. *Computers & Structures* 47 (3): 441–450.

Strebelle, S. (2000). *Sequential simulation drawing structures from training images*. PhD Thesis. California: Stanford University.

Strebelle, S. (2002). Conditional simulation of complex geological structures using multiple-point statistics. *Mathematical Geology* 34 (1): 1–21.

Strebelle, S. (2012). Multiple-point geostatistics: From theory to practice. In: *Expanded Abstract Collection from Ninth International Geostatistics Congress*, 65. Oslo, Norway: Chevron, Norwegian Computing Center.

Strebelle, S. and Levy, M. (2008). Using multiple-point statistics to build geologically realistic reservoir models: The mps/fdm workflow. *Geological Society, London, Special Publications* 309 (1): 67–74.

Strebelle, S., Payrazyan, K., and Caers, J. (2002). Modeling of a deepwater turbidite reservoir conditional to seismic data using multiple-point geostatistics. In: *SPE Annual Technical Conference and Exhibition*. San Antonio, Texas: Society of Petroleum Engineers September 2002, paper SPE-77425-MS.

Strebelle, S.B. and Journel, A.G. (2001). Reservoir modeling using multiple-point statistics. In: *SPE Annual Technical Conference and Exhibition*. New Orleans, Louisiana: Society of Petroleum Engineers September 2001, paper SPE-71324-MS.

Sweeney, C., Flynn, J., Nuernberger, B. et al. (2015). Efficient computation of absolute pose for gravity-aware augmented reality. In: *2015 IEEE International Symposium on Mixed and Augmented Reality (ISMAR)*, 19–24. Japan: Fukuoka.

Tanskanen, P., Kolev, K., Meier, L. et al. (2013). Live metric 3D reconstruction on mobile phones. In: *2013 IEEE International Conference on Computer Vision (ICCV)*, 65–72. Sydney: Australia.

Tavani, S., Corradetti, A., Granado, P. et al. (2019). Smartphone: An alternative to ground control points for orienting virtual outcrop models and assessing their quality. *Geosphere* 15 (6): 2043–2052.

Trinks, I., Clegg, P., McCaffrey, K. et al. (2005). Mapping and analysing virtual outcrops. *Visual Geosciences* 10 (1): 13–19.

Van Wagoner, J. (1988). An overview of the fundamentals of sequence stratigraphy and key definitions. In: *Sea Level Changes: An Integrated Approach*, 39–46. Special Publications of SEPM.

Viseur, S., Roudaut, R., Bertozzi, R. et al. (2014). 3D interactive geological interpretations on digital outcrops using a touch pad. In: *Vertical Geology Conference (VGC)*. Lausanne, Switzerland.

Vollset, J. and Doré, A.G. (1984). *A Revised Triassic and Jurassic Lithostratigraphic Nomenclature for the Norwegian North Sea*. Norwegian Petroleum Directorate (NPD): Oljedirektoratet.

Wolniewicz, P. (2014). SedMob: A mobile application for creating sedimentary logs in the field. *Computers and Geosciences* 66 (0): 211–218.

Wong, P. and Shibli, S. (2003). Modeling a fluvial reservoir with multipoint statistics and principal components. *Developments in Petroleum Science* 51: 175–185.

6

Image Analysis Algorithms for Semiautomatic Lineament Detection in Geological Outcrops

Silvia Mittempergher[1] *and Andrea Bistacchi*[2]

[1] *Dipartimento di Scienze Chimiche e Geologiche, Università degli Studi di Modena e Reggio Emilia, Italy*
[2] *Dipartimento di Scienze dell'Ambiente e della Terra, Università degli Studi di Milano Bicocca, Milano, Italy*

Abstract

Digital Outcrop Models (DOMs) are increasingly applied to obtain large, statistically valuable datasets of 3D geological data. However, extracting the traces of geological features from outcrop images is still a time-consuming and subjective process, limiting the production of very large datasets and the reproducibility of the results. We are currently developing the DOMStudio workflow, a tool for speeding up and making more objective the detection of lineaments in outcrop images by combining image analysis with high-resolution image texturing on photogrammetric DOMs. In this chapter, we test and compare four algorithms for lineament detection not yet routinely used in geology-oriented applications: (1) a two-classes image segmentation based on Markov random fields, (2) the Difference of Gaussian filter, (3) the symmetry of image phases determined by log-Gabor wavelets, and (4) the congruency of image phases as obtained by convolving the images with complex shearlets. We introduce and discuss pre- and postprocessing workflows for extracting thinned, 1-pixel-thick lineaments useful to be converted in vector format. All the described algorithms leverage the image analysis capabilities of Matlab® libraries. We applied the algorithms to detect structural lineaments (faults, stylolites, veins, and fractures) in three images from DOMs collected in different geological settings: fractured sandstone, fractured limestone, and pseudotachylyte-bearing faults in granodiorite. We finally discuss the pros and cons of the four workflows in relation to the spectral signatures of lineaments against wall rocks in the examined outcrops.

6.1 Introduction

Different remote sensing technologies, including photogrammetry and LiDAR, allow datasets to be collected that can be used to create 3D digital representations of outcrop surfaces, called Digital Outcrop Models (DOMs) (e.g. Bellian et al., 2005; Chapter 2 by Bistacchi et al., 2021). Thanks to the improvements in Structure From Motion (SFM) algorithms in the last few decades (Ullman and Brenner, 1979; Snavely et al., 2006), photogrammetry is now routinely applied to obtain 3D models of outcrops with accuracy comparable or even higher than that of laser scanner systems (e.g. Bistacchi et al., 2015; Wilkinson et al., 2016). DOM datasets are initially obtained as photorealistic point clouds, which are datasets composed of millions of points with XYZ coordinates and RGB color. Point clouds can be analyzed for extracting the orientation of planar discontinuities in the rock mass with algorithms recognizing facets or other morphological features of the outcrop surfaces (e.g. Baker et al., 2008; Ferrero et al., 2009; Viseur et al., 2007). However, many geological features relevant for structural analysis, such as faults or veins cemented by hydrothermal minerals, might not produce any 3D morphological expressions on the outcrop surface. These geological structures can be identified only by exploiting the RGB attributes of the point cloud or high-resolution pictures of the outcrop surfaces (Thiele et al., 2015, 2017). A typical example of this situation is the study of geological structures exposed on smooth glacier-polished outcrops (e.g. Bistacchi et al., 2011). In these cases, photogrammetry SFM datasets should be preferred with respect to a laser scanner, since point clouds produced by photogrammetry are directly derived from outcrop pictures, thus establishing a biunivocal relation between pixels on the pictures and the 3D geometry represented as a point cloud. Images of the outcrop can be mapped or 'textured' onto a triangulated surface interpolating the point cloud (applying a technology originally developed for movies and videogames (Catmull, 1974), building a textured triangulated surface DOM (TS-DOM,

3D Digital Geological Models: From Terrestrial Outcrops to Planetary Surfaces, First Edition.
Edited by Andrea Bistacchi, Matteo Massironi, and Sophie Viseur.

Figure 6.1) (see Chapter 2), or they can be projected on to a reference plane to obtain an orthophotograph. The images texturing the TS-DOM, or the orthophotographs, can then be interpreted manually to outline the structures of interest (e.g., Corradetti et al., 2018; Martinelli et al., 2020) or with the aid of image processing algorithms (e.g. Vasuki et al., 2014; Thiele et al., 2017; Seers and Hodgetts, 2016). In this contribution, we briefly introduce our workflow to obtain 3D lineament traces from photogrammetric DOMs, the DOMStudioImage toolbox. We then compare and discuss extensively the efficiency of four image analysis algorithms in highlighting linear features on three geological outcrops having markedly different appearances in terms of how lineaments stand out compared with the background. The outcomes of this comparison are relevant also to the design of image analysis algorithms for the analysis of any outcrop images, also not connected to a DOM.

6.2 The DOMStudioImage Toolbox

The DOMStudioTS workflow is designed to interpret TS-DOMs obtained from photogrammetric surveys. The structural interpretation on a TS-DOM is carried out in terms of traces, which are the intersections of "geological" surfaces (fractures, faults, bedding, etc.) with the outcrop surface. This method is well suited for outcrops with smooth surfaces and 3D morphology, where the outcrop morphology is not controlled by fractures (i.e. facets are not the expression of the geological structures of interest). The key point of our approach is that a biunivocal correspondence between points on the high-resolution outcrop images and points on the triangulated surface is established, so that the traces extracted from high-resolution outcrop images can be projected on the surface and converted to 3D traces (see Bistacchi et al., 2015, and Chapter 2 by Bistacchi et al., 2021, for a detailed explanation).

Photogrammetry software based on SFM algorithms calculates, starting from a redundant set of images of the object to be modelled, the orientation parameters of each image, i.e. the camera positions in the 3D space (x, y, z coordinates of the point cloud) and the internal geometry and optics of the camera. The camera orientation parameters define for each image the projection transformation between the real-world 3D space and the 2D image plane, allowing to transform the x, y, z coordinates of an object in the 3D space to the corresponding 2D position in the image plane and vice versa. Camera parameters also allow images for lens distortion to be corrected, a step performed by the photogrammetry software. Once images are undistorted, a simple pinhole transformation can be applied to project image points into the 3D point cloud (Bistacchi et al., 2015).

Each vertex of the TS-DOM therefore corresponds to a point on one (or several) optically undistorted images, so that high-resolution images can be textured on the mesh (Figure 6.1). For obvious geometrical effects, different portions of the surface of 3D objects are best represented in different images. The TS-DOM is therefore subdivided into portions, each one associated with an image chosen among those used for building the point cloud. A subset of optically undistorted images shot nearly perpendicular to each portion of the mesh is used for texturing the mesh and for tracing the lineaments. Additionally, we prefer to use only the central parts of the images, which are nearly perpendicular to the outcrop surface, neglecting the parts whose line of sight makes an angle higher than 15° with the normal to the surface. Each image is therefore associated with a clipping mask, which highlights only the parts of the images effectively projected on the TS-DOM (Figure 6.2).

The optically undistorted images and the associated clipping masks are the input of the DOMStudioImage toolbox, which is the module of the DOMStudioTS workflow aimed at extracting (automatically as far as possible) the lineament traces from the images using image processing tools implemented in MATLAB®, following an approach inspired by that of Vasuki et al. (2014) and Seers and Hodgetts (2016).

6.3 Lineament Detection Workflow

6.3.1 Image Preprocessing: Conversion to Grayscale and Adaptive Histogram Equalization

Most lineament detection algorithms work on grayscale images, so outcrop images collected during photogrammetric surveys have to be converted to grayscale. Standard conversion to grayscale is attained as a linear combination of the R, G, and B bands (e.g. Solomon and Breckon, 2011):

$$\text{Gray value} = 0.2989\text{R} + 0.5870\text{G} + 0.1140\text{B}$$

An alternative method consists in converting the RGB image to the perceptual HSV color space (Hue, Saturation, Value), and using only the Value (or brightness) band for lineament detection (e.g. Mather and Koch, 2011). For outcrop images with uneven shadowing, we prefer this second option.

Depending on the conditions of illumination of the outcrop surface, it can be useful to apply histogram equalization to enhance and homogenize the contrast of the image set. In pictures taken in full sunlight, better results are obtained by applying adaptive equalization. Adaptive histogram equalization operates separately in small portions of the image (tiles). Using the adaptive equalization algorithm "`adapthisteq`" included in the

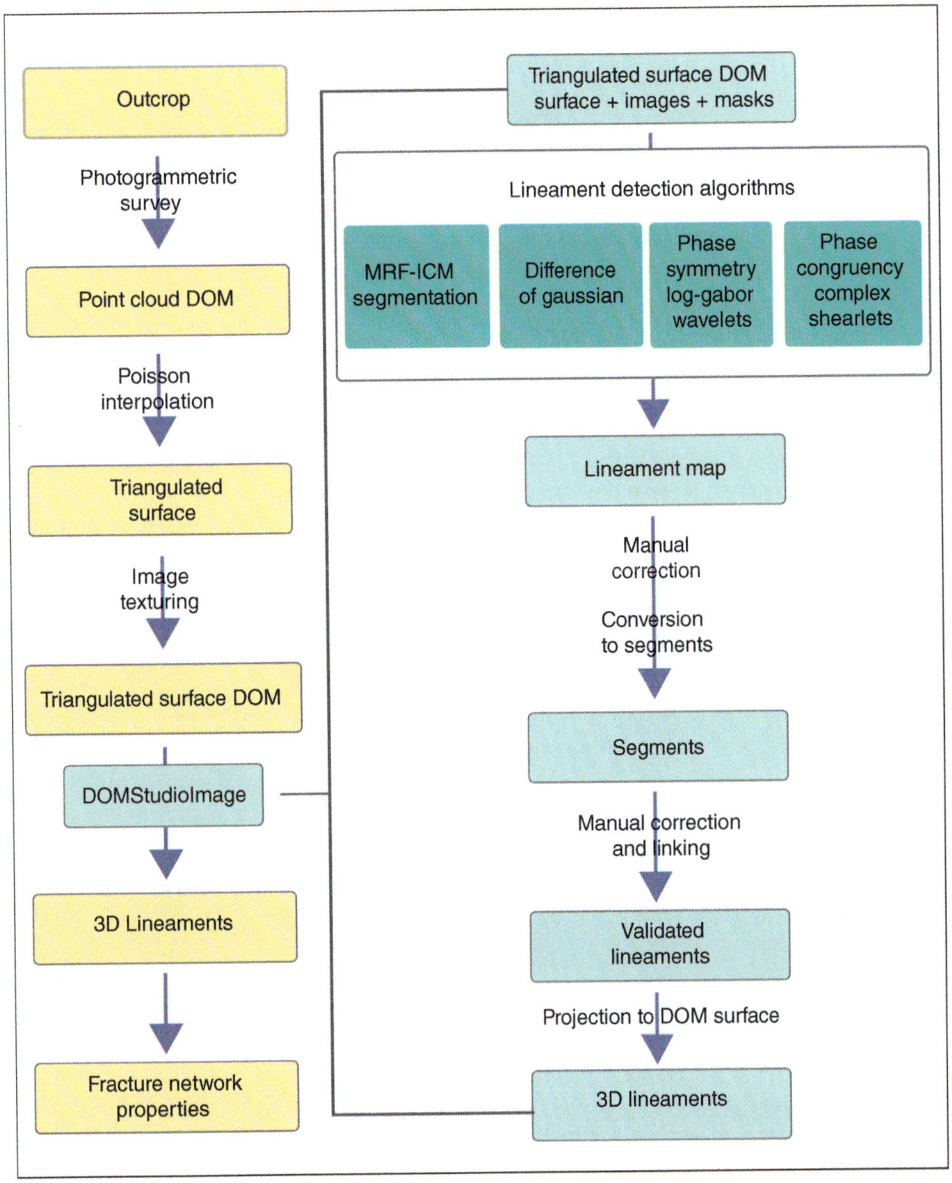

Figure 6.1 The DOMStudioTS workflow for building and interpreting photogrammetric DOMs. In dark green, the DOMStudioImage toolbox, discussed in this chapter.

MATLAB® Image Processing Toolbox, it is possible to set the number of tiles to be used for the adaptive equalization (Figure 6.3). After (adaptive) equalization and conversion to grayscale, the image is ready to be processed for extracting lineaments by applying one of the lineament extraction methods described in the next section.

6.3.2 Lineament Detection Algorithms

Geological lineaments appear in outcrop images as elongated areas where pixels have color or gray value different from that of the wall rock. To highlight lineaments with respect to the background, image analysis algorithms follow two main approaches: (i) image segmentation or (ii) lineament enhancement, by applying specifically designed filterbanks. In this work, we tested one lineament detection method based on segmentation, the Markov Random Field ICM segmentation (here MRF-ICM), and three methods based on different filterbanks, namely (i) the Difference of Gaussian (DoG), (ii) the Phase Symmetry (PhSym) (Kovesi, 1997), and (iii) the Complex Shearlet Phase Congruency (CSPhCon) (Reisenhofer et al., 2016).

6.3.3 MRF-ICM: Markov Random Field ICM Segmentation

When lineaments have a strong contrast with the background and the background is homogeneous, image

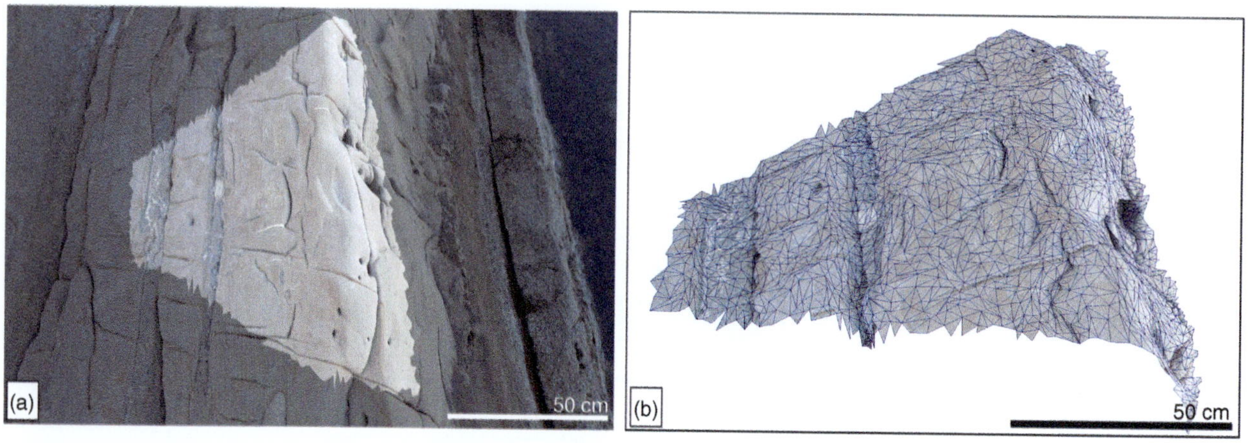

Figure 6.2 The link between field pictures and the DOM triangulated surface. (a) Original field image of an outcrop of fractured sandstone. The shaded area is not selected for texturing because the angle of view is oblique and would cause image distortion when projected on the surface. (b) The light area in (a) as it appears when projected on the corresponding part of the triangulated DOM surface.

Figure 6.3 Image of Outcrop 1 (a), converted to grayscale and equalized using a 20 ×20 arrangement of tiles (b).

segmentation can be an effective and simple method to highlight lineaments. Image segmentation is the process by which an image is subdivided into its constituent regions or objects (e.g. Solomon and Breckon, 2011). This is attained by assigning each pixel to a class, in the simplest case identifying it as foreground (in our case, the lineaments) or background (the wall rock). The simplest way to segment an image is to apply an intensity threshold, which can be determined empirically or with automatic methods, for instance finding one or more thresholds that minimize the within-class variance (Otsu's method, 1979). Intensity thresholding is a rather rough approach to segmentation, because the spatial relationships between pixels is not considered. More refined segmentation techniques are based on region growing (or splitting) and consider not only the variability of gray levels within each class, but also the spatial relationships of pixels in a neighborhood (Solomon and Breckon, 2011).

Markov Random Field (MRF) segmentation is a powerful and sophisticated segmentation method that determines how well a pixel belongs to a particular class by modelling both the variability of gray levels within each class and its dependency upon the classification of the pixels in its neighborhood, hence imposing spatial continuity in the output segmented image (Solomon and Breckon, 2011). In our case, even if in general MRF is able to classify several classes, we choose to segment just two classes: lineaments and background (Figure 6.4). To obtain the MRF segmentation, we use the Iterated Conditional Modes (ICM) algorithm (Besag, 1986), which works iteratively, attributing pixels to different classes following an energy minimization approach. To detect large-scale features only, avoiding small branches or irregularities in lineament traces, before running the MRF-ICM algorithm, it is advisable to apply a Gauss filter to smooth the image and remove noise. The optimal half-width of the Gauss

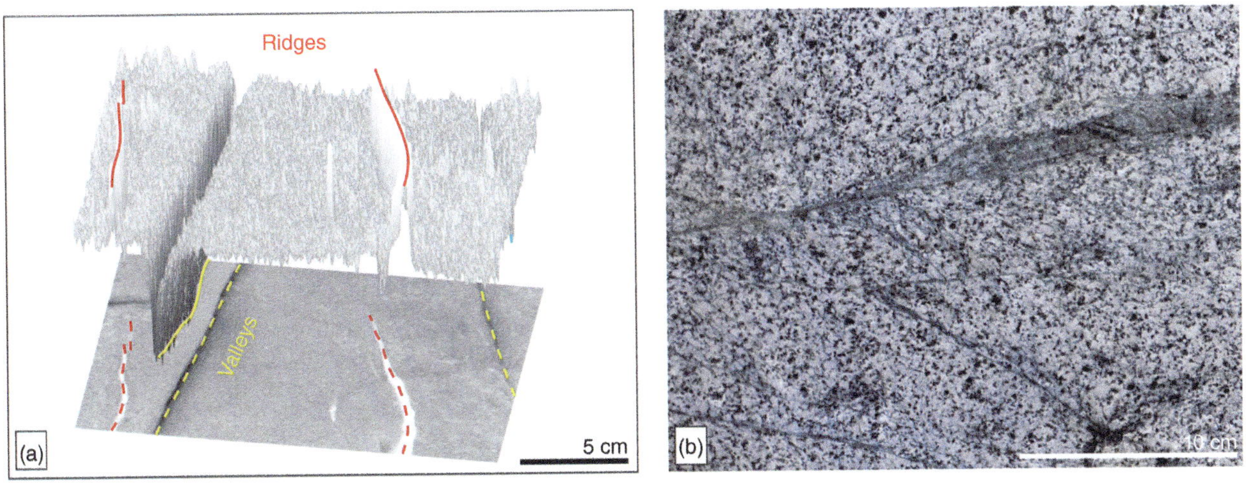

Figure 6.4 Example of an image of Outcrop 2 processed with the MRF-ICM segmentation algorithm. (a) Original outcrop image. (b) The binary image obtained by image segmentation with the two-classes MRF-ICM segmentation algorithm. Lineaments are represented as the foreground (white).

filter kernel and its standard deviation value have to be empirically calibrated for each set of images, taken on different outcrops and in different conditions, making the output of the workflow a bit arbitrary.

6.3.4 DoG: Difference of Gaussian Filter

Image segmentation is not optimal for detecting lineaments when they have variable gray levels across the image, as it happens frequently in outcrop images, due to uneven illumination, to variable width or color of the lineaments themselves, or to variations in color or texture of the wall rock. In these cases, lineaments in grayscale images can be modelled as ridges and valleys, which, in computer vision, are defined as an alignment of points where the intensity of gray levels reach either a local maximum (ridge) or a local minimum (valley; Figure 6.5a). Ridge detection algorithms are designed to highlight the crest lines of valleys and ridges and are mostly based on pattern recognition and filtering methods (Lindberg, 1998).

The simplest ridge detection filters are based on the second derivative (Laplacian) of the image, which assumes

Figure 6.5 Details of lineaments in outcrop images. (a) Detail of the picture in Figure 6.2a, converted to grayscale, showing an open fracture (dark) and a calcite vein (white). When gray levels are considered, calcite veins can be modelled as ridges and dark fractures as valleys. (b) Detail of a granodiorite outcrop cut by hydrothermally altered green fault traces. Fault traces have variable width and color along the strike, and the dark biotite grains in granodiorite form a very heterogeneous background.

very high or very low values along ridges or valleys respectively (Figure 6.5). However, the second derivative is very sensitive to noise, which appears as local peaks or sinks; hence a Gaussian smoothing of the image is necessary before applying second derivative operators. Laplacian and Gaussian filters are both linear filters and thus can be combined to improve computational performance by convolving the image with a Laplacian of Gaussian filter. Moreover, the Laplacian of Gaussian filter can be approximated with the difference of two Gaussian kernels (e.g. Solomon and Breckon, 2011), obtaining the Difference of a Gaussian (DoG) filter. This is a well-known and easily implemented edge detection algorithm, used also in standard image processing software, which consists in convolving the image with a filter obtained as the difference between two Gaussian kernels having different standard deviations.

The goal of the DoG filter is generally to highlight edges (i.e. sharp boundaries between areas with different gray levels). However, by choosing a sufficiently wide kernel size and suitable standard deviations of the two Gaussian filters, it is possible to detect ridges instead of edges (e.g. Magnier et al., 2011). The results shown in Figure 6.6 have been obtained by using two Gaussian filters having standard deviations of 2 and 10, and a kernel size of 15 pixel.

6.3.5 PhSym: Phase Symmetry Line Detection

Often outcrop images include lineaments having variable width and locally variable color and gray value contrast with the background. In other words, the lineaments appear continuous to the human eye, but have objectively variable characters along their length. In these cases, multiscale and multidirectional filterbanks need to be applied (e.g. Vasuki et al., 2014; Magnier et al., 2011). Gray values of lineaments that appear as ridges in images are symmetric along profiles sampled along all orientations other than parallel to the lineament itself. On the other hand, profiles sampled along the axis of symmetry of a lineament tend to correspond to a sequence of local maxima or minima. Lineaments can be detected by performing a frequency analysis of the image and correspond to the pixels where all the frequency components are either at a maximum or a minimum. We used the PhSym phase symmetry algorithm, implemented by Kovesi (2000) and successfully applied to fracture detection in rock outcrops by Vasuki et al. (2014). In PhSym the frequency analysis is performed by convolving the image with a set of multiscale complex-valued Log-Gabor wavelets, which are suitable for performing frequency analysis on images because they can measure the localized frequency response of the image (unlike, for instance, the Fourier transform, which is not localized). By varying the scale of the wavelets, it is possible to detect lineaments of different widths. The symmetry or asymmetry of phase is computed locally, so the detection of lineaments is not sensitive to variations in image illumination or contrast, and it is not necessary to equalize images before processing. The output of PhSym is an indexed image, where pixel values correspond to the degree of symmetry at that point, normalized between 0 (no symmetry) and 1 (maximum of symmetry) (Figure 6.7). Because of the scale and contrast independency of the phase symmetry calculated by PhSym, several indexed images covering an outcrop can be thresholded by using the same value for all images, irrespective of local illumination.

A limit of PhSym is that, being very sensitive, it also detects extremely tiny and short lines, resulting in a noisy output image (Figure 6.7b). This can be solved by filtering the lineaments below a threshold length, which must be established depending on lineament size and image

Figure 6.6 Output of the DoG filter on an image of Outcrop 1. (a) The original outcrop image. (b) Output of the DoG with a kernel of 15 pixel and standard deviation of 10 and 2. Color coding indicates the probability of a pixel to be a ridge, with 1 being the pixels with the highest probability to belong to a ridge.

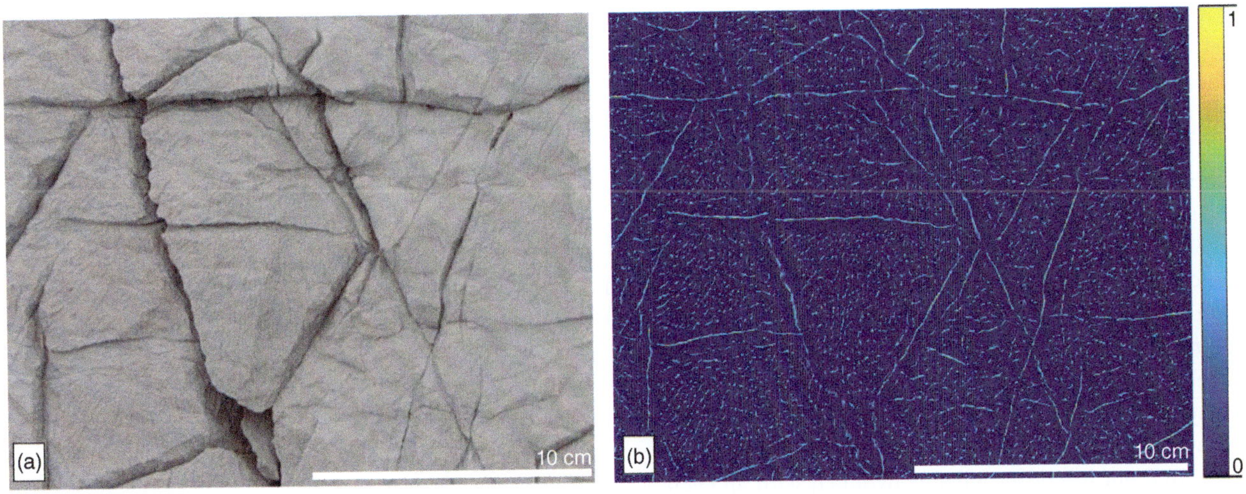

Figure 6.7 Detail of the output of the PhSym line detection algorithm. (a) Detail of the image of Outcrop 2, showing lineaments having different brightness contrast. (b) Image of the phase symmetry on the area represented in (a). Phase symmetry can detect very tiny details independently of their contrast. The smaller ones, which represent noise, must be filtered in a following analysis step.

resolution. Another limit, compared with simpler image enhancement filters such as the DoG, is that PhSym has several parameters to be calibrated, according to local conditions of the image set in order to obtain the desired output and this makes the analysis a bit arbitrary.

6.3.6 CSPhCon: Complex Shearlet Phase Congruency Ridge Detector

When outcrops have an inhomogeneous background, as in Figure 6.5b (tonalite peppered with biotite grains), filters should be designed to separate more or less continuous elongated lineaments from the black spots in the background. 2D wavelets are widely used in image analysis because they represent a generalization of the Fourier transform by using a basis function that allows considering both location and frequency (with a tradeoff between location and frequency resolution that reflects the uncertainty principle in signal processing and image analysis). Classical wavelet systems are isotropic operators (Figure 6.8a), so they might not be effective in detecting anisotropic features like ridges, and particularly when trying to separate isotropic and anisotropic features. There have been several attempts to build directional wavelet-like systems

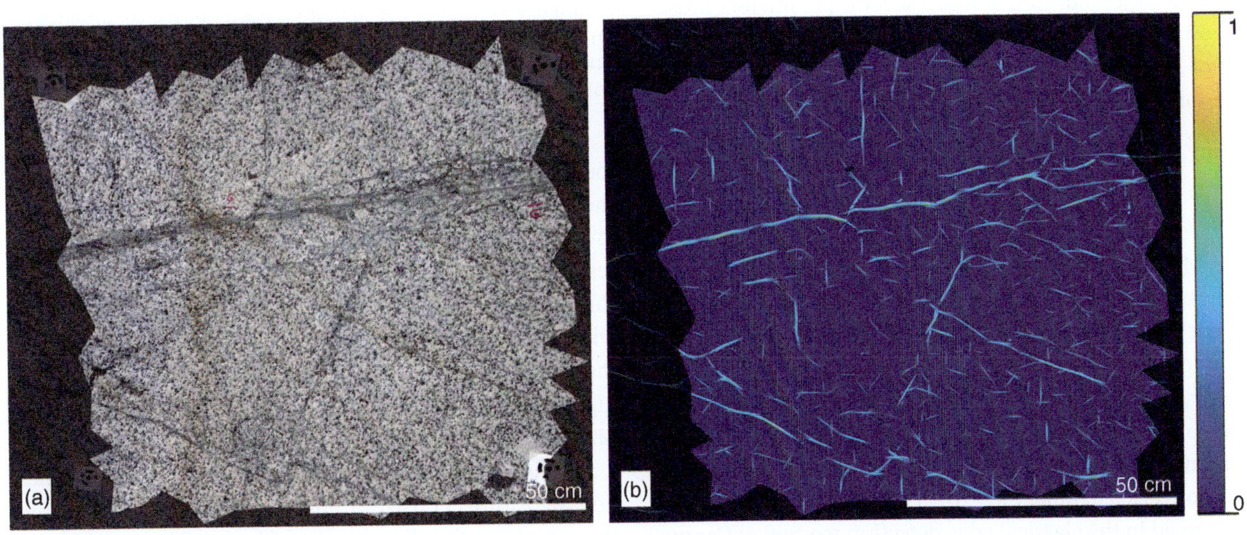

Figure 6.8 (a) Isotropic tiling scheme for a wavelet. (b) Anisotropic tiling scheme for a shearlet. (c) Dilations of the even-symmetric wavelet that, in the shearlet implementation, is scaled anisotropically according to tiling scheme B. (d) Dilations of the odd-symmetric wavelet that, in the shearlet implementation, is scaled anisotropically according to tiling scheme B. (e) Even-symmetric wavelet/shearlet and its interaction with an idealized ridge in 2D. (f) Odd-symmetric wavelet/shearlet and its interaction with an idealized ridge in 2D. *Source*: Modified from Reisenhofer et al., 2016.

to overcome this limitation, namely curvelets (Candés and Donoho, 2000), contourlets (Do and Vetterli, 2005) or shearlet systems (Figure 6.8b) (Kutyniok and Labate, 2012). Among these, shearlets have been demonstrated to be very effective operators for detecting anisotropic features (Kutyniok and Lim, 2010).

Shearlets can be used to improve the performances of wavelet-based image analysis methods, such as, for instance, the Phase Congruency algorithm implemented by Kovesi (1997). The Phase Congruency algorithm, in a way similar to the PhSym algorithm described above, is based on the frequency analysis of the image through a pair of even symmetric and odd symmetric Log-Gabor wavelets. Ridges (or valleys) in the input image are identified based on the different responses when convoluted with an even symmetric wavelet (inflection points) or an odd symmetric wavelet (maximum and minimum points) (Figures 6.8e, f). Reisenhofer et al. (2016) updated the Phase Congruency algorithm by using a pair of even symmetric and odd symmetric shearlets instead of wavelets (Complex Shearlet Phase Congruency, CSPhCon). The amplitude and the scales of a shearlet system can be calibrated to detect lineaments of different thicknesses. CSPhCon has the advantage of using an anisotropic tiling of frequency domains, which is suitable for distinguishing anisotropic features (e.g. fracture lineaments) from isotropic features (e.g. dark biotite grains in an igneous rock), and the sensitivity to anisotropy can be tuned depending on the images to be analyzed. CSPhCon is implemented in the SymFD Matlab® toolbox developed by Reisenhofer et al. (2019, www.math.uni-bremen.de/cda/software.html#SymFD). As for the Phase Symmetry method, in CSPhCon several parameters must be calibrated for building filterbanks suitable for the analysis of images collected on different outcrops and under different conditions, thus making the results of the analysis slightly arbitrary (Figure 6.9).

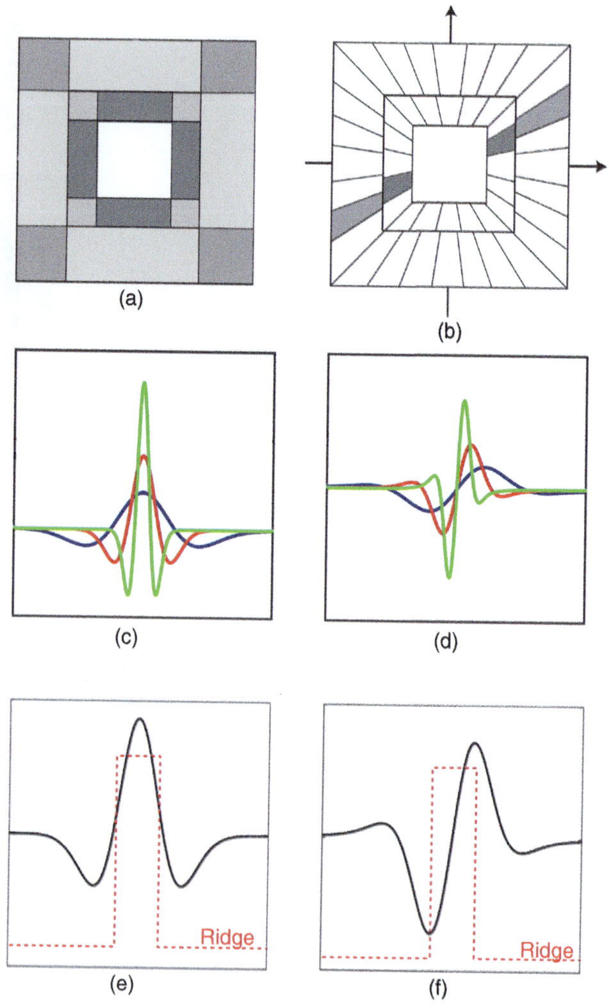

Figure 6.9 Output of the complex shearlet phase congruency on an image of Outcrop 3. (a) Outcrop 3 image, showing tiny lineaments in a strongly textured wall rock. (b) Output of the complex shearlet phase congruency algorithm.

6.3.7 Lineament Thinning and Skeletonization

The MRF-ICM segmentation algorithm yields binary images of white areas interpreted as lineaments (value 1) against a black background (value 0). The other lineament detection methods based on filterbanks yield indexed images, where the pixel index expresses the probability of a pixel to belong to a lineament (lineament map). The lineament map can be treated as a grayscale image, which needs to be converted to binary to extract the lineament traces. As introduced in the classical edge detection method of Canny (1986), before converting the grayscale lineament maps to a binary by applying an intensity threshold, it is advisable to clean the lineament map by performing a non-maxima suppression, i.e. suppressing all the pixels that are not local maxima. This operation reduces the thickness of the ribbons that contain the lineaments, excluding also low-intensity branches that have low probability to represent actual lineaments. After non-maxima suppression, thresholding is performed by using hysteresis: all pixels with gray value below a lower threshold are kept only if they are connected with pixels having gray value above a higher threshold. This algorithm has been implemented in Matlab® by Kovesi (2000).

At this point, the outputs of all the lineament extraction algorithms consist of binary images of white ribbons that contain the lineaments. These ribbons must be reduced to 1 pixel-thick curves, which will be successively converted into vectorial polylines. From here, the processing is the same for the three lineament extraction algorithms.

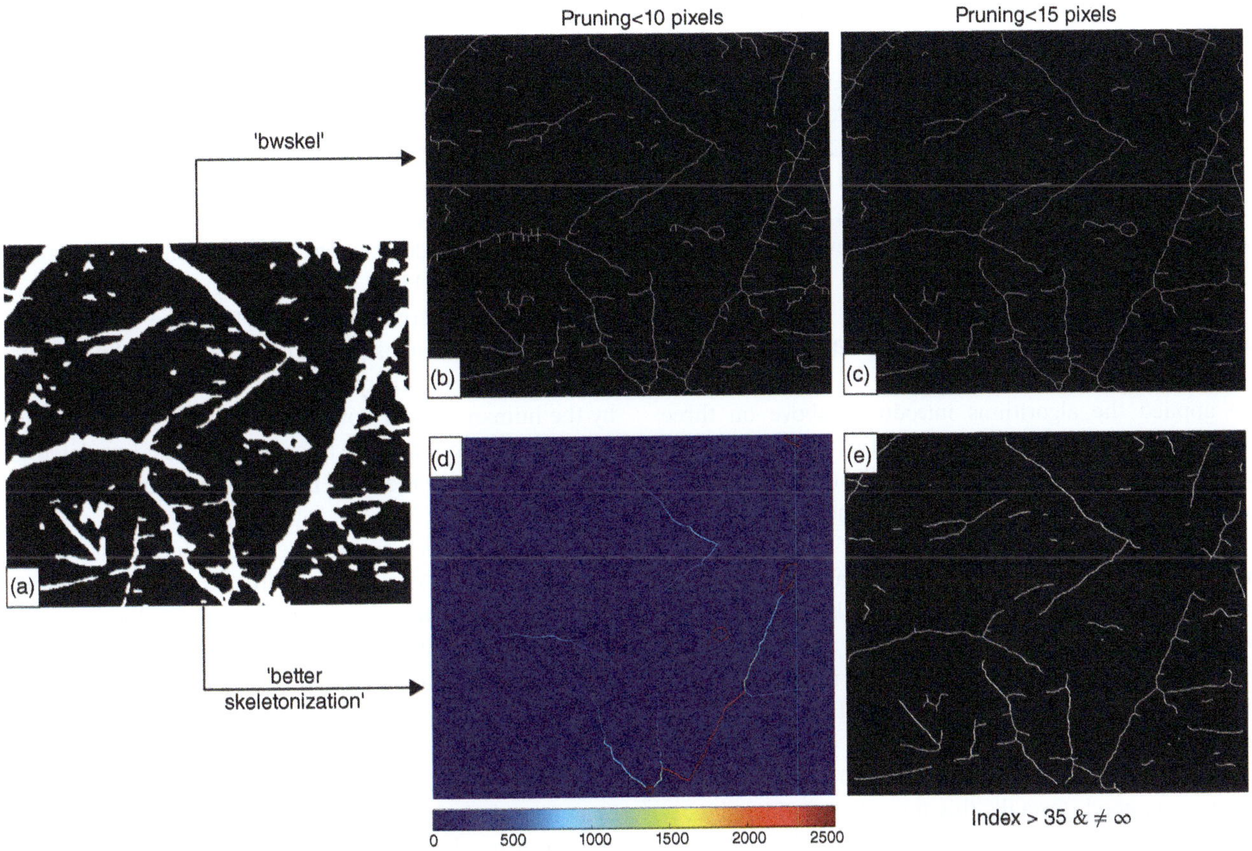

Figure 6.10 Examples of skeletonization workflows applied to a detail of the output (a) of the MRF-ICM algorithm on Outcrop 2. (b) and (c) Output of the Matlab® built-in "`bwskel`" algorithm, with pruning applied to segments longer than 10 pixels and 15 pixels, respectively. (d) Index map of the skeleton as output of the "better skeletonization" algorithm by Howe (2020). Indexes are proportional to the probability to be a part of the skeleton, i.e. are proportional to the length of the outline from which they originated. Loops have a dark-red color corresponding to an infinite value. (e) Skeleton obtained by thresholding (d) and eliminating loops.

The process of simplifying 2D shapes to a "thin" 1D curve is called skeletonization and is widely used to simplify objects in image processing (see Saha et al., 2017, for a review). For our applications to lineament extraction, one of the main problems is the fact that quite often skeletonization produces many more branches than desired, deriving from small protrusions in the object outline (Figure 6.10b). Unwanted branches can be reduced by a postprocessing step called skeleton pruning. We tested two methods of skeletonization and pruning, based on different approaches: the algorithm "`bwskel`" included in the Matlab® Image Processing Toolbox (www.mathworks.com/products/image.html and the "better skeletonization" tool implemented in Matlab by Howe (2021) based on an algorithm described by Telea and van Wijk (2002).

The "`bwskel`" included in the "`bwmorph`" toolbox of Matlab is based on the process of morphological thinning, which peels off the boundary pixels layer by layer, identifying points whose removal does not affect the object's topology. The algorithm includes an option to prune the skeleton by removing branches shorter than a given threshold. Branches are common, especially for the outputs of the MRF-ICM algorithm (Figure 6.10), and tests have to be performed to each set of images to find the correct length threshold (Figure 6.10b, c). If ribbons with very different thicknesses are present, the same threshold might not work for the whole image set.

The "better skeletonization" algorithm is based on a distance transform of the object's boundary. This method keeps trace of the boundary from which the skeleton is derived; therefore it is possible to threshold the image in order to remove the contributions from small protrusions in the outline. Loops have Inf (infinite) value in the indexed image, so it is simple to remove loops from the skeleton, if needed (Figure 6.10d, e). Pruning through thresholding of the indexed image appears to be less scale-sensitive than thresholding the branches based on their final length.

In our outcrop images, both skeletonization algorithms perform well when lineament maps are composed of thin and straight ribbons, as happens for the lineament maps from the DoG, PhSym, and CSPhCon algorithms. The lineament maps obtained from MRF-ICM are the most critical for the skeletonization processes, because the ribbons including the lineaments have very irregular outlines and variable thicknesses (Figure 6.10a). To uniform our investigation, we applied the "better skeletonization" algorithm to all the lineament maps, with a threshold of 35 pixels to all the image processing workflows.

6.4 Results on Geological Images

We applied the algorithms introduced above on three outcrop images from DOMs collected in different geological settings. Outcrop 1 is a cliff of calcareous sandstones belonging to the Dinaric Flysch of the Island of Pag (Croatia) (e.g., Babič and Zupanič, 1996; Mittempergher et al., 2019). The sandstone beds are a series of amalgamated turbidites about 30–50 cm thick and are cut by two sets of bed-perpendicular fractures, sometimes filled by white calcite (Figures 6.3 and 6.6). Fractures are well identifiable in field pictures because they are slightly eroded and lined by thin layers of iron oxides. The wall rock is very homogeneous and the contrast with darker fractures is good. Overall, this must be considered as a very easy environment to perform automatic lineament interpretation.

Outcrop 2 is a portion of a bed surface of limestone of the island of Pag (Croatia) belonging to the Late Cretaceous Gornji Humac Formation (e.g. Babič and Zupanič, 1996; Mittempergher et al., 2019). The rudist-bearing limestone is crosscut by several sets of tiny veins and stylolites, which are preferential sites for karst dissolution and appear as centimeter-scale grooves on the bed surfaces (Figure 6.4). The wall rock is fairly homogeneous and has a good contrast with the lineaments. However, due to different dissolution levels, grooves have variable width and depth (Figure 6.7). We consider this an intermediate difficulty outcrop, representative of many situations encountered in nature, particularly on carbonate rocks.

Outcrop 3 is a glacier-polished rock surface in granodiorite of the Adamello Massif (Italian Southern Alps). The outcrop is located within the Gole Larghe Fault Zone (Di Toro and Pennacchioni, 2005), an exhumed fault zone including thousands of small-scale faults cemented by greenish hydrothermal minerals and black pseudotachylytes. Because of the diffuse presence of mm-sized biotite in the tonalite, the wall rock has a spotty appearance, and lineaments (particularly the black pseudotachylyte veins that are the target for lineament tracing) are difficult to segment with respect to biotite grains with automatic image analysis. Moreover, pseudotachylyte veins and secondary fractures are sometimes very thin ("hairline fractures") and difficult to identify if image resolution is not sufficiently high, so in our analysis we used images at 0.2 mm/pixel, and this makes the automatic processing of the images computationally intensive and the need for efficient automatic analysis even greater.

In Outcrop 3, the difficulty in differentiating biotite grains from pseudotachylyte faults depends on (i) both having the same black color and (ii) the biotite grains being generally larger (2–10 mm) than the thickness of pseudotachylyte veins (1–5 mm). Paradoxically, biotite grains and pseudotachylyte veins are easily recognized by the human eye but not by automatic methods, and we think that the key to solve this problem must be related to the spatial organization and morphology of the lineaments (Figure 6.9).

The three outcrop images have been interpreted manually to define a "desired" and "meaningful" output and to be compared with the output of the tested algorithms. Lineaments were traced using the free painting program Krita (https://krita.org/en/), by adding a raster layer with the same resolution and size above the original outcrop images. The manually interpreted ground truth image has been compared pixelwise with the outputs of the four lineament detection algorithms. The performance of each algorithm has been evaluated in terms of precision and recall, assuming that the manually traced image contains all the lineaments (positives) in the image. Recall (or sensitivity) is the fraction of positive examples correctly predicted by the algorithm under evaluation, and is expressed as the ratio of true positives (pixels labelled correctly as lineaments by the algorithm) and the sum of false negatives and true positives (equal to the pixels labelled as lineaments by the operator). A recall indicates how many of the automatically identified lineaments are relevant. Precision is defined as the ratio between the true positives and the sum of true and false positives (how many of the pixels identified as lineaments by the operator are also interpreted correctly by the algorithm) (Ting, 2011). Precision indicates how many of the selected lineaments are relevant. Precision and recall have been calculated by comparing binary images of 1-pixel thick lineaments. To account for small-scale differences between the manual and automatic interpretations, pixels interpreted as positives in automatically generated lineament maps are considered as true positives if they fall in a buffer area of 10 pixels from the manually traced lineament map. Introducing a buffer is necessary because, given the high resolution of the images, each lineament is several tens of pixels thick, and any trace within it can be considered a good approximation for the lineament itself. In the case of a very noisy output, i.e. of many false positives, this simple buffering might overestimate the recall. However, in that case the precision will be low, and will cause rejection of the solution. The images used for the output evaluation and the related Matlab script is freely available at the following doi: 10.5281/zenodo.3954283.

We applied the four algorithms described in the previous paragraph to each outcrop image, and report the results in Table 6.1 and Figure 6.11. In Table 6.1 the results of the PhSym algorithm are reported with and without filtering lineaments shorter than 20 pixels.

Recall and precision are in all cases below 0.8, meaning that, also in the best conditions, only 80% of the pixels pertaining to lineaments were correctly identified. In the case of very noisy outputs, recall might have an anomalously high value, as in the case of the DoG algorithm applied to Outcrop 3. In this case the algorithm returned a number of lineaments that was higher than the ones really present in the image (lineaments detected by the operator). Precision for the same analysis is a mere 20%, meaning that six lineaments out of seven returned by this algorithm are false positives.

Acceptable outputs are thus those that have at the same time high precision and high recall values (e.g. DoG for Outcrop 1 and CSPhCon for Outcrops 1 and 2). The results are discussed in detail in the following section.

6.5 Discussion

We tested four image analysis approaches to detect lineaments in outcrops showing markedly different contrast between lineaments and background. The three outcrops present increasing challenges for automatic lineament detection methods from Outcrop 1 (fractures in sandstone) to Outcrop 3 (pseudotachylyte faults in tonalite). Our tests suggest that not all methods are suitable for analyzing every kind of outcrop. However, after a correct analysis of the characteristics of the outcrop, it is possible to obtain fairly good results by choosing an appropriate image analysis method. The outcomes of the four lineament detection algorithms are visually summarized in Figure 6.11, and the resulting recall and precision are reported in Table 6.1.

The MRF-ICM segmentation algorithm, being based on image segmentation, works well when lineaments are clearly distinguishable from the background, thanks to their color or gray level. The best compromise between recall and precision is indeed obtained for Outcrop 2, where lineaments are much darker than the wall rock. In this case, however, the MRF-ICM algorithm fails to detect thin lineaments having only moderate contrast with the background. In the other two outcrops, the MRF-ICM algorithm gives a relatively high recall, i.e. it detects many of the lineaments of interest, but the precision is low, because other features are erroneously identified as lineaments, and the traces of lineaments are not always well located within the white ribbons identified as lineaments. Some of these issues (particularly those related to thin lineaments) can be solved by increasing the spatial resolution of images, but others represent more fundamental limits of the method (e.g. the difficulty in separating black pseudotachylytes from black biotite grains in Outcrop 3).

The DoG algorithm is effective in detecting lineaments in Outcrops 1 and 2, giving both high recall and precision. However, the DoG algorithm performs very badly in Outcrop 3, where it attains a very low precision. This occurs because the DoG algorithm is spatially isotropic, thus, in Outcrop 3, it also highlights dark minerals of the wall rock. In this respect, the result is quite similar to the output form MRF-ICM, even if the algorithm is completely different. In the presence of well-defined lineaments (Outcrops 1 and 2), having high contrast with the background, the DoG algorithm performs very well and it might be recommended in practical applications, especially considering that it is a very simple and computationally fast algorithm.

In Outcrops 1 and 2, the PhSym algorithm performs well in terms of recall, i.e. it detects many of the lineaments of interest, but with low precision, because it detects also very tiny non-significant lineaments, due to its scale independence. The precision improves by filtering the lineaments by length (keeping only lineaments longer than 20 pixels), but the precision is still lower than that of the DoG algorithm. In Outcrop 3, the PhSym algorithm does not work well in terms of recall or of precision. The very local response of the phase symmetry algorithm cannot distinguish the dark biotite grains of the tonalite from the pseudotachylyte veins that represent the lineaments to be detected.

The CSPhCon algorithm is precise in identifying lineaments in Outcrops 1 and 2, but with a relatively low recall. This means that not all the lineaments of interest are identified, but their trace is accurate and there are few false positives. In Outcrop 3, recall and precision have similar values around 0.3. This means that in the presence of a strong pattern in the wall rock, this method also fails to produce a precise lineament map, i.e. there is a significant proportion of false positives.

To summarize, Table 6.1 shows that accurate results from automatic image interpretations can be obtained only in Outcrops 1 and 2. In Outcrop 3, all algorithms give precision and recall below 0.5 (or anomalous values above 1). As it clearly appears from Figure 6.11, lineament traces obtained from automatic image interpretation are generally discontinuous, especially at intersection points, and need to be connected, either manually or automatically, in order

Figure 6.11 Comparison of the lineament traces obtained on the three outcrop images by applying the four algorithms described above.

Table 6.1 Summary of the performances expressed as precision and recall of each line detection algorithm, as applied to Outcrops 1, 2, and 3. See the discussion in the main text.

	Outcrop 1		Outcrop 2		Outcrop 3	
	Precision	Recall	Precision	Recall	Precision	Recall
MRF-ICM	0.3	0.6	0.5	0.4	0.2	0.3
DoG	0.6	0.7	0.5	0.6	0.2	1.4
PhSym	0.2	0.8	0.3	0.8	0.2	0.4
PhSym lineaments > 20 px	0.4	0.7	0.4	0.6	0.2	0.3
CSPhCon	0.7	0.4	0.7	0.5	0.3	0.3

to obtain geologically meaningful fault or fracture traces. Further developments of the workflow will be aimed at connecting lineament segments by using semiautomatic, supervised methods.

Regarding the two goals of automatic interpretation of DOMs: (i) cutting time required for interpretation and (ii) obtaining less subjective and more reproducible results, we must admit that the first goal is easily achieved, but the second one still represents an open problem, since the image analysis algorithm, the parameters of the algorithm, and the skeletonization algorithm and parameters still depend on subjective choices.

6.6 Conclusions

We are currently developing a workflow for interpreting photogrammetric DOMs for structural geology purposes. Photogrammetric DOMs can be textured with full-resolution outcrop pictures, given the one-to-one correspondence between the images and the 3D model built from the images themselves. In the same way, structural lineaments identified on images can be projected on the DOM surfaces, obtaining three-dimensional traces suitable for 3D structural analysis. Here, we compared four algorithms that can be used for extracting lineaments from images in a semiautomatic way.

Methods for identifying lineaments in outcrop images are based either on image segmentation or on image filtering. Here we tested a segmentation-based algorithm (based on Markov Random Fields implemented with the Iterated Conditional Modes algorithm) and three algorithms based on filters (the difference of Gaussian, the phase symmetry method implemented by log-Gabor wavelets, and the phase congruency method implemented using complex-valued shearlets). Our results suggest that:

1. When lineaments are well distinguishable from the background (e.g. Outcrops 1 and 2), relatively simple algorithms like the Difference of Gaussian filtering (DoG) outperform more sophisticated methods.
2. When lineaments are poorly distinguishable from the background, only frequency-based methods sensitive also to directionality (PhSym and CSPhCon) are effective in extracting lineaments with a reduced number of false positives.
3. Lineaments extracted automatically tend to be segmented and include false positives: manual interpretation and cleaning is still needed for obtaining geologically meaningful lineament traces.
4. This kind of automatic analysis is fast and results in huge datasets, but its outcome is still subjective owing to choices in the image analysis and skeletonization algorithms.

References

Babič, L. and Zupanič, J. (1996). Coastal Dinaric flysch belt: paleotransport model for the Pazin Basin, and the role of a foreland uplift. *Natura Croatica* 5: 317–327.

Baker, B.R., Gessner, K., Holden, E.-J., and Squelch, A.P. (2008). Automatic detection of anisotropic features on rock surfaces. *Geosphere* 4 (2): 418. https://doi.org/10.1130/GES00145.1.

Bellian, J.A., Kerans, C., and Jennette, D.C. (2005). Digital outcrop models: Applications of terrestrial scanning LiDAR technology in stratigraphic modeling. *Journal of*

Sedimentary Research 75 (2): 166–176. https://doi.org/https://doi.org/10.2110/jsr.2005.013.

Besag, J.E. (1986). On the statistical analysis of dirty pictures. *Journal of the Royal Statistical Society, Series B* 48 (3): 259–302. JSTOR 2345426.

Bistacchi, A., Griffith, W.A., Smith, S.A.F. et al. (2011). Fault roughness at seismogenic depths from LiDAR and photogrammetric analysis. *Pure and Applied Geophysics* 168: 2345–2363. https://doi.org/10.1007/s00024-011-0301-7.

Bistacchi, A., Balsamo, F., Storti, F. et al. (2015). Photogrammetric digital outcrop reconstruction, visualization with textured surfaces, and three-dimensional structural analysis and modeling: Innovative methodologies applied to fault-related dolomitization (Vajont Limestone, Southern Alps, Italy). *Geosphere* 11 (6): 2031–2048. https://doi.org/10.1130/GES01005.1.

Bistacchi, A., Mittempergher, S., and Martinelli, M. (2021). Digital Outcrop Model reconstruction and interpretation. In: *3D Digital Geological Models: From Terrestrial Outcrops to Planetary Surfaces*. Wiley, this volume.

Canny, J. (1986). A computational approach to edge detection. *IEEE Trans. Pattern Analysis and Machine Intelligence* 8 (6): 679–698.

Catmull, E.E. (1974). *A subdivision algorithm for computer display of curved surfaces*. University of Utah.

Corradetti, A., Tavani, S., Parente, M. et al. (2018). Distribution and arrest of vertical through-going joints in a seismic-scale carbonate platform exposure (Sorrento peninsula, Italy): Insights from integrating field survey and digital outcrop model. *Journal of Structural Geology* 108: 121–136. http://dx.doi.org/https://doi.org/10.1016/j.jsg.2017.09.009.

Candes, E. and Donoho, D.L. (1999). *Curvelets: A surprisingly effective nonadaptive representation for objects with edges*. Department of Statistics: Stanford University.

Di Toro, G. and Pennacchioni, G. (2005). Fault plane processes and mesoscopic structure of a strong-type seismogenic fault in tonalites (Adamello batholith, Southern Alps). *Tectonophysics* 402: 55–80. https://doi.org/https://doi.org/10.1016/j.tecto.2004.12.036.

Do, M.N. and Vetterli, M. (2005). The contourlet transform: An efficient directional multiresolution image representation. *IEEE Transactions on Image Processing* 14: 2091–2106. https://doi.org/10.1109/TIP.2005.859376.

Ferrero, A.M., Forlani, G., Roncella, R., and Voyat, H.I. (2009). Advanced geostructural survey methods applied to rock mass characterization. *Rock Mechanics and Rock Engineering* 42 (4): 631–665. https://doi.org/10.1007/s00603-008-0010-4.

Howe, N. (2021). Better Skeletonization. https://www.mathworks.com/matlabcentral/fileexchange/11123-better-skeletonization, MATLAB Central File Exchange. Retrieved August 24, 2021.

Kovesi, P. (1997). Symmetry and asymmetry from local phase. *Proceedings of the Tenth Australian Joint Conference on Artificial Intelligence* 2–4.

Kovesi, P. (2000). MATLAB and Octave Functions for Computer Vision and Image Processing. Available from: http://www.peterkovesi.com/matlabfns/.

Kutyniok, G. and Labate, D. (2012). Introduction to Shearlets. In: *Shearlets. Applied and Numerical Harmonic Analysis* (ed. G. Kutyniok and D. Labate). Birkhäuser, Boston: https://doi.org/10.1007/978-0-8176-8316-0_1.

Kutyniok, G. and Lim, W.Q. (2010). Image separation using wavelets and shearlets. In: *Curves and Surfaces, 2010. Lecture Notes in Computer Science*, vol. 6920 (ed. J.D. Boissonnat et al.). Berlin, Heidelberg: Springer https://doi.org/10.1007/978-3-642-27413-8_26.

Lindberg, T. (1998). Edge detection and ridge detection with automatic scale selection. *International Journal of Computer Vision* 30 (2): 117–154.

Magnier, B., Montesinos, P., and Diep, D. (2011). Ridges and valleys detection in images using difference of rotating half smoothing filters. In: *Lecture Notes in Computer Science (Including Subseries Lecture Notes in Artificial Intelligence and Lecture Notes in Bioinformatics), 6915 LNCS*, 261–272. https://doi.org/10.1007/978-3-642-23687-7_24.

Martinelli, M., Bistacchi, A., Mitempergher, S. et al. (2020). Fault damage zone characterization combining scan-line and scan-area analysis on a km-scale Digital Outcrop Model: The Qala Fault in Gozo. *Malta. Journal of Structural Geology* 140: 104144. https://doi.org/https://doi.org/10.1016/j.jsg.2020.104144.

Mather, P.M. and Koch, M. (2011). *Computer Processing of Remotely-Sensed Images: An Introduction*, Fourth Edition, 434 pp. Ltd: John Wiley & Sons doi: 10.1002/9780470666517.

Mittempergher, S., Succo, A., Bistacchi, A. et al. (2019). Geological and structural map of the southeastern Pag Island, Croatia: Field constraints on the Cretaceous–Eocene evolution of the Dinarides foreland. *Geological Field Trips and Maps* 11: 1–19. https://doi.org/10.3301/GFT.2019.06.

Otsu, N. (1979). A threshold selection method from gray-level istograms. *IEEE Transactions on Systems, Man, and Cybernetics* 9 (1): 62–66.

Reisenhofer, R., Kiefer, J., and King, E.J. (2016). Shearlet-based detection of flame fronts. *Experiments in Fluids* 57 (3): https://doi.org/10.1007/s00348-016-2128-6.

Reisenhofer, R. and King, E.J. (2019). Edge, ridge and blob detection with symmetric molecules. *SIAM Journal of Imaging Sciences* 12: 1585–1626. doi: 10.1137/19M1240861.

Saha, P.k., Borgefors, G., and Sanniti di Baja, G. (2017). Chapter 1 – Skeletonization and its applications – a review. In: *Skeletonization*, 3–42. Academic Press https://doi.org/10.1016/B978-0-08-101291-8.00002-X.

Seers, T.D. and Hodgetts, D. (2016). Extraction of three-dimensional fracture trace maps from calibrated image sequences. *Geosphere* 12 (4): 1323–1340. https://doi.org/10.1130/GES01276.1.

Solomon, C. and Breckon, T. (2011). *Fundamentals of Digital Image Processing: A Practical Approach with Examples in Matlab*. John Wiley 344 pp. ISBN: 978-0-470-84472-4.

Snavely, N., Seitz, S., and Szeliski, R. (2006). PhotoTourism: Exploring photo collections in 3D. In: *SIGGRAPH Conference Proceedings*, 835–846. https://doi.org/10.1145/1141911.1141964.

Telea, A. and van Wijk, J.J. (2002). An augmented fast marching method for computing skeletons and centerlines. In: *Proceedings of VisSym* (ed. D. Ebert, P. Brunet and I. Navazo), 251–258.

Thiele, S.T., Grose, L., Samsu, A. et al. (2017). Rapid, semi-automatic fracture and contact mapping for point clouds, images and geophysical data. *Solid Earth* 8 (6): 1241–1253. https://doi.org/10.5194/se-8-1241-2017.

Thiele, S.T., Micklethwaite, S., Bourke, P. et al. (2015). Insights into the mechanics of en-échelon sigmoidal vein formation using ultra-high resolution photogrammetry and computed tomography. *Journal of Structural Geology* 77: 27–44. https://doi.org/10.1016/j.jsg.2015.05.006.

Ting, K.M. (2011). Precision and recall. In: *Encyclopedia of Machine Learning, Springer* (ed. C. Sammut and G.I. Webb). Boston, MA.

Ullman, S. and Brenner, S. (1979). The interpretation of structure from motion. *Royal Society London B* 203: 405–426. https://doi.org/10.1098/rspb.1979.0006.

Vasuki, Y., Holden, E.J., Kovesi, P., and Micklethwaite, S. (2014). Semi-automatic mapping of geological structures using UAV-based photogrammetric data: An image analysis approach. *Computers and Geosciences* 69 (November 2016): 22–32. https://doi.org/10.1016/j.cageo.2014.04.012.

Viseur, S., Richet, R., Borgomano, J., and Adams, E. (2007). Semi-automated detections of geological features from DOM – The Gresse-en-Vercors Cliff. In: *In 69th European Association of Geoscientists and Engineers Conference and Exhibition 2007: Securing the Future*. Incorporating SPE EUROPEC 2007, vol. 1, 88–92.

Wilkinson, M.W., Jones, R.R., Woods, C.E. et al. (2016). A comparison of terrestrial laser scanning and structure-from-motion photogrammetry as methods for digital outcrop acquisition. *Geosphere* 12: 1865–1880. https://doi.org/10.1130/GES01342.1.

Part II

Morphometric Analysis Across Different Scales and Planets

7

Mapping Coastal Erosion of a Mediterranean Cliff with a Boat-Borne Laser Scanner: Performance, Processing, and Cliff Erosion Rate

Jérémy Giuliano[1,2,3], Thomas J. B. Dewez[4], Thomas Lebourg[2], Vincent Godard[3], Mélody Prémaillon[4], and Nathalie Marçot[4]

[1] GeoConseil, Risk and Geological Consulting, Le Val 83143, France
[2] Université de Nice Sophia Antipolis, CNRS, GéoAzur UMR 7329, Observatoire de la Côte d'Azur, Valbone 06560, France
[3] Aix-Marseille Université, CNRS CEREGE UMR 7330, Aix-en-Provence 13545, France
[4] BRGM, 45060 Orléans, 34000 Montpellier and 13009, Marseille, France

Abstract

Terrestrial laser scanners (TLS) are well known for providing an efficient means to monitor coastal cliff erosion. Cliffs along micro-tidal coasts, however, have often escaped quantification because the narrow or absent coastal platforms do not offer stable and embracing vantage points. To circumvent this issue, mobile laser scanning surveys from a boat can be used. We present a case study from Provence-Alpes-Côte-d'Azur (Mediterranean coast-southern France) to quantify cliff erosion in such micro-tidal environments. Three surveys were subcontracted over a period of 17 months to monitor a 3.5-km-long cliff of Carry-le-Rouet (15 km west of Marseille). Data quality was checked independently using man-made planar walls positioned above the cliff face, to retrieve survey precision and change detection thresholds. A boat-borne mobile LiDAR system was capable of describing planar features with a precision of 3–4 cm (epoch 1 and 2) and improved to 2.6 cm (epoch 3) with point densities around 100 points/m^2. Absolute positioning accuracy varied between 0.1 cm and 0.3 cm. Because the coastline is very sinuous, we describe a method to unfold the point clouds using a continuous analytical surface made of vertical planes joined by arcs of cylinders and perform the analysis in 2.5D. Change was detected using a unique, conservative, threshold of 14 cm (99% quantile estimated on the plane change) on grids of 10×10 cm pixels. Integrated over the entire cliff face, the average annual cliff recession rate at Carry-le-Rouet is 1.1 cm/year. In 17 months, erosion was three times more effective in sandstone and marl layers than in calcarenites and conglomerates. Erosion varies vertically with erosion three times more effectively in the lower 25 m of the cliffs than above. Despite imperfections, boat-borne laser scanning systems are capable of delivering meaningful erosion data even in this low erosion context.

7.1 Introduction

In recent years, a wide range of portable laser scanners has been developed and significant operational improvements make them a choice equipment for field measurements (Lichti et al., 2000a, 2000b; Ingensand et al., 2003; Girardeau-Montaut 2005; Heritage and Hetherington, 2007; Dewez, et al. 2017). Among the main geomorphological applications, the study of gravitational processes such as rock falls or landslides (Kemeny and Post, 2003; Bitelli et al., 2004; Teza et al., 2007; Dunning et al., 2009; Dewez et al., 2007; Abellán et al., 2010; Dewez et al., 2013) and evaluation of associated geological hazards (Hunter et al., 2003; Abellán et al., 2006; Jaboyedoff et al., 2009; Furlani et al., 2014) have strongly benefited from these methodological developments. In the particular case of an inaccessible cliff face, the possibility to monitor spatio-temporal changes of complex surfaces is important for the study of erosion, in order to map the location, dimension, and epoch of occurrence of collapse events (Lim et al., 2005, 2010; Rosser et al., 2005; Young and Ashford, 2006, 2007; Dewez et al., 2007, 2013; Collins and Sitar, 2008; Olsen et al., 2009; Young et al., 2009, 2010; Rohmer and Dewez, 2013; Kuhn and Prüfer, 2014). One of the applications is to interpret the collapse magnitude in terms of erosion activity to appreciate the stage of cliff evolution. For example, an increase in rock falls erosion can "pre-date" the occurrence of larger events (Rosser et al., 2007), whereas a decreasing activity may correspond to a temporal "stabilization" (Perdrazzini et al., 2010).

3D Digital Geological Models: From Terrestrial Outcrops to Planetary Surfaces, First Edition.
Edited by Andrea Bistacchi, Matteo Massironi, and Sophie Viseur.

Terrestrial laser scanners (TLS) are particularly suited for the survey of steep cliffs with high resolution and the possibility to deal with complex geometries (Ingensand, 2006; Young et al, 2010). Moreover, this technique is best adapted to vertical terrain configurations compared to airborne laser scanners. Indeed, for an aerial technique, high-incidence angles between vertical cliff faces and a laser beam are responsible for inaccuracy and the lack of vertical measures (Baltsavias, 1999; Young et al., 2010; Earlie et al., 2014). For that reason, coastal cliff studies are generally performed with terrestrial laser scanners looking at the cliff face from the beach/platform at the foot of the cliff. This setup requires, however, installing a tripod in front of the cliff, which is not always feasible and especially not along a micro-tidal coastline where a dry coastal platform is either narrow or non-existent. We can note also that TLS is not always economical over several kilometer-long sections, because of the effort needed to assemble all stations together in a rigid reference frame (Michoud et al., 2014).

For these reasons, mobile laser scanning is an attractive technology that allows a long section to be scanned with high precision (Ellum and El-Sheimy, 2002; Barber et al., 2008; Jaakkola et al., 2008; Kukko et al., 2012; Glennie et al., 2013). These systems have been adapted on boat configurations in order to detect topographic changes in fluvial environments (Alho et al., 2009; Vaaja et al., 2011, 2013) and more recently in coastal environments (Michoud et al., 2014). This latter application, over 30-km-long coastal cliffs in northern Europe (Normandy, France), has demonstrated its capability to map geomorphological features smaller than 0.5 m^2 and to detect rock fall volumes larger than 1 m^3 between two acquisitions (in 2012 and 2013). At the time of performing this work, back in 2012, Uncrewed Airborne Vehicles (UAVs) were not allowed to fly, and nor were structure-from-motion (SFM) with multiview stereo (MVS) photogrammetric techniques broadly available as they are today in 2021. Despite this disruptive evolution, the processing concepts and findings described in this work remain valid.

Here we discuss the example of a 3.5-km-long section of cliff in Provence, along the French Mediterranean coast, near the town of Carry-le-Rouet (15 km west of Marseille). Along this part of the coastline, potential scanning stations are scarce, given the micro-tidal regime. Using different mobile mapping systems for time series measurements will become increasingly frequent in the future, given the cost, specificity of the equipment, commercial market regulations, and rapid technological improvements of mobile mapping equipment. Therefore, despite it not being ideal for research purposes, we describe how these different survey datasets were treated and how system performance can be assessed. Our work is targeted at end-user engineers who are not survey specialists and nevertheless need to independently check the data they paid for.

Repeated surveys were thus performed with a similar mobile boat-borne laser scanner system. In this study, we address both generic issues related to the survey technique and allude to site-specific information concerning the geomorphic processes at work. First, the positional performance of the different measurement systems and their ability to be combined to produce temporal sequences of point clouds are described: (i) single survey point survey precision; (ii) repeatability of the measures between successive surveys; and (iii) inter-point spacing (i.e. the inverse of point density) at each survey epoch. The method employed is particularly adapted in the absence of an external model (e.g. without a terrestrial laser reference). Second, given the sinuous plan-form morphology of the cliff, 3D point clouds cannot be readily compared and brought in 2.5D Geographic Information System (GIS) analysis environments for further contextual analyses. We describe a transformation method based on continuous analytical surfaces made of vertical planes and arcs of a cylinder to unfold the coastline. The resulting surface is then compatible with cliff-face-oriented orthophotography and 2D geological sections drawn in GIS software. Third, this processing workflow is applied to the case study of Carry-le-Rouet, where three boat-borne surveys captured cliff erosion between February 2011 and July 2012. Finally, we discuss the boat-borne laser capacity and some first-order geo-morphological findings.

7.2 Test Site and Study Setting

The case study is a section of the Côte-Bleue in the Provence region (France), along an East-West 3.5-km-long sea cliff near the town of Carry-le-Rouet (43°19N, 5°09E). This part of the coast is sinuous and composed of a succession of five headlands (Figure 7.1a). Cliff height is between 2 and 56 m (relative to the NGF69 terrestrial datum, 1 m DEM, Litto3D®V1 IGN-SHOM). The cliff exposes various lithological formations from the Upper Oligocene to Lower Miocene that includes marine marly–sandy-calcareous successions with bed thicknesses ranging from 0.3 to 5 m and occasional occurrences of lenticular conglomerates (Demory et al., 2011). The series is made of subhorizontal to gently South-East dipping beds (2 to 10°) (Figure 7.1b). The cliff profiles present both vertical and inclined faces with well-developed abrasion notches at the foot.

The period of the strongest waves (windstorm from SW and SE) is concentrated in winter, with waves breaking on the coast between 0.5 and more than 2 m in height (e.g.

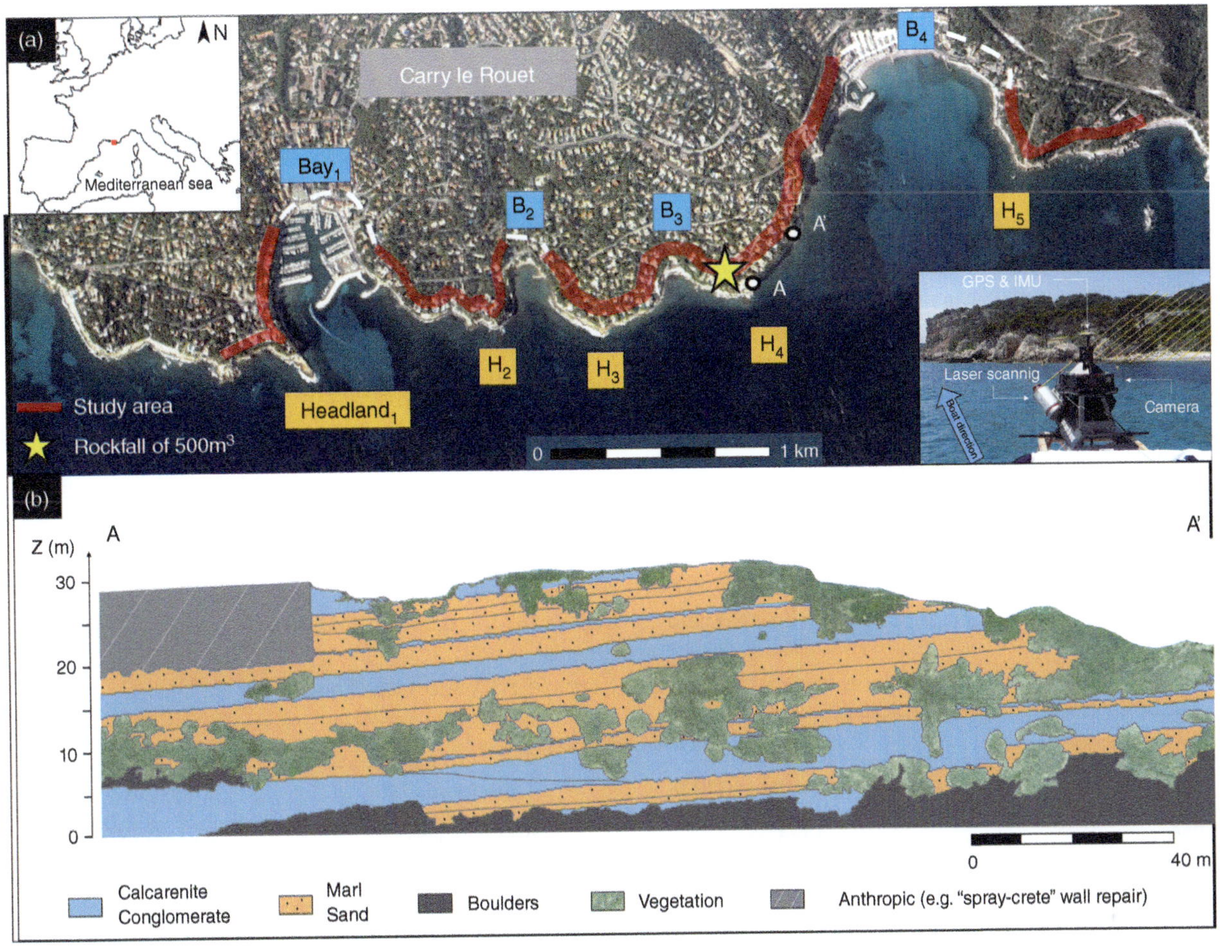

Figure 7.1 Study site of Carry-le-Rouet along the micro-tidal Mediterranean coastline (15 km west of Marseille, southern France). (a) Map view of the sinuous coastline composed of five headlands and four bays and study area; (b) Front view of the section AA' and geological interpretation of the Oligo-Miocene sedimentary stack drawn from orthophotography. The "spray-crete" wall repair corresponds to the northern end of a rockfall section of about 500 m³ which occurred on 6 February 2008.

Millot, 1990; Ciavola et al., 2007; Sabatier et al., 2009). Rainfall is mainly concentrated in autumn in the form of torrential rain (Lionnello et al., 2006). These rains can correspond to more than 200 mm of water falling only in 24 h (Marseille-Marignane weather station, series of 1923–2018).

The rocky shore dynamics along the French Mediterranean cliffs has not been much studied and the respective contributions of marine versus continental contributions in the Mediterranean area are not constrained (Furlani et al., 2014). Furthermore, this part of the French coast is not prone to spectacular and/or frequent cliff collapse events, such as those of chalk cliffs of northern Europe (e.g. Duperret et al., 2002; Costa et al. 2004; Regard et al. 2012; Dewez et al., 2013), but risk-wise, the vicinity of the cliff is subjected to increasing residential, touristic, and economic pressures. On 6 February 2008, a rock fall of about 500 m³ occurred in the central part of our study area (yellow star in Figure 7.1a), precisely in the garden of a villa, leaving less than 2 m between the cliff edge and a swimming pool.

As a consequence, cliff collapse risks have become a major concern for the local populations and authority and a cliff collapse hazard quantification was desired in order to characterize the erosion setting. We also need to evaluate whether this rock fall corresponds to a rare, isolated, or frequent event. We thus set out to apply the probabilistic analysis framework proposed by Dewez et al (2013) to start assessing the cliff collapse hazard.

7.3 Datasets

The temporal cliff evolution study in Carry-le-Rouet started with a first survey on 28 February 2011 (epoch 1), with an experimental mobile boat-borne LiDAR mapping system developed by the now-defunct surveying company

ATM3D. The following two other surveys, 30 November/1 December 2011 (epoch 2) and 7 July 2012 (epoch 3), were acquired by FUGRO-GEOIDE using two different setups: the Boatmap® system (BMS) and the Drivemap® system (DMS) respectively.

7.3.1 Description of a Mobile Mapping System

While each of the three LiDAR mobile mapping systems implemented different hardware specifications (Table 7.1), the instrumental concept is similar. A rigid frame on the boat contains (Figure 7.1a) a laser scanner, a pair of GNSS antennae, and an Inertial Measurement Unit (IMU). The accuracy of the mobile boat positioning system determines the overall performance of the survey.

7.3.2 Point Clouds and Registration

All three-point clouds were processed by the FUGRO-GEOIDE survey company in Montpellier (France) and delivered as X, Y, Z coordinates expressed in the national Lambert93 reference frame, along with a raw LiDAR point intensity value. No other additional data were made available. In total, the complete dataset of all three scans contains 73.2 million points (Table 7.2). The contractor performed a multipass point cloud registration process using both physical targets and point cloud matching with the ICP algorithm (e.g. Besl and McKay, 1992). Although locally variable, the quality of multipass registration could not be assessed because the point pass identifier (whether the west-to-east or east-to-west pass) was not specified in the file, and only a single merged point cloud was delivered to us.

7.3.3 Orthophotography

A series of cliff face photographs were taken in December 2012 and shot with an industrial shutter-free full frame Imperx *16 Mpix* camera fitted with a 50 mm lens (Table 7.1).

Table 7.1 Technical specifications of the three boat mobile mapping systems (ATM3D, Boat-Map® and Drive-Map®).

	BOAT MOBILE MAPPING SYSTEM		
Epoch	**1 *(February 2011)***	**2 *(November 2011)***	**3 *(July 2012)***
SYSTEM	**ATM3D**	**BOAT-MAP®**	**DRIVE-MAP ®**
Number of points	$10.6*10^6$	$18.7*10^6$	$43.9*10^6$
	1. Positioning sensors (GNSS & IMU)		
GNSS brand	*NOVATEL*	*TRIMBLE*	*TOPCON*
GNSS sampling frequency [Hz]	1	1	20
Brand (IMU)	*LandINS (IXSEA)*	*FSAS (NOVATEL)*	*TOPCON*
IMU sampling frequency [Hz]	200	200	100
Rolling/pitch precision [°]	0.005	0.008	0.02
Bend precision	0.010	0.023	0.04
	2. Mapping sensors (LASER scanner)		
LASER scanner brand	*LMS Z390i (RIEGL)*	*IIris HD ER (OPTECH)*	*VQ 250 (RIEGL)*
Aperture angle [°]	80	40	360
Measurement rate [pts.s^{-1}]	11000	10000	300000
Distance range [m]	1-400	3–1800	1.5–200
Angular resolution [°]	0.002	0.00115	0.0018
σ angular measured [°]	0.001	0.00458	0.001
σ distance measured [mm]	4	7mm–100m	5
	3. Camera		
Camera brand	-	*Imperx-16M-G(IMPERX)*	-
Focale length [mm]	-	35	-
Sensor length [mm]	-	36.07*24.05	-
Pixel size [Mpx]	-	4872*3248	-

Table 7.2 Synthesis of point clouds quality. (a) Single epoch quality of the three epochs (statistical point precision and modal value); (b) Inter epoch quality of the three epochs (point precision, repeatability and threshold distance). *The 95 and 68.3 quantiles would be the equivalent of classical 2-sigma and 1-sigma thresholds if the distribution were truly Gaussian.

(a) Single epoch quality of three point clouds: point precision [cm]

Epoch		*1 (February 2011)*	*2 (November 2011)*	*3 (July 2012)*
N walls [pts]		*7630*	*8968*	*23232*
Quantile of	***99.9***	8.8	8.8	4.3
absolute	***95.0 (2σ*)***	3.1	3.8	2.6
distance	***68.3 (1σ*)***	1.4	2.2	1.1
[cm]	***50.0***	1.0	1.6	0.7
Modal value [cm]		0.1	1.1	0.1

(b) Inter-epoch quality of three point clouds (see)

1. Sum of the squared point precision [cm] +

	Epoch 1	Epoch 2	Epoch 3
Epoch 1	-	-	-
Epoch 2	17.6	-	-
Epoch 3	13.1	13.1	-

2. Repeatability distance [cm] =

	Epoch 1	Epoch 2	Epoch 3
Epoch 1	-	-	-
Epoch 2	0.29	-	-
Epoch 3	0.31	0.13	-

3. Threshold distance [cm]

	Epoch 1	Epoch 2	Epoch 3
Epoch 1	-	-	-
Epoch 2	17.9	-	-
Epoch 3	13.4	13.2	-

These photographs were then orthorectified by the contractor as 2-cm-per-pixel continuous orthophotography. These orthophotos were used for segmenting the point clouds (e.g. on/off cliff points and cliff-growing shrubs) and producing litho-stratigraphic sections of cliff rocks (b).

7.4 Point Cloud: Quality Assessment

Point clouds delivered by subcontractors, or received from an unknown source, often lack a clear description of the data quality. Here we describe an *ad hoc* procedure to retrieve point cloud precision, repeatability, and effective point density empirically. The procedure relies on the fortunate presence of well-distributed man-made walls close to the cliff top, which means at a similar distance and incidence to the LiDAR position on the cliff. Modelled as best-fit planes, these walls serve to empirically derive point precision, accuracy, and point density on the surface close enough to the cliff that statistics from modelling these planes can be extrapolated to the cliff with a reasonable degree of confidence.

7.4.1 Validation Metrics and Procedure

Five walls are visible in all three scans. Walls are 4 m to 16 m in length and 2 m to 4 m in height. Point cloud precision of a survey corresponds to the amalgamated residuals to the best fit planes (e.g. Monserrat and Croseto, 2008) (Figure 7.2a). Repeatability, or accuracy, assesses whether the successive survey campaigns located the walls in the same absolute location. It is assessed as the most-likely distance between the best-fit planes of epoch n with that of epoch $n + 1$ (e.g. Besl and McKay, 1992; Michoud et al., 2014) (Figure 7.2b). Because several authors describe errors as not being normally distributed (e.g. Höhle and Höhle, 2009), classical error propagation cannot be worked out as the sum of residual variances. The threshold value above which cloud-to-cloud differences are not simply due to chance has been worked out in a different fashion. Because the selected walls have remained unchanged between epochs, any cloud-to-cloud difference observed is noise coming from imprecise measurements. We pick a threshold distance as the empirical quantile at 99.9% (Dewez et al., 2013). Above this value, cloud-to-cloud distance has as little as a 1 chance in 1,000 to be noise.

The sum of the squared-point precisions and repeatability error gives the shortest distance of the erosion detection, which we call the threshold distance (7.2c). Finally, inter-point spacing (or point density) was also computed to optimize grid pixel sizes required in a later step and check the delivery for compliance with the survey specifications (e.g. Soudarissanane et al., 2011; Argüelles-Fraga et al., 2013).

To model the planar shape of the walls from XYZ Lambert93 coordinates, a Principal Component Analysis (PCA) was performed. This geometrical transform is preferred to a linear regression solution because all X, Y, and Z values

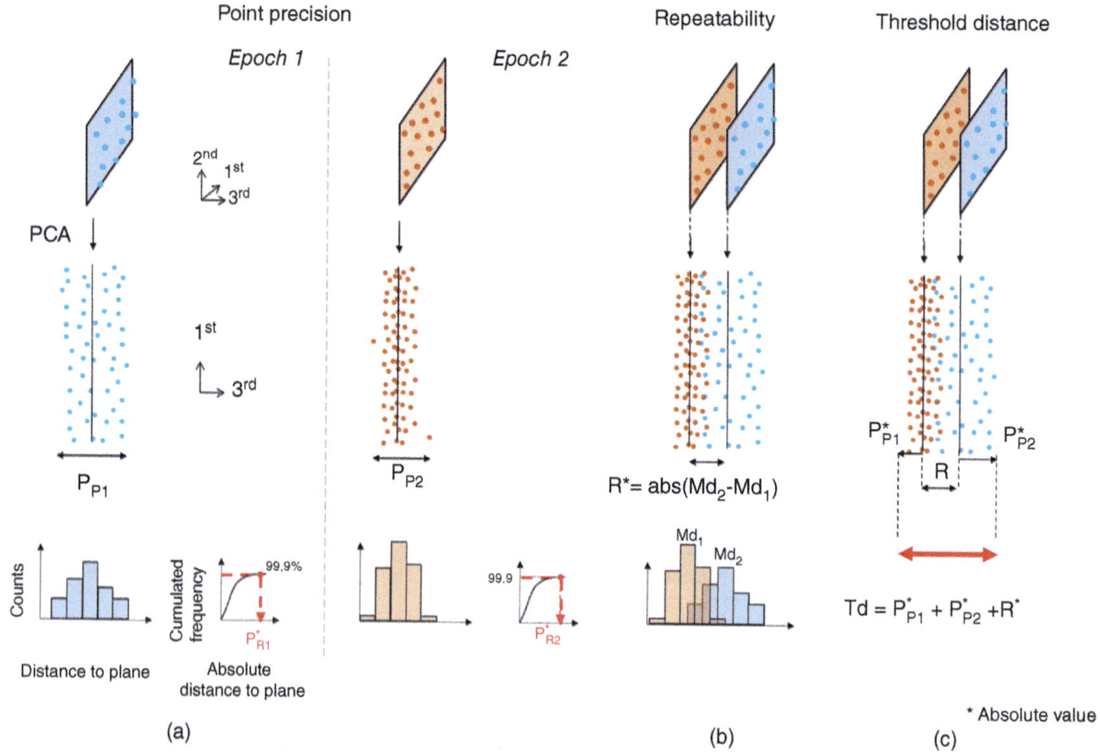

Figure 7.2 Quality assessment principle applied to a planar object. Point precision (PP), repeatability distance error (R), and threshold distance (Td) assessment through the third component (3rd) of the Principal Component Analysis applied to diachronic wall measurements (examples for epoch 1 and 2). The 3rd component corresponds to the distance to the wall plan model. (a) Point precision (PP1 and PP2) values are obtained as the 99.9% quantile of the cumulated frequency of the absolute 3rd component. (b) Repeatability error (R) corresponds to the absolute difference of the median of the 3rd component values (Md1 and Md2). (c) Threshold distance (Td) is the total uncertainty of the detection between two epochs.

are known with some degree of error. In this case, the principal component analysis retrieves three components, by decreasing order of variability. The first and second components are wall co-planar coordinates; the third component is orthogonal to this best-fit plane and describes the dispersion on either side of the plane (Figure 7.2). We should note that wall point cloud samples were segmented well away from the wall edges to avoid contamination by outliers. Unwanted residual distribution deviations could exist if the walls were not true planes. The walls were indeed true planes as residuals plotted versus wall-parallel coordinates did not display coherent high-order polynomials trends.

7.4.2 Point Precision for a Single Survey (Pp)

The point to plane distance distribution is often not normally distributed (Höhle and Höhle, 2009). This means that using Gaussian-based metrics will not reflect the actual empirical error distribution. To counteract this issue, Höhle and Hohle (2009) used robust quantile-based descriptors that matched the empirical distributions. This means describing the central value of a distribution with the median, as opposed to the mean, and describing the dispersion with quantiles of the absolute distance to the median, as opposed to the standard deviation.

Here we use the 99.9% quantile of the absolute deviation to the median as the point precision descriptor (Figure 7.2a and Figure 7.3). This quality threshold corresponds to a much smaller proportion of points than the usual 5% of a 2-sigma threshold (Table 7.2), but minimizes the proportion of meaningless differences given the sample size of millions of points.

7.4.3 Repeatability (R)

Repeatability describes whether successive surveys position walls in the same location. It was determined by measuring the distance between the point cloud of epoch 2 with modelled planes of epoch 1. The median of this distribution delivers the value of repeatability (Table 7.2).

7.4.4 Threshold Distance to Detect Erosion (Td)

The remaining cliff pixels exhibit a range of distances that needs thresholding to map eroded blocks. Short distances indicate that the cliff has not changed much while larger values depict erosion more confidently. Picking the appropriate threshold is again a matter of robust statistics. The detection threshold between two-point clouds depends on their accuracy, and is obtained by the sum of absolute point

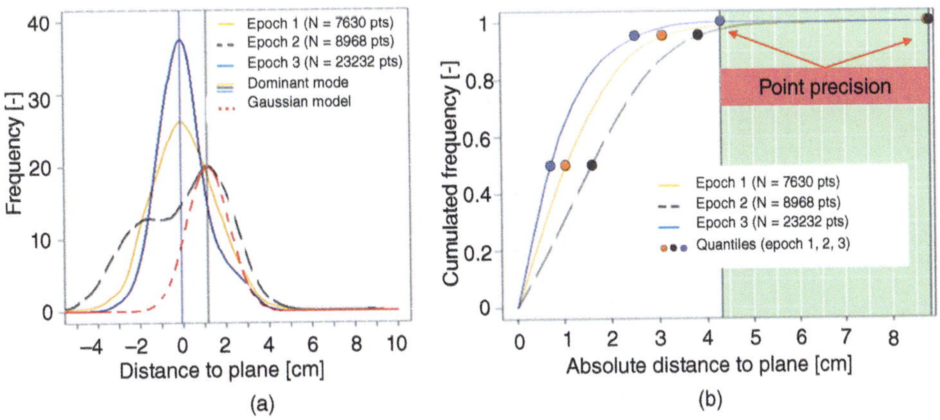

Figure 7.3 Point precision (PP) assessment. (a) Frequency density (bandwidth of 3 mm) of PCA third component deviation to the median. Dominant modes (maxima of density curves) are marked by vertical lines. The red short-dashed curve represents a Gaussian distribution centered on the dominant mode of the epoch 2. (b) Cumulated frequency of absolute deviation from the median. The plain dots represent the 50%, 95%, and 99.9% quantiles of each survey. Epoch 2 is more skewed to the right than epochs 1 and 3.

precisions and absolute repeatability distances (Figure 7.2c and Table 7.2).

7.4.5 Inter-point Spacing Estimation

In a later preprocessing stage, the three-point clouds are projected into 2.5D grids to enable interactions with orthophotography and further explore erosion dynamics. The choice of an optimal grid pixel size is a recurrent issue in GIS. Here, we solve it by computing the nearest-neighbor Euclidean distance statistics in wall-coplanar coordinates (Figure 7.4). This solution is designed for users with little computing power. An alternative would be to compute the volume density in a large enough neighborhood (e.g. 0.5 m). This option would have the advantage of accounting for possible point density biases linked with LiDAR line-of-sight obliquity. Our option nevertheless is equally valid since cliff face and sampling walls have similar orientations. Any line-of-sight bias on the wall would affect point density on the cliff in the same manner. Because the inter-point spacing is obviously variable in space, a minimum representative inter-point spacing can be obtained by the median values (Michoud et al., 2014), which minimize influences of outliers (Höhle and Höhle, 2009).

7.5 LiDAR Data Processing

The scientific problems we wish to solve concern the total mass balance and discrete block erosion over the entire cliff face. In order to do so, local point-wise erosion can be assessed with 3D cloud-to-cloud distances (e.g. Lague et al. 2013), but mass balance or scar inventories still require computing an integrated estimate of erosion, either through meshing or gridding. Further, we would like to explain erosion with various methods readily implemented in 2.5D GIS and currently lacking in full-3D software. The data processing workflow therefore consists of a succession of four steps: (i) project 3D point clouds into an analytical 2.5D surface (unfolding); (ii) calculate inter-epoch distances (point clouds comparison); (iii) classify and check for point visibility to assess erosion; and (iv) threshold cloud-to-cloud distances to retrieve only meaningful erosion (Figure 7.5).

7.5.1 3D to 2.5D Projection Method

The first step consists in projecting 3D points into an unfolded 2.5D cliff surface. Most terrestrial laser scanning studies in rocky coast environments are performed along straight or semistraight cliff segments (Lim et al., 2005, 2010; Dewez et al., 2007, 2013; Collins and Sitar, 2008; Abellán et al., 2010; Young et al., 2010; Stock et al., 2011; Kuhn and Prüfer, 2014; Michoud et al., 2014) and all of them admit, sometimes implicitly, that when the cliffs become truly 3D, processing is much more difficult.

Full 3D comparisons, like the M3C2 method of Lague et al. 2013, require computing local plane normals over a given neighborhood, but plane normals may misbehave at places with sharp gradients. The characteristic dimensions of cliff smoothness is also spatially variable (see Feldmann et al., 2018). On the other hand, it is well known that planar cliff configurations close to 2D or 2.5D allow the determination of eroded volume through DoD (Digital Elevation Model of Difference; see Rosser et al., 2005; Abellán et al., 2010; O'Neal and Pizzuto, 2011; Bernard, et al., 2021). However, in 3D complex configurations, DoD will poorly depict strongly oblique surfaces. The coast of Carry-le-Rouet is very sinuous, with a succession of headlands and bays and needs to be simplified (see Figure 7.6a).

To simplify the visualization and processing of a 3D configuration in 2.5D, we detail a method to unfold point

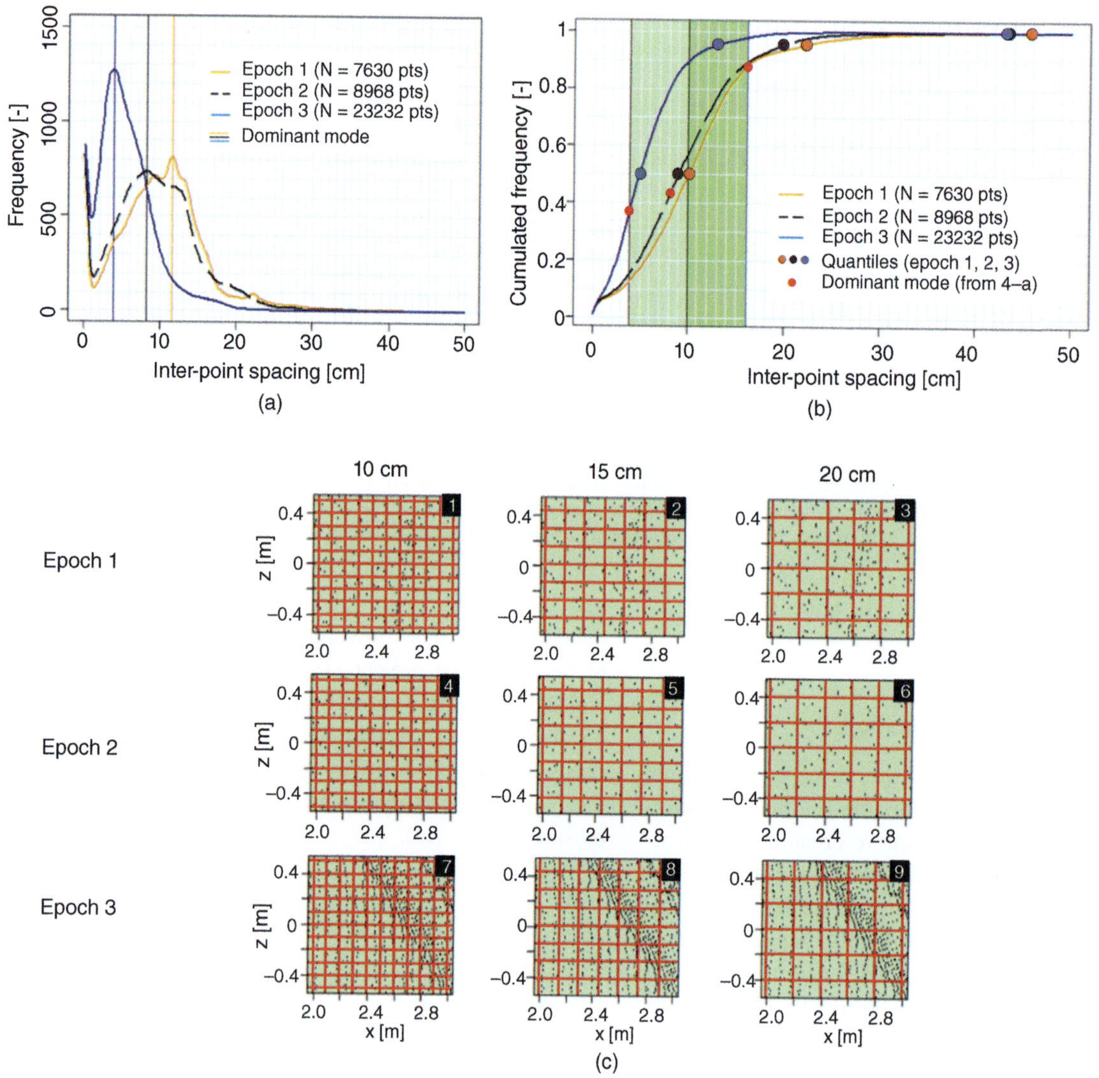

Figure 7.4 Inter-point assessment determined by a projection of 2.5D grids for the three epochs. (a) Frequency density of inter-point Euclidian spacing in a wall-coplanar reference frame (bandwidth = 0.5 mm). Dominant modes are defined by vertical lines. (b) Cumulated frequency of the inter-point spacing. (c) Example of a different grid cell size (10, 15, 20 cm) applied to point clouds from a wall extracted for the three epochs.

clouds (Figure 7.6). The numerical method consists in rotated XY co-ordinates onto a vertical surface made either of contiguous plane sections or arcs of cylinders. Elevation coordinates, Z, remain unchanged. One of two types of transformations are applied: (i) an angular rotation for planar sections (Figure 7.6c1) or (ii) an angular distance calculation for cylindrical sections (Figure 7.6c2):

1. For each rectilinear section, unfolded co-ordinates (X_U, Y_U) of a subcloud are obtained by applying a rotation of α to the initial points (X, Y), which corresponds to the angle between the rectilinear section and the re-projected straight line (Figure 7.6c1).
2. For each circular section defined by a circle O, unfolded co-ordinates X_U and Y_U of the subcloud are obtained separately (Figure 7.7):

- X_U coordinate corresponds to the angular distance of the points from the initial arc section Figure 7.7a):

$$X_U = (R \times Y) \times US[2\pi] \tag{7.1}$$

where R is the radius of the circle O, Y is the angle between the starting section and point to project, and US is the unfolding sense (by convention $US = 1$ for the inland section at sea, and $US = -1$ for the bay section).

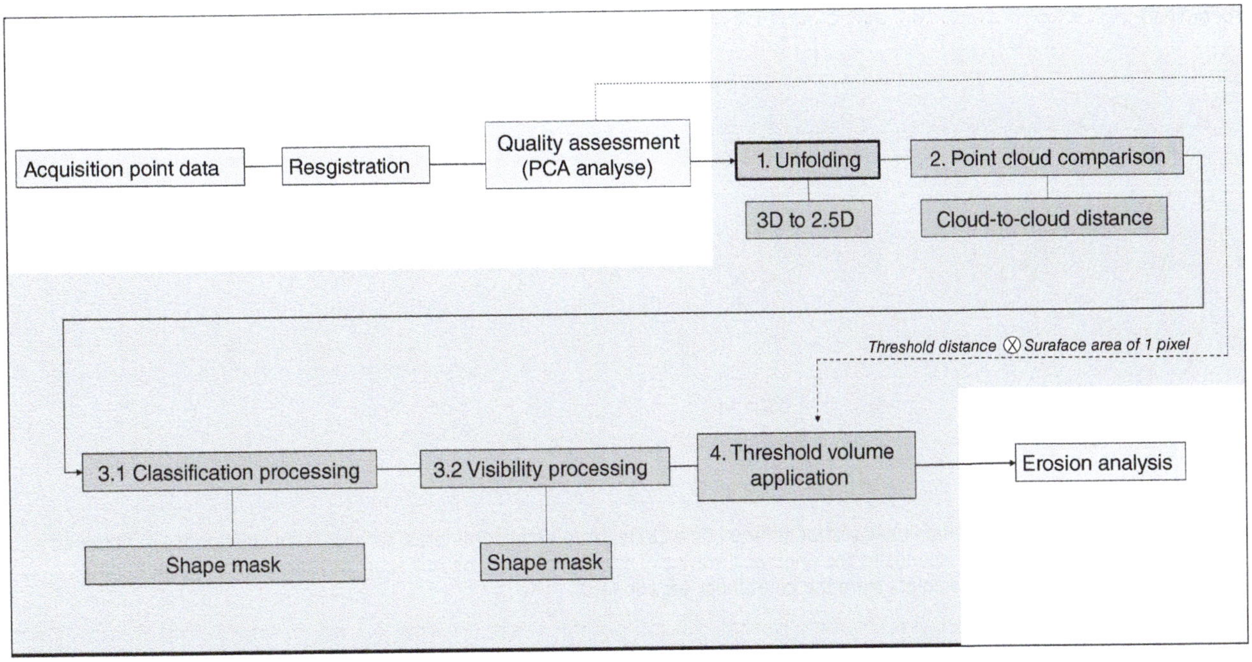

Figure 7.5 Flow diagram of LiDAR data processing (from data acquisition to erosion analysis). Grey polygon indicates LiDAR data processing based on the unfolding method.

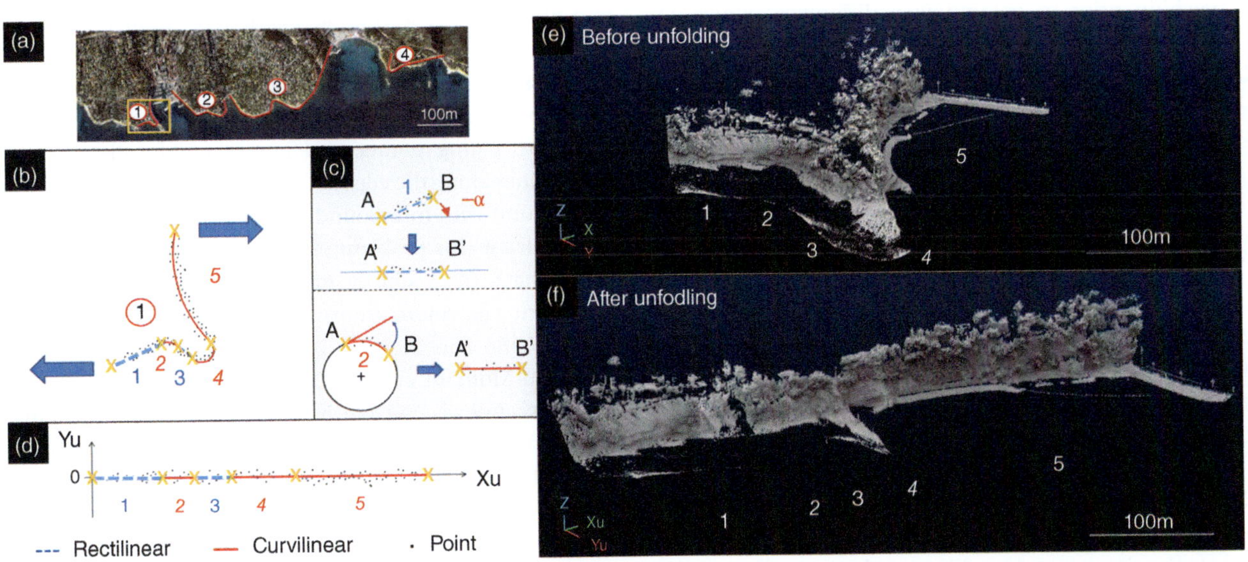

Figure 7.6 Projection method applied to a complex sinuous point cloud, an example of area 1. (a) Map view of the general trend of coastline defined through GIS. Map projection is defined in the same system as point clouds. (b) Focus of sinuous area 1. General trend and point cloud are segmented into linear (dashed blue line) and curvilinear sections (solid red curve). (c) Two types of transformation are applied to the sub-point cloud, (c.1) angular rotation for rectilinear and (c.2) angular distance calculation for curvilinear section. (d) Point cloud unfolded and re-projected in unfolded coordinates along a straight line. (e) and (f) View of point cloud before and after unfolding.

- Y_U coordinate corresponds to the distance of the points from the curviline, positive or negative (Figure 7.7b):

$$Y_U = (L - R) \times US \tag{7.2}$$

with L the distance between the point to unfold (X, Y) and the center of the circle O.

The same transformation scheme was applied to orthorectify photographs shot in November 2011. For complementary information see the Appendix.

7.5.2 Point Clouds Comparison Method

After projecting all point clouds into the 2.5D space, the second step is to quantify changes in range due to cliff erosion.

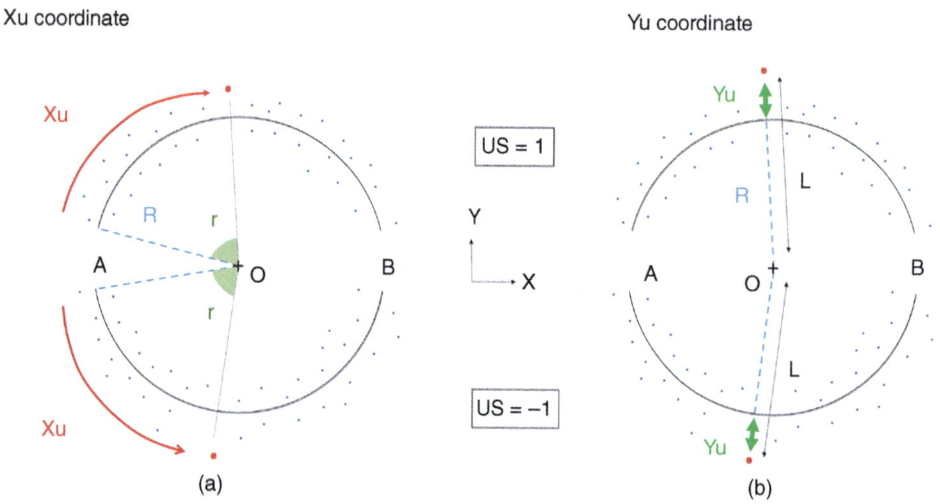

Figure 7.7 Unfolding method of curvilinear section defined by a circle. Unfolded coordinates XU and YU of subclouds are obtained separately. (a) XU coordinate corresponds to the angular distance of the points from the initial arc section (A). (b) YU coordinate corresponds to the distance of the points from the curvilinear section (AB).

Many authors have argued that a 3D cloud distance comparison was possible and desirable (Girardeau-Montault et al., 2005; Lague et al., 2013). This is opposed to gridding or meshing prior to computing a difference, which is amenable to loss of information or ghost volume creation. Hence, we computed a cloud-to-cloud distance between projected point clouds with *Cloud Compare* software before gridding the result.

7.5.3 Point Clouds Segmentation and Visibility Solution

Among all the points measured during the survey, many correspond to irrelevant features: vegetation, houses, and man-made features, tourists standing in front of the cliff, etc. The third processing step consists in segmenting the point clouds into object classes.

7.5.3.1 Classification Method

Based on the American Society for Photogrammetry and Remote Sensing LAS 1.4 labelling standard, a semi-manual classification of different features was performed.We distinguish the cliff (class identifier 2), all types of vegetation (tree, shrub, and grass), which are combined into a single unit (class 3), and two newly created classes: the external part of the scene (class 64) and the cliff foreground (class 65). This latter class will be further processed in a second stage because it is associated with occlusion problems. This classification was performed manually using orthophotos in a GIS (Figure 7.8). Mapping produced contiguous polygons from which the value of the class was transferred to the point clouds (Figure 7.9).

7.5.3.2 Visibility Solving Method (Shadow Effects)

Finally, because laser scanners acquire 3D data with a line-of-sight strategy, full-3D objects are likely to contain unseen faces during acquisition. To limit the extent of these shadows, the survey was performed in at least two passes. Shadows are detrimental for our purpose as cloud-to-cloud comparisons can only compute distances between existing points, and hence may locally present large values along the edges of hidden areas, unduly joining background with foreground points. The classical filtering approach for this problem employs the calculation of the depth maps (or visibility maps) from a known scanner viewpoint (for more information see Murakami et al., 1999; Vögtle and Steinle., 2004; Vosselman et al., 2004; Giradeau-Montaut et al., 2005). In our case, however, the mobile scanner is constantly on the move and so tagging a specific position with its corresponding viewpoint is impossible per se. Additionally, the subcontractors did not provide the space/time scanner trajectory. We therefore had to develop a different approach to solve this visibility problem.

Shadows in the cliff plane were computed from a raster mask of point presence (Figure 7.10a). Mask values of each epoch were labelled as a power of 2. Therefore, the sum of visibility masks produces a unique identifying code that determines simultaneously when and how many times points were visible (Figure 7.10b). This visibility mask is then combined with the masks of classified objects (cliff, vegetation, and foreground) to retain only the topographic changes of cliff portions. Processing quality depends on the choice of pixel size, and this choice is constrained by the representative inter-point spacing discussed above (cf. Section 7.4.5).

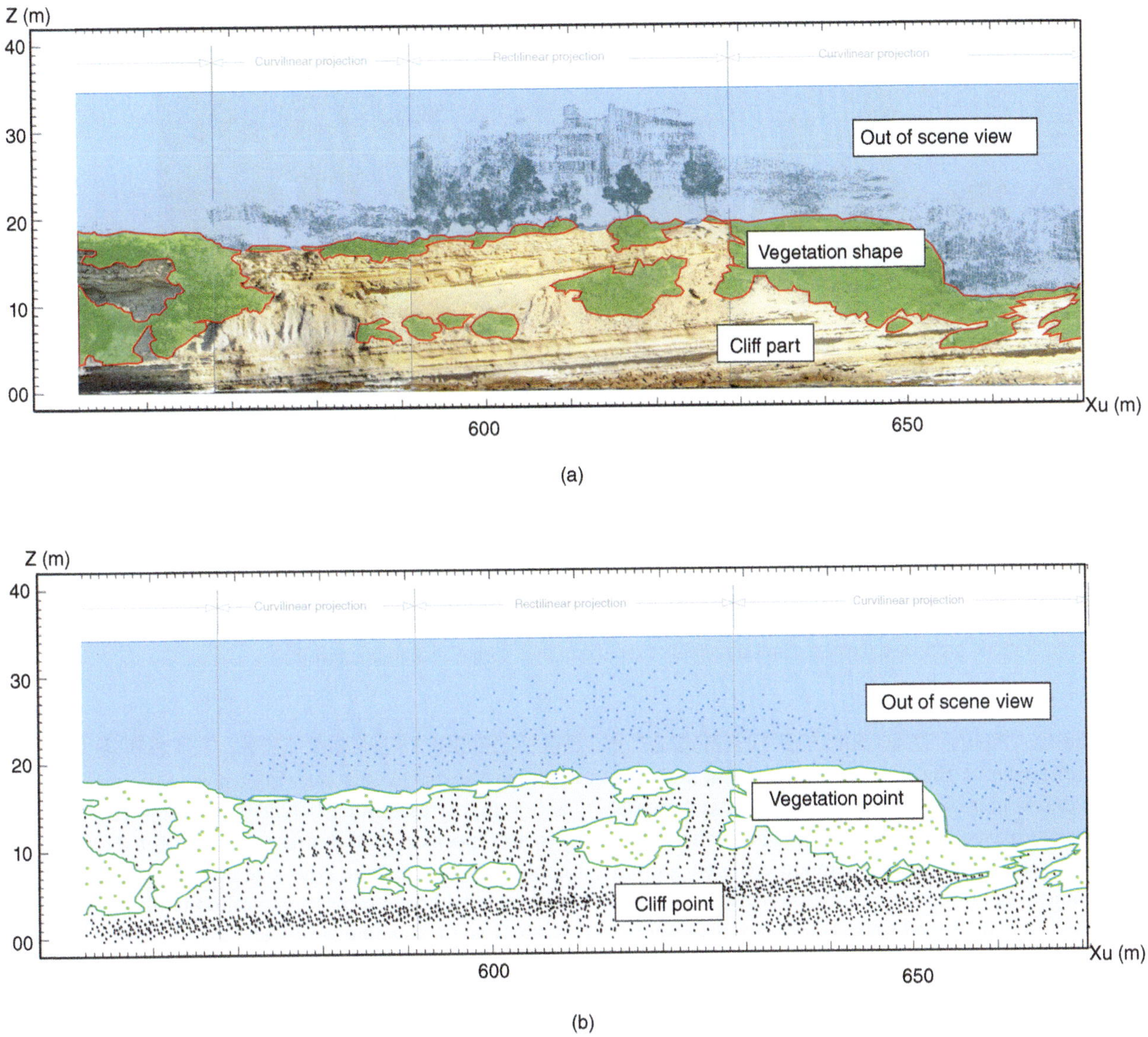

Figure 7.8 Semi-automatic classification by mask application to an unfolded point cloud. (a) Example of a section of an orthophotography digitalized by three polygons that correspond to "out of scene view," "vegetation," and "cliff." (b) Mask segmentation applied by polygons overlay on the front view (Xu, Z) of an unfolded point cloud.

7.5.4 Threshold Volume and Erosion Estimation

Finally, the remaining cliff pixels exhibit a range of distances that needs thresholding to map eroded blocks. The minimum volume of erosion detected is thus the result of the combination between the threshold distance and the surface area of one pixel size (cf. Sections 7.3.3 and 7.3.4 respectively).

In order to quantify the erosion volume and study rockfall scar properties, we need to estimate the eroded surface area. As such, surface area and volume of an estimation of eroded material is not possible with cloud-to-cloud comparisons. Some degree of integration is necessary. Cloud-to-cloud distances of unfolded point clouds are gridded into a 2.5D, 10 cm × 10 cm, Digital Surface Model of erosion (DSMe; see Figure 7.11). At this stage of the process, DSMe is acceptable because unfolded points clouds are almost planar (Dewez et al., 2013). This threshold is applied to quantify the total eroded volume.

7.6 Results

7.6.1 Quality Assessment

The point precision is determined as the distribution of the absolute distance between best-fit wall points and

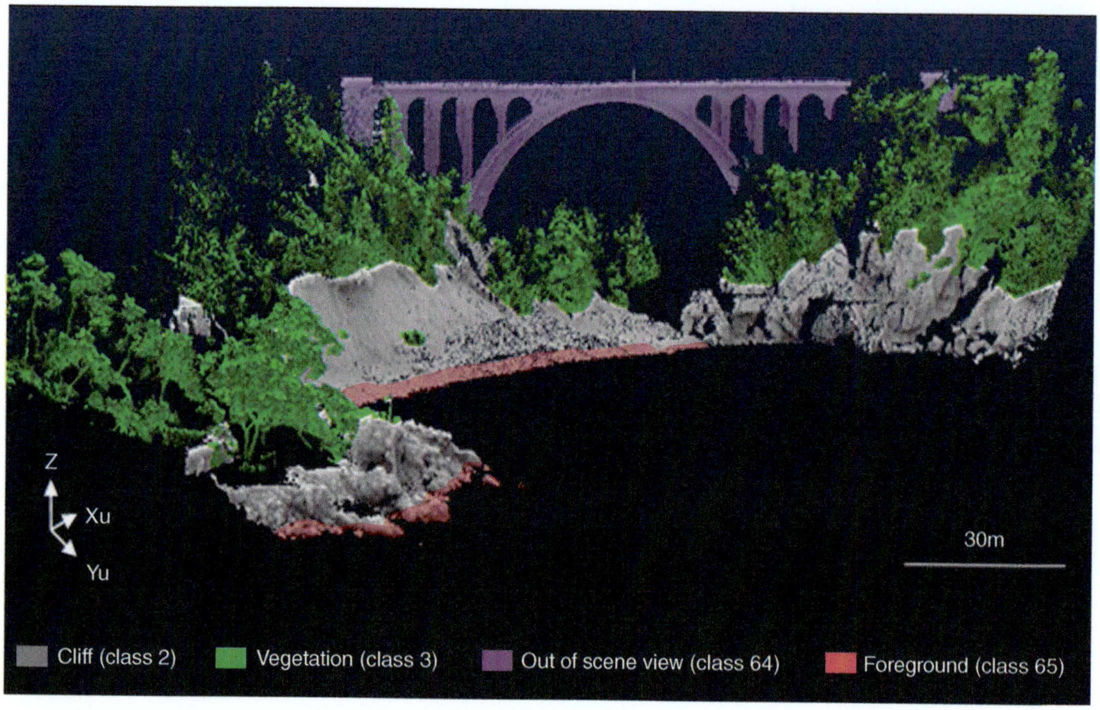

Figure 7.9 Point cloud processed (unfolded and classified). Example of the eastern section of epoch 3.

their median (Figures 7.2a and 7.3). Distributions of absolute distances to the median are unimodal and centered around 0.1 cm for epochs 1 and 3. Measurements do not seem biased. At epoch 2, however, the distribution is bimodal with the first mode centered around 1.1 cm, indicating an internal multipass registration issue (Figure 7.3). Statistical results of the precision are presented in Table 7.2. Classical point measurement precisions in the 95% quantile (2-sigma) are 2.6 cm (epoch 3) and between 3 and 4 cm (epoch 1 and epoch 2), whereas the 99.9% quantile are 4.3 cm (epoch 3) and 8.8 cm (epoch 1 and epoch 2).

Looking at the absolute difference to the median between each distribution of distance to the wall plan (Figure 7.2b and Table 7.2), the repeatability error of the three epochs comes to 1 mm (between epochs 2 and 3) and presents a shift of 3 mm (between epochs 1 and 2 and epochs 1 and 3).

The inter-point spacing is obviously variable in space. Its distributions turn out to be unimodal and asymmetric for all three scans (Figure 7.4). They all present similar features, but the distribution mode of epoch 3 corresponds to a denser point spacing. The most frequent inter-point spacing is centered around 16 cm (88% quantile), 8 cm (44% quantile), and 4 cm (38% quantile) for the epochs 1, 2, and 3 respectively. The 1-sigma values (68.3% quantile) are 12, 11.5, and 7.1 cm, which correspond to a density of 69, 76, 178 pts/m^2 (epochs 1, 2, and 3). A minimum representative inter-point spacing is obtained by the median values. In this case inter-point spacing is 10, 9, and 5 cm, corresponding to a density of 100, 123, and 400 pts/m^2 (epochs 1, 2, and 3 respectively). For gridding, we choose an intermediate pixel size of 10 cm and adopted a nearest neighbor interpolation method to fill pixels. Pixels without points remained unfilled.

Detection thresholds for the 99.9% cloud-to-cloud distance quantile from reference walls range between 13.1 cm and 18.6 cm (Table 7.2, part 2). To apply a consistent detection through time, we thus pick a compromise value of 14 cm as the minimum distance to consider when erosion really occurred between epoch 1 and 3.

Given the point cloud characteristics and processing described, the minimum volume potentially detectable is 0.0014 m^3 (i.e. 1.4 liters). No smaller volumes can be detected. This is not to say that erosion scars as small as these make sense from a geomorphological point of view. In fact, the scar inventory completeness was tested with Stepp's (1972, cited and used by Dewez et al., 2013) method. The probability distribution function of scar volume scales as a regular power law between 0.006 m^3 and 0.5 m^3. Volumes smaller than 0.006 m^3 are not completely counted because of measurements and processing limitations. For larger volumes, erosion did not occur in the 17 month time frame (precisely 492 days) between the first and last survey.

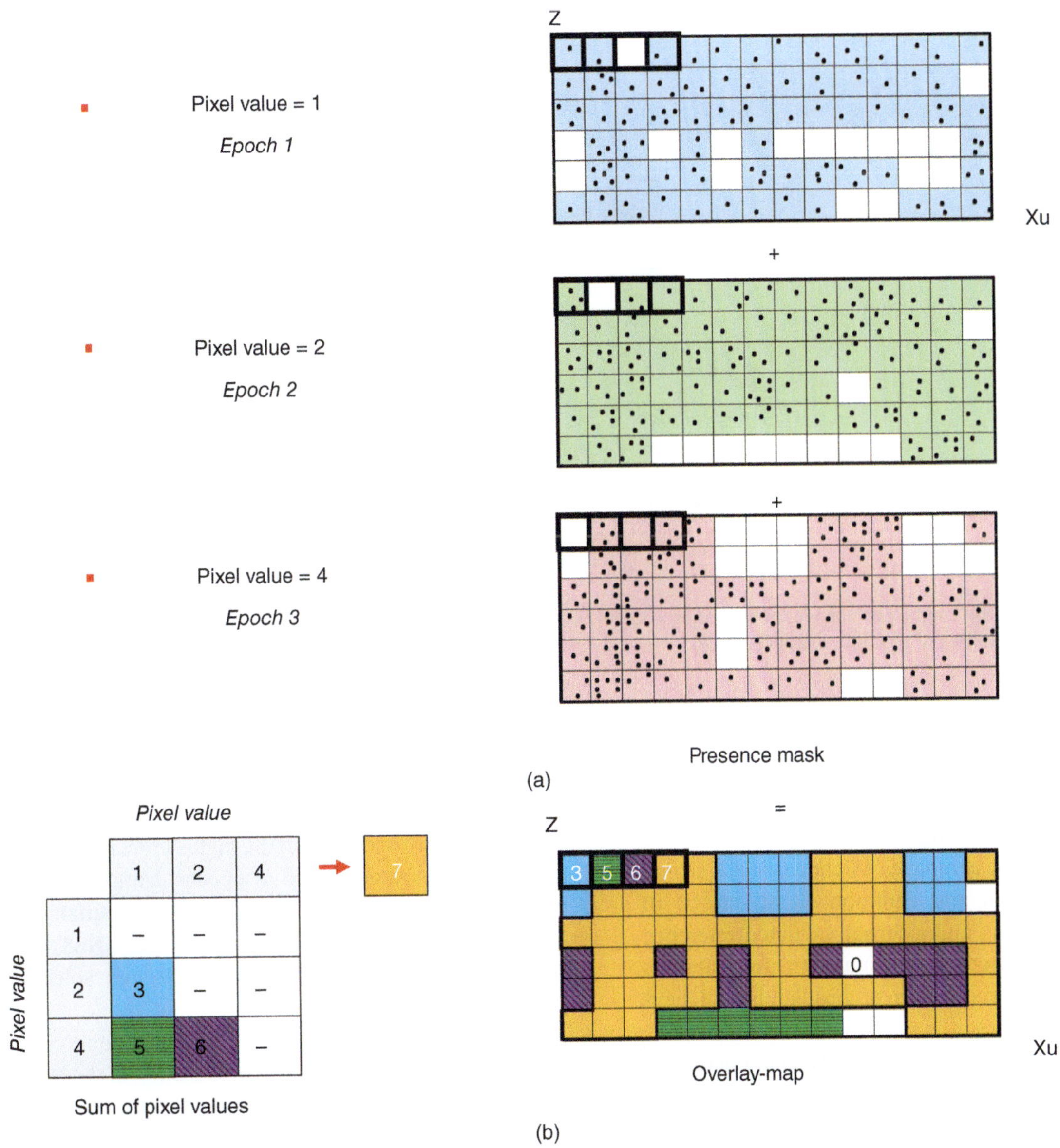

Figure 7.10 Overlay-map generation in two steps. (a) Point presence mask creation for the three epochs. The pixel label is incremented at each epoch. (b) Overlay map generated by the sum of the point presence mask. The nominal pixel value shows distinct visibility between the three epochs.

7.6.2 Erosion Estimation Between Epochs 1 and 3

Erosion scars show up as scattered patches of erosion values (Figure 7.11). The minimum volume of erosion detected is 1.4 liters (i.e. one pixel of 100 cm^2 with an erosion detection threshold of 14 cm). The total surface of cliff exposed is 30,141 m^2, where erosion affected 1,107 m^2 (only 3.67% of the exposed cliff). To further compute erosion scar properties, the grids were turned into polygon layers with a GIS software, where 14,925 distinct scars make up for a total volume of 323.30 m^3. In space, erosion is scattered over the cliff face (Figure 7.11). The largest collapses occurred at "Cap de la Vierge" (cape H_5 with a total of 150 m^3) and involved the basal sandy marl units (Figure 7.11a). Relative to the inventory, a "substantial" collapse (4 m^3) also occurred just below the repairs of the largest-known erosion scar at the area near H_4 on the "Barqueroute" cliff. Apart from these large events, smaller ones are scattered over the cliff face.

7.7 Discussion

Although mobile laser scanning systems have been used in various other contexts (Alho et al., 2009; Ellum and

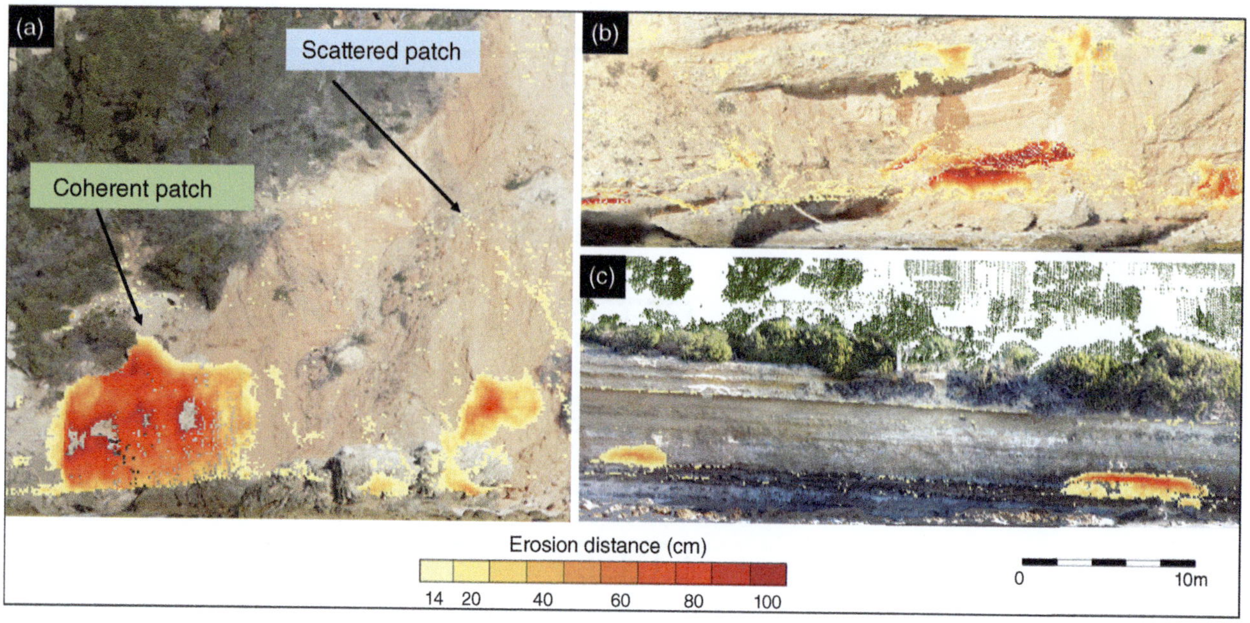

Figure 7.11 Examples of the Digital Surface Model of erosion (DSMe) overlaid on orthophotography. Erosion scars show up as coherent and scattered patches of absolute distance values.

El-Sheimy, 2002; Barber et. al., 2008; Jaakkola et al., 2008; Kukko et al., 2012; Glennie et al., 2013; Vaaja et al., 2011, 2013), there are very few published studies that used boat-borne mobile laser acquisitions to detect coastal cliff erosion along semi-straight cliff segments (Michoud et al., 2014). Our work completes Michoud et al's (2014) work for two practical and common situations: (i) in the absence of previous and external validated reference data (e.g. terrestrial laser scanning) and (ii) in a sinuous configuration with a succession of headlands and bays (Figure 7.1).

In the absence of validated and robust reference data, modelling point deviation from planar features visible in all scans provides quality assessment in terms of internal consistency (i.e. the ability to accurately represent an object with known geometry, e.g. López et al., 2014). From this we learned that:

1. Despite being rocked and rolled by sea waves (20–30 cm of amplitude), boat-borne mobile laser systems are capable of describing, in our case, planar features with a point range precision of 4.3 cm (epoch 3) to 8.8 cm (epochs 1 and 2) at the 99.9% quantile (Table 7.2) at distances between 50 m to 300 m from the cliff face. The order of magnitude of precision to be expected with these devices comes between 1/1,000 and 1/7,500 of the distance to the cliff.
2. Internal mis-registration may exist between successive passes in front of the cliff during the same survey. Here, epoch 2 was affected by a mis-registration reaching about 3 cm between passes. Looking at the residual distribution to a planar feature may help identify and quantify this defect. Obviously such co-registration errors strongly limit the performance of the mapping system, in a particular context of low erosion rates. All efforts should thrive to reach the highest precision. Even though laser scanners used at epochs 2 and 3 appear to be of the same ranging precision, the instantaneous orientation solution (from combined IMU/GNSS) was at fault and was not compensated through software post-processing.
3. Because a repeatability error-between 1 and 3 mm-could be reached (Table 7.2), global compensation of the entire surveys is achievable with millimeter accuracy. Consequently, the threshold distances are mostly influenced by point precision values (Figure 7.2c). The rather poor erosion detection thresholds, compared to terrestrial laser scanning, comprised between 13.1 and 18.6 cm (Table 7.2) are affected by the poor overall precision of epoch 1 and the internal bias of epoch 2. Nevertheless, this finding is globally in line with the performances described by Michoud et al. (2014).
4. One should note that our statistical accuracy assessment criterion, the 99.9% quantile of cloud-to-cloud differences, suggested by Dewez et al. (2013), is more conservative than in many other studies. For example, in the case where the median absolute deviation threshold is used (Michoud et al., 2014), every point contained a larger deviation than the threshold. The erosion may not always be detected as such. We preferred a conservative approach in a context of low erosion.

In a sinuous coastal configuration, point clouds need to be simplified in order to examine the erosion with various methods readily implemented in 2.5D GIS, such as the common DoD approach (Digital Elevation Model of Difference, as described by Rosser et al., 2005; Abellán et al., 2010; O'Neal and Pizzuto, 2011). Therefore, the key to the data processing workflow (Figure 6.2) is the unfolding method (Figures 7.6 and Figure 7.3). We are, of course, aware of the full-3D M3C2 algorithm developed by Lague et al. (2013), revised by Bernard et al. (2021), and implemented in *Cloud Compare* (Girardeau-Montaut et al., 2005), but computing cloud-to-cloud distances is never the end of the geomorphological analysis. Erosion needs to be compared with other sources of information (e.g. lithological maps), manually segmented or integrated over some meshing scheme. For these operations, 2.5D GIS software remain standard tools for many users and allow, for example, point clouds with orthophotos to be classified. Although manual processing is error prone and time consuming, early tests with *Canupo* (Brodu and Lague, 2012) were not conclusive to segment vegetation of an epoch 1 point cloud, possibly due to the high level of noise and comparatively to the scale of short cliff-dwelling vegetation.

Finally, the goal of this work was to generate a rock fall scar inventory from the Mediterranean coast of France, which had received little attention for cliff retreat so far. This work was prompted by a strong societal demand. The cliff collapse risk there is high, not because of the collapse hazard, but rather for the high-value assets exposed both along the top and foot of the cliff (houses, beaches, etc., see Figure 7.1a). This work initiated a quantitative assessment of collapse hazard. With a volume of erosion detection limit of 1.4 liters (one pixel of 100 cm^2 and a threshold erosion of 14 cm), three first-order collapse hazard readings were computed: (i) the overall annual cliff recession rate, (ii) the cliff face erosion map, and (iii) vertical and along-shore synthetic erosion profiles:

1. The observed average retreat rate equates to 1.1 cm/year (total eroded volume over 17 months divided by the exposed cliff surface area), which is one order of magnitude smaller than in many coastal cliff settings discussed in the literature (see a broad literature compilation in Premaillon et al., 2018). Macrotidal chalk cliffs of the English Channel erode at 10–50 cm/year (Costa et al., 2004; Regard et al., 2012; Dewez et al., 2013), mesotidal hard sedimentary cliffs of NE England (Rosser et al. 2013) or SW France (Premaillon et al., 2021) at 1–5 cm/year, and mesotidal marly-limestone cliffs of the Atlantic coast of Portugal erode faster than 10 cm/year (Marques et al., 2006). Furthermore, such centimeter-scale retreat rates are in line with changes we measured between the 1920s and 2010 using vertical aerial photography over the entire coastline of Provence–Alpes–Cote d'Azur (Giuliano, 2015), and to tree root dendrochronological cliff erosion measurements (Lopez-Saez et al., 2018). This retreat rate is also compatible with the rare and occasional events collated in the last decade.
2. The erosion scar inventory contains 14,925 distinct patches with erosion thicker than 14 cm. The total erosion affected 1,107 m^2 of the exposed cliff (3.67%) with a low total volume (about 320 m^3 in 17 months) in comparison with the Normandy cliff erosion (e.g. 100,000 m^3 in 12 months observed by Michoud et al., 2014), although this massive short-term volume is affected by a huge coastal landslide in Dieppe (see also Jacob et al. 2018). To further explore the relationship between erosion and rock types, we follow a normalization method proposed by Van Westen (1997) and used in De Guidi and Scudero (2013) whereby the relative response of each lithology to erosion can be ranked and quantified. From the observed erosion map, marls and sands are more prone to erosion (60% chances to be eroded) compared to calcarenites and conglomerates (20% each).
3. Alongshore (Figure 7.12.a), the scar volumes and surfaces are unevenly distributed and no direct response can easily be picked out with respect to relative headland/bay morphology. This lack of fingerprint can be due to the stochastic nature of rock falls and to the short time span of our observations. We do not regard this statement as conclusive. Finally, the vertical profile of erosion (Figure 7.12b) shows that erosion is focused on the lower part of the cliff, between 0 and 10 m elevation. The secondary erosion peak between 20 and 25 m elevation is caused by a single rock fall, east of "Cap de la Vierge" (headland H_5). Visual examination of scars are compatible with the suggestion of Rosser et al. (2013) or Prémaillon et al. (2021) that collapses do propagate, progressing from a basal notch at sea-level upward. Therefore the Carry-le-Rouet cliffs evolve according to common processes but at a rather slow pace over our 17 months survey interval.

7.8 Conclusion

This paper describes methods to deal with 3D point cloud acquisitions from a boat-borne laser scanner along sinuous cliffs. Man-made wall planes conveniently located along

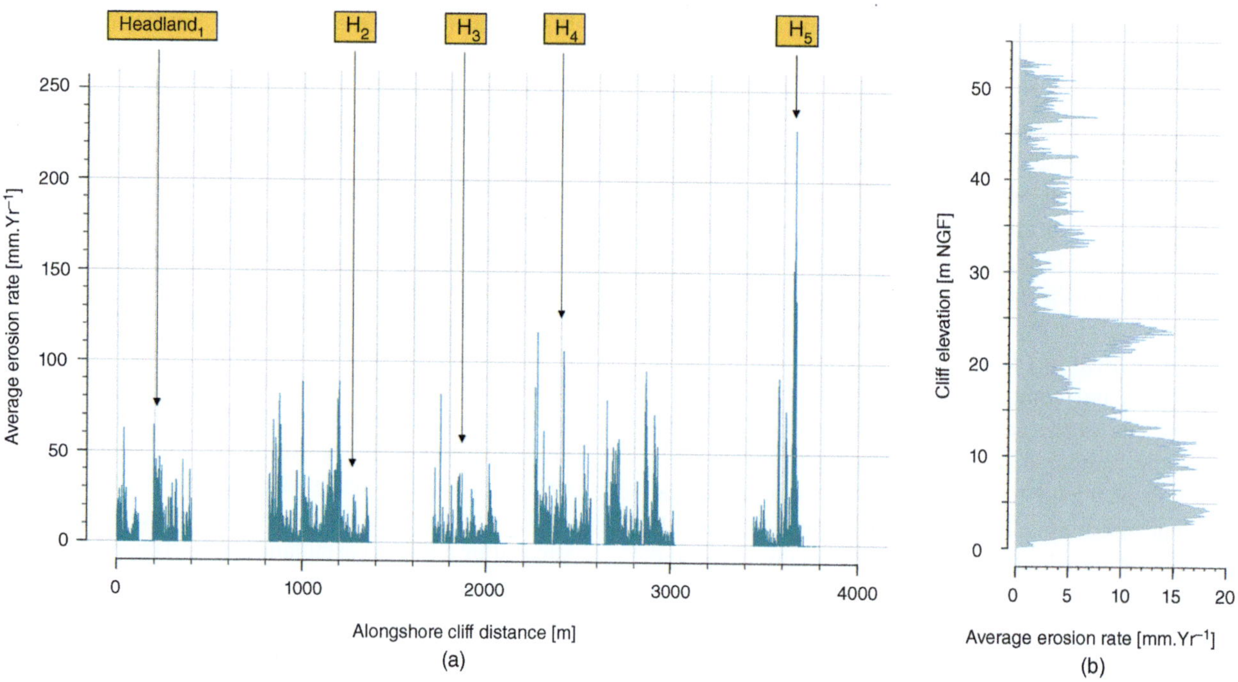

Figure 7.12 Average erosion rates integrated along-shore (a) and vertically (b). (a) All headlands flagged from west to east. The easternmost headland, H5 ("Cap de la Vierge"), locally focused erosion with three large events. (b) The vertically integrated profile shows a basal section, up to 15 m in height, affected by erosion rates above 10 mm/year while the upper part of the cliff only reaches 5 mm/year.

the top of the coastal cliff were fitted with robust planar models to empirically retrieve precision and accuracy of the surveyed point clouds. Although full-3D change detection is now possible with recently available software, many contextual interpretations still cannot be fully performed with such software. This is why we describe a method to unfold sinuous coastline point clouds onto analytical surfaces made of vertical plane segments and cylinder arcs (an unfolding method). Converting 3D to 2.5D leverages the rich GIS tools set for manual mapping, orthophoto interpretation, and conventional geomorphological analysis.

Applied to Carry-le-Rouet over a time span of 17 months, the average recession rate comes to 1.1 cm/year, when retaining only points with a significant difference threshold larger than the 99.9% quantile. Such an average recession rate lies at the lower end of cliff recession rates studies world-wide. Erosion was three times more effective in sand/marls than in calcarenites or conglomerates and was mostly focused in the lower part of the cliff. The model of cliff instability propagation from its base upward is also observed in Carry-le-Rouet, even though further work is needed to describe it in more detail. At this short time span, cliff erosion is thus dominated by marine processes over continental processes.

Acknowledgments

Jérémy Giuliano's PhD project (2011–2015) and Mélody Prémaillon's Master project (2014) were funded by Region PACA in the framework of project Valse (*Vulnérabilité et Adapation pour les Sociétés face aux Erosions de Falaise Côtières en Région Provence Alpes Côte d'Azur*). Mobile laser scanning was funded by Region PACA and DREAL-PACA to BRGM. Thomas Dewez was partially funded by the BRGM DEV-ESCARP project. The authors thank Julien Lassauque for his R programing contribution.

Appendix. Script for Unfolding Point Clouds (R)

Point cloud unfolding was coded in R (R Core Team, 2018) through the custom-made Unfolding Point Cloud function. This function can quickly treat any cloud in XYZ ASCII format. The script requires two inputs: (i) a point cloud to process that contains the information of each sector and (ii) parameters for coastline trends (either planar or curvilinear) (Figure 7.A-1). Angle values are obtained by the complex number method, through the Complex package in R (Becker et al., 1988). The output comes in ASCII with both initial and unfolded coordinates.

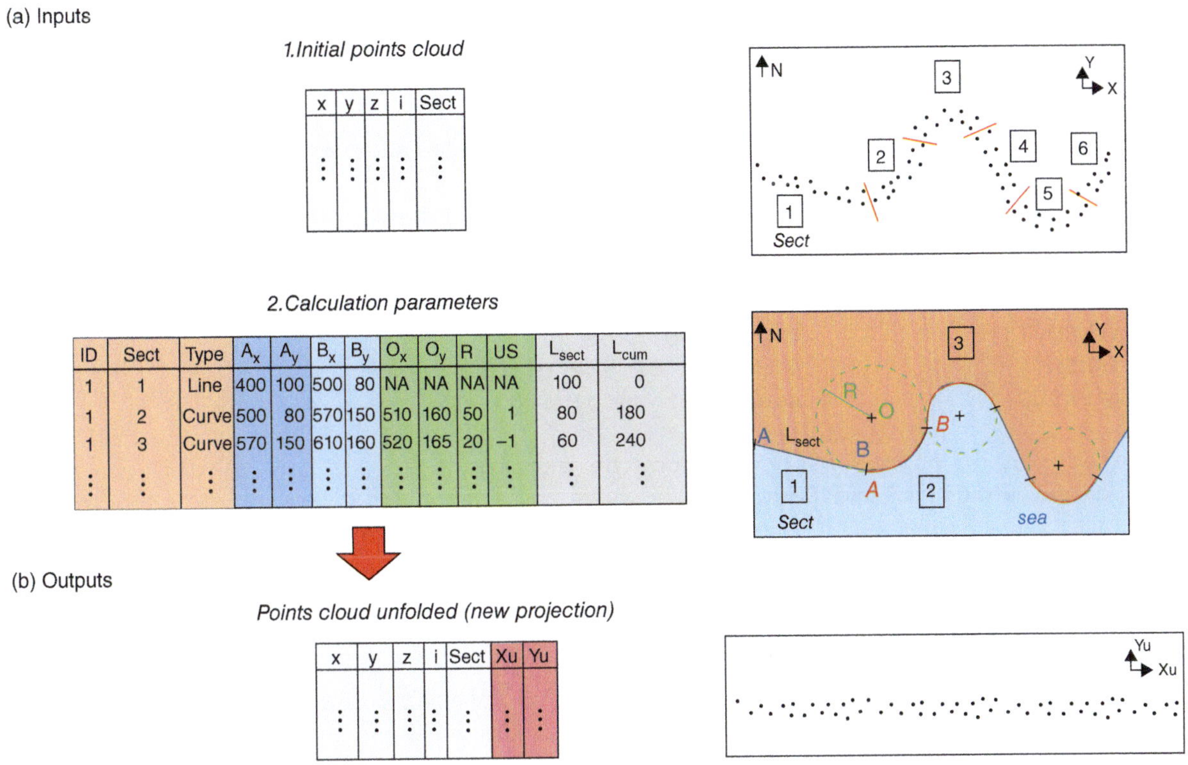

Figure 7.A-1 Sketch of the unfolding procedure. The initial point cloud is fed as text files comprising xyz coordinates, a series of scalar fields and a sector identifier (Sect), The corresponding map view is drawned on the right. A calculation parameters table details: sector identifier (Sect), the type of sector (Line or Curve), planview coordinates of sector origin (A_x, A_y) and end (B_x, B_y), circle origin (O_x, O_y) in case of "Curve" sector type, circle radius (R) in case of curve type, unfolding sense (US) 1 or −1, sector length (L_{sect}) and curvilinear distance from the coastal stretch origin (L_{cum}). Outputs produce a text file containing original point coordinates (xyz) and corresponding unfolded coordinates planimetric coordinates (Xu, Yu), Xu being the curvilinear distance from origin, positive to the right, and Yu, a landward cross-shore distance with positive Yu pointing landward. Note that z coordinates remain unchanged in the unfolding process and is not replicated.

```
#
### A- Data input ###
# A1- Points cloud (Figure 6.A-1.a.1)
infile.pc <- file.choose()  # open initial points cloud to be unfolded
pcloud <- as.data.frame(scan(file=infile.pc, what=list(x=numeric(),y=numeric(),
          z=numeric(), Rr=numeric(),G=numeric(),B=numeric(),i=numeric(),
          Sect=numeric()), flush=T, skip=1))
#
# A2- Parameters (Figure 6.A-1.a.2)
infile.pm <- file.choose()# open parameters file for unfolding operations (lin-
ear and curvilinaer sections)
pmark<-as.data.frame(scan(file=infile.pm,what=list(ID=0, Sect=0,Type="",Ax=0.0,
         Ay=0.0, Bx=0.0, By=0.0, Ox=0.0, Oy=0.0, R=0.0, lsect=0.0, lcum=0.0,
         US=0.0),skip=1))
pcloud$XU<-NA                    # to store x unfolded
pcloud$YU<-NA                    # to store y unfolded
#
### B- Loop for unfolding ###
for (j in 1:length(pmark$Sect)){
# B1- Type of section
  type <- as.character(pmark$Type[j])
```

```
  #
  # B2- Selection of parameters related to the section (Figure 6.A-1.a.2)
 xA <-pmark$Ax[j]              # X coordinate of start of the section
  yA <-pmark$Ay[j]              # y coordinate ...
  xB <-pmark$Bx[j]              # x coordinate of end of the section
  yB <-pmark$By[j]              # y coordinate ...
  lcum <-pmark$lcum[j]          # lcum : cumulated length at the beginning
                                  of the section
  lsect <-pmark$lsect[j]        # lsect : section length
  xO <-pmark$Ox[j]              # x coordinate of the center of the circle
  yO <-pmark$Oy[j]              # y coordinate ...
  R  <-pmark$R[j]               # radius of the circle
  US <- pmark$US[j]            # sense for unfolding
#
  # B3- Selection of points belonging to the section
  lgs <- pcloud[which(pcloud$Sect==j),]
  #
# B4- Unfolding
  if (type=="line"){
    lgUnf <- unfol.line(lgs, xA, yA, xB, yB, lcum)
  } else if (type == "curve"){
    lgUnf <- unfol.curve(lgs, xA, yA, xO, yO, R, US, lcum)
  } else {
  }
  #
  # B5- Data storing
  pcloud$XU[which(pcloud$Sect==j)] <- lgUnf$XU
  pcloud$YU[which(pcloud$Sect==j)] <- lgUnf$YU
} #
### C- Data output (Figure 6.A-1.b) ###
#
output.file <- paste(dirname(infile.pc), "/unfolded24.txt", sep="")
write.table(pcloud, output.file, row.names=F)
#
#
unfol.line <- function(lgs, xA, yA, xB, yB, lcum){
  # Translation of origin
  lgUnf <- data.frame(XU=lgs$x - xA, YU=lgs$y - yA)
  # Rotation
  alpha <- Arg(complex(real=(xB-xA), imaginary=(yB-yA)))
  rot <- matrix(c(cos(alpha), sin(alpha), -sin(alpha), cos(alpha)),ncol=2,
          nrow=2, byrow=T)
  temp <- t(rot %*% t(cbind(lgUnf$XU, lgUnf$YU)))
  # Translation back to lcum
  lgUnf$XU <- temp[,1] + lcum
  lgUnf$YU <- temp[,2]
  return(lgUnf)
}
#
##END##
```

References

Abellán, A., Vilaplana, J.M., and Martínez, J. (2006). Application of a long-range terrestrial laser scanner to a detailed rockfall study at Vall de Núria (Eastern Pyrenees, Spain). *Engineering Geology* 88 (3–4): 136–148.

Abellán, A., Calvet, J., Vilaplana, J.M., and Blanchard, J. (2010). Detection and spatial prediction of rockfalls by means of terrestrial laser scanner monitoring. *Geomorphology* 119: 162–171.

Alho, P., Kukko, A., Hyyppä, H. et al. (2009). Application of boat-based laser scanning for river survey. *Earth Surface Processes and Landforms* 34: 1831–1838.

Argüelles-Fraga, R., Ordóñez, C., García-Cortés, S., and Roca-Pardiñas, J. (2013). Measurement planning for circular cross-section tunnels using terrestrial laser scanning. *Automation in Construction* 31: 1–9.

Baltsavias, E.P. (1999). Airborne laser scanning: Basic relations and formulas. *ISPRS Journal of Photogrammetry and Remote Sensing* 54: 199–221.

Barber, D., Mills, J., and Smith-Voysey, S. (2008). Geometric validation of a ground-based mobile laser scanning system. *ISPRS Journal of Photogrammetry and Remote Sensing* 63: 128–141.

Becker, R.A., Chambers, J.M., and Wilks, A.R. (1988). *The New S Language*, Wadsworth & Brooks/Cole.

Bernard, T.G., Lague, D., and Steer, P. (2021). Beyond 2D landslide inventories and their rollover: Synoptic 3D inventories and volume from repeat LiDAR data. *Earth Surface Dynamics* 9: 1013–1044.

Besl, P.J. and McKay, N.D. (1992). A method for registration of 3-D shapes. *IEEE Transactions on Pattern Analysis and Machine Intelligence* 14 (2): 239–256.

Bitelli, G., Dubbini, M., and Zanutta, A. (2004). Terrestrial laser scanning and digital photogrammetry techniques to monitor landslide bodies. In: *Proceedings of XX ISPRS Congress* (ed. O. Altan), 246–251. Turkey. Commission V, WG V/2: O.Istanbul.

Brodu, M. and Lague, D. (2012). 3D terrestrial LiDAR data classification of complex natural scenes using a multi-scale dimensionality criterion: Application in geomorphology. *ISPRS Journal of Photogrammetry and Remote Sensing* 68: 121–134.

Ciavola, P., Armaroli, C., Chiggiato, J. et al. (2007). Impact of storms along the coastline of Emilia-Romagna: The morphological signature on the Ravenna coastline (Italy). *Journal of Coastal Research, SI* 50: 540–544.

Collins, B. and Sitar, N. (2008). Processes of coastal bluff erosion in weakly lithified sands, Pacifica, California, USA. *Geomorphology* 97 (3): 483–501.

Costa, S., Delahaye, D., Freiré-Diaz, S. et al. (2004). Quantification of the Normandy and Picardy chalk cliff retreat by photogrammetric analysis. In: *Coastal Chalk Cliff Instability*, vol. 20 (ed. R.N. Mortimore and A. Duperret), 139–148. London, Engineering Geology. Special Issue: Geological Society.

De Guidi, G. and Scudero, S. (2013). Landslide susceptibility assessment in the Peloritani Mts. (Sicily, Italy) and clues for tectonic control of relief processes. *Natural Hazard and Earth System Sciences* 13: 949–963.

Demory, F., Conesa, G., Oudet, J. et al. (2011). Magnetostratigraphy and paleoenvironments, in shallow-water carbonates: the Oligocene-Miocene sediments in the northern margin of the Liguro-Provençal basin (West Marseille, southeastern France). *Bulletin de la Société Géologique de France* 182 (1): 37–55.

Dewez, T.J.B., Rohmer, J., and Closset, L. (2007). Laser survey and mechanical modeling of chalky sea cliff collapse in Normandy, France. In: *Proceedings of Landslides and Climate Change, Challenges and Solutions* (ed. R. McInnes, J. Jakeways, H. Fairbanks and E. Mathie), 281–288. England: *Isle of Wight*.

Dewez, T.J.B., Rohmer, J., Regard, V., and Cnudde, C. (2013). Probabilistic coastal cliff collapse hazard from repeated terrestrial laser surveys: Case study from Mesnil Val (Normandy, northern France). In: *Proceedings of the 12th International Coastal Symposium (Plymouth, England), Journal of Coastal Research*. Special Issue, vol. 65 (ed. D. Conley, G. Masselink, P. Russel and T. O'Hare), 702–707.

Dewez, T.J.B., Yart, S., Thuon, Y. et al. (2017). Towards cavity-collapse hazard maps with Zeb-Revo handheld laser scanner point clouds. *Photogram Rec.* 32: 354–376. https://doi.org/10/gmfbzp.

Dunning, S.A., Massey, C.R., and Rosser, N.J. (2009). Structural and geomorphological controls on landslides in the Bhutan Himalaya using Terrestrial Laser Scanning. *Geomorphology* 103: 17–29.

Duperret, A., Genter, A., Mortimore, R.N. et al. (2002). Coastal rock cliff erosion by collapse at Puys, France: The role of impervious marl seams within chalk of NW Europe. *Journal of Coastal Research* 18: 52–61.

Earlie, C.S., Masselink, G., Russell, P.E., and Shail, R.K. (2014). Application of airborne LiDAR to investigate rates of recession in rocky coast environments. *Journal of Coast Conservation* 15 pp., doi: 10.1007/s11852-014-0340-1.

Ellum, C. and El-Sheimy, N. (2002). Land-based mobile mapping systems. *Photogrammetric Engineering and Remote Sensing* 68 (1): 13–17.

Feldmann, B., Lévy, C., Dewez, T.J.B., and Marçot, N. (2018). Méthode de comparaison de nuages de points par scanner laser mobile pour caractériser les éboulements de falaise côtières. *Revue XYZ* 156: 57–62.

Furlani, S., Pappalardo, M., Gomez-Pujol, L., and Chelli, A. (2014). The rock coast of the Mediterranean and Black Seas. In: *Rock Coast Geomorphology: A Global Synthesis*, vol. 40 (ed. D.M. Kennedy, W.J. Stephenson and L.A. Naylor), 89–123. London Memoirs: Geological Society.

Girardeau-Montaut, D., Roux, M., Marc, R., and Thibault, G. (2005). Change detection on points cloud data acquired with a ground laser scanner. *International Archives of Photogrammetry, Remote Sensing and Spatial Information Sciences* 36 (3): 30–35.

Giuliano, J. (2015). Érosion des falaises de la région Provence-Alpes-Côte d'Azur:Évolution et origine de la morphologie côtière en Méditerranée: Télédétection, géochronologie, géomorphologie, PhD Thesis, Université Nice Sophia Antipolis.

Glennie, C., Brooks, B., Ericksen, T. et al. (2013). Compact multipurpose mobile laser scanning system–initial tests and results. *Remote Sensing* 5: 521–538.

Heritage, G. and Hetherington, D. (2007). Towards a protocol for laser scanning in fluvial geomorphology. *Earth Surface Processes and Landforms* 32: 66–74.

Höhle, J. and Höhle, M. (2009). Accuracy assessment of digital elevation models by means of robust statistical methods. *ISPRS Journal of Photogrammetry and Remote Sensing* 64: 398–406.

Hunter, G., Pinkerton, H., Airey, R., and Calvari, S. (2003). The application of a long-range laser scanner for monitoring volcanic activity on Mount Etna. *Journal of Volcanolology and Geothermal Research* 123: 203–210.

Ingensand, H. (2006). Metrological aspects in terrestrial laser-scanning technology. In: *Proceedings of 3rd IAG/12th FIG Symposium* (ed. A. Chrzanowski and R. Wilkins). Germany: Baden, 22–24 May2006.

Ingensand, H., Ryf, A., and Schulz, T. (2003). Performances and experiences in terrestrial lasers canning. In: *Proceedings of Optical 3D-Measurement Techniques VI* (ed. A. Grün and H. Kahmen). Zurich 22–25 September 2003.

Jaakkola, A., Hyyppä, J., Hyyppä, H., and Kukko, A. (2008). Retrieval algorithms for road surface modelling using laser-based mobile mapping. *Sensors* 8: 5238–5249.

Jaboyedoff, M., Demers, D., Locat, J. et al. (2009). Use of terrestrial laser scanning for the characterization of retrogressive landslides in sensitive clay and rotational landslides in river banks. *Canadian Geotechnical Journal* 46: 1379–1390.

Jacob, T., Samyn, K., Bitri, A. et al. (2018). Mapping sand and clay-filled depressions on a coastal chalk clifftop using gravity and seismic tomography refraction for landslide hazard assessment, in Normandy, France. *Engineering Geology* 246: 262–276.

Kemeny, J. and Post, R. (2003). Estimating three-dimensional rock discontinuity orientation from digital images of fracture traces. *Computers and Geosciences* 29: 65–77.

Kuhn, D. and Prüfer, S. (2014). Coastal cliff monitoring and analysis of mass wasting processes with the application of terrestrial laser scanning: A case study of Rügen, Germany. *Geomorphology* 213: 153–165.

Kukko, A., Kaartinen, H., Hyyppä, J., and Chen, Y. (2012). Multiplatform mobile laser scanning: Ssability and performance. *Sensors* 12: 11712–11733.

Lague, D., Brodu, N., and Leroux, J. (2013). Accurate 3D comparison of complex topography with terrestrial laser scanner: Application to the Rangitikei canyon (NZ). *ISPRS Journal of Photogrammetry and Remote Sensing* 82: 10–26.

Lichti, D.D., Steward, M.P., Tsakiri, M., and Snow, J. (2000a). Calibration and testing of a terrestrial laser scanner. *International Archives of Photogrammetry, Remote Sensing and Spatial Information Sciences* 33 (B5): 485–492.

Lichti, D.D., Steward, M.P., Tsakiri, M., and Snow, J. (2000b). Benchmark tests on a three-dimensional Laser scanning system. *Geomatics Research Australia* 72: 1–24.

Lim, M., Petley, D.N., Rosser, N.J. et al. (2005). Combined digital photogrammetry and time-of-flight laser scanning for monitoring cliff evolution. *The Photogrammetric Record* 20 (110): 109–129.

Lim, M., Rosser, N.J., Allisson, R.J., and Petley, D.N. (2010). Erosional processes in the hard rock coastal cliffs at Staithes, North Yorkshire. *Geomorphology* 114: 12–21.

López, J. and Frances, F. (2010). Influence of the North Atlantic oscillation and the Western Mediterranean oscillation in the maximum flow events in Spain. In: *In International Workshop Advances in Statistical Hydrology, May 23–25, 2010*. Taormina: Italy.

Lopez-Saez, J., Corona, C., Morel, P. et al. (2018). Quantification of cliff retreat in coastal Quaternary sediments using anatomical changes in exposed tree roots. *Earth Surface Processes and Landforms* 43: 2983–2997.

Marques, F.M.S.F. (2006). Rates, patterns and timing of cliff retreat. A case study on the west of Portugal. *Zeitschrift für Geomorphologie-Supplementbande* 144: 231–257.

Michoud, C., Carrea, D., Costa, S. et al. (2014). Landslide detection and monitoring capability of boat-based mobile laser scanning along Dieppe coastal cliffs, Normandy. *Landslides, Technical Note*. doi: https://doi.org/10.1007/s10346-014-0542-5.

Millot, C. (1990). The Gulf of Lion's hydrodynamics. *Continental Shelf Research* 10 (9–11): 885–894.

Monserrat, O. and Crosetto, M. (2008). Deformation measurement using terrestrial laser scanning data and least square 3D surface matching. *Photogrammetric Engineering and Remote Sensing* 63: 142–154.

Murakami, H., Nakagawa, K., Hasegawa, H. et al. (1999). Change detection of buildings using an airborne laser scanner. *ISPRS Journal of Photogrammetry and Remote Sensing* 54: 148–152.

O'Neal, M.A. and Pizzuto, J.E. (2011). The rates and spatial patterns of annual riverbank erosion revealed through terrestrial laser-scanner surveys of the South River, Virginia. *Earth Surface Processes and Landforms* 36 (5): 695–701.

Olsen, M., Johnstone, E., Driscoll, N. et al. (2009). Terrestrial laser scanning of extended cliff sections in dynamic environments: Parameter analysis. *Journal of Surveying Engineering* 135: 161–169.

Pedrazzini, A., Oppikofer, T., Jaboyedoff, M. et al. (2010). Assessment of rockslide and rockfall problems in an active quarry: Case study of the Arvel quarry (western Switzerland). In: *European Rock Mechanics Symposium*, 593–596. (EUROCK 2010), Lausanne, Switzerland: CRC Press, Leiden.

Prémaillon, M., Regard, V., Dewez, T.J.B., and Auda, Y. (2018). GlobR2C2 (Global Recession Rates of Coastal Cliffs): A global relational database to investigate coastal rocky cliff erosion rate variations. *Earth Surface Dynamics* 6: 651–668. https://doi.org/10/gd3hng.

Prémaillon, M., Dewez, T.J.B., Regard, V. et al. (2021). Conceptual model of fracture-limited sea cliff erosion: Erosion of the seaward tilted flyschs of Socoa, Basque Country, France. *Earth Surface Processes and Landforms* 46: 2690–2709. https://doi.org/10/gmrc9p.

R Core Team (2018). R: A language and environment for statistical commuting. R Foundation for Statistical Computing, Vienna, Austria. ISB 3-900051-07-0, URL http://www.R-project.org.

Regard, V., Dewez, T., Bourlès, D.L. et al. (2012). Late Holocene seacliff retreat recorded by 10Be profiles across a coastal platform: Theory and example from the English Channel. *Quaternary Geochronology* 11: 87–97.

Rohmer, J. and Dewez, T.J.B. (2013). On the deviation of extreme sea cliff instabilities from the power law frequency volume distribution: Practical implications for coastal management. In: *Proceedings of the 12th International Coastal Symposium (Plymouth, England), Journal of Coastal Research*. Special Issue 65 (ed. D. Conley, G. Masselink, P. Russel and T. O'Hare), 1698–1703.

Rosser, N.J., Petley, D.N., Lim, M. et al. (2005). Terrestrial laser scanning for monitoring the process of hard rock coastal cliff erosion. *Quarterly Journal of Engineering Geology and Hydrgeology* 38: 363–375.

Rosser, N.J., Lim, N., Petley, D.N. et al. (2007). Patterns of precursory rockfall prior to slope failure. *Journal of Geophysical Research* 112: F04014.

Rosser, N.J., Brain, M.J., Petley, D.N. et al. (2013). Coastline retreat via progressive failure of rocky coastal cliffs. *Geology* 41: 939–942.

Sabatier, P., Samat, O., Ullmann, A., and Suanez, S. (2009). Connecting large-scale coastal behaviour with coastal management of the Rhône delta. *Geomorphology* 107: 79–89.

Schwarz, K.P., Chapman, M.A., Cannon, M.W., and Gong, P. (1993). An integrated INS/GPS approach to the georeferencing of remotely-sensed sata. *Photogrammetric Engineering and Remote Sensing* 59 (11): 1667–1674.

Soudarissanane, S., Lindenbergh, R., Menenti, M., and Teunissen, P. (2011). Scanning geometry: Influencing factor on the quality of terrestrial laser scanning points. *ISPRS Journal of Photogrammetry and Remote Sensing* 66: 389–399.

Stepp, J.C. (1972). Analysis of completeness of the earthquake sample in the Puget Sound area and its effect on statistical estimates of earthquake hazard. In: *Proceedings of the First Microzonation Conference*, 897–909. Seattle, USA.

Stock, G.M., Bawden, G.W., Green, J.K. et al. (2011). High-resolution three-dimensional imaging and analysis of rock falls in Yosemite Valley, California. *Geosphere* 7 (2): 573–581.

Teza, G., Galgaro, A., Zaltron, N., and Genevois, R. (2007). Terrestrial laser scanner to detect landslide displacement fields: a new approach. *International Journal of Remote Sensing* 28: 3425–3446.

Vaaja, M., Hyyppä, J., Kukko, A. et al. (2011). Mapping topography changes and elevation accuracies using a mobile laser scanner. *Remote Sensing* 3: 587–600.

Vaaja, M., Kukko, A., Kaartinen, H. et al. (2013). Data processing and quality evaluation of a boat-based mobile laser scanning system. *Sensors* 13: 12497–12515.

van Westen, C. (1997). Statistical landslide hazard analysis, Application guide, ILWIS 2.1 for Windows. ITC, Enschede, The Netherlands.

Vögtle, T. and Steinle, E. (2004). Detection and recognition of changes in building geometry derived from multitemporal laser scanning data. *International Archives of Photogrammetry, Remote Sensing and Spatial Information Sciences* 35 (B2): 428–433.

Vosselman, G., Gorte, B.G.H., and Sithol, G. (2004). Change detection for updating medium scale maps using laser altimetry. *International Archives of Photogrammetry, Remote Sensing and Spatial Information Sciences* 35 (B3): 207–212.

Young, A.P. and Ashford, S.A. (2006). Application of airborne LIDAR for seacliff volumetric change and beach sediment contributions. *Journal of Coastal Research* 22: 307–318.

Young, A.P. and Ashford, S.A. (2007). Quantifying sub-regional seacliff erosion using mobile terrestrial LiDAR. *Shore Beach* 75: 38–43.

Young, A.P., Flick, R.E., Gutierrez, R., and Guza, R.T. (2009). Comparison of short-term seacliff retreat measurement methods in Del Mar, California. *Geomorphology* 112: 318–323.

Young, A.P., Olsen, M.J., Driscoll, N. et al. (2010). Comparison of airborne and terrestrial LIDAR estimates of seacliff erosion in southern California. *Photogrammetric Engineering and Remote Sensing* 76: 421–427.

8

A DEM-Based Volume Extraction Approach: From Micro-Scale Weathering Forms to Planetary Lava Tubes

Riccardo Pozzobon[1,2], Claudio Mazzoli[1], Silvia Salvini[1], Francesco Sauro[3], Matteo Massironi[1], and Tommaso Santagata[4]

[1]*Department of Geosciences, University of Padova, Italy*
[2]*INAF-OAPD, Italy*
[3]*Department of Biological, Geological and Environmental Sciences, Italian Institute of Speleology, University of Bologna, Italy*
[4]*VIGEA – Virtual Geographic Agency, Italy*

Abstract

We show how a GIS-based approach on 3D morphologies can be used to analyze volume variations from the microscopic scale on rock samples to large collapse pits on Earth and Mars.

The microscale analyses were performed on scans acquired by a confocal laser scanning microscope from carbonate rock plates dissolved by immersion in slightly acidic solutions. Each studied sample underwent an increasing number of immersions aiming to calculate the recession rate of such stones commonly used in cultural heritage when exposed to acidic rain. We achieved this by creating a synthetic reference surface and calculating the difference in height with the scanned sample surface.

The same approach was applied in planetary remote sensing to evaluate the actual volume of collapsed conduit sections of Earth and Martian lava tubes from DEMs. Lava tubes can be up to tens of kilometers long on Earth and up to hundreds of kilometers on Mars but a numerical estimate of the collapse volumes (and thus of the voids) was never attempted. The creation of a synthetic surface on top of collapses best approximating the pristine topography and calculation of the volumes in between allows comparing the total volume of the collapsed sections in distinct planetary bodies.

8.1 Introduction

Volume measurements on DEMs (digital elevation models) are fundamental to apply a quantitative approach in several studies in the field of geology, geomorphology, rock mechanics, and geochemistry. In literature there are a number of methods that have been applied to these problematics (Tseng et al., 2013; Grohmann and Sawakuchi, 2013; Martha et al., 2010; Jakobsson et al, 2007; Nichols et al., 2006' Kalmar et al, 1995); nonetheless a general approach applicable in all different fields and at different scales is still missing. All of these works, however, deal either with multitemporal datasets measuring their variation (landslides, landscape modification) in the presence of a defined reference system or calculating volume of water bodies that are constrained to shorelines or basins that, by their nature, are developed between the DEM and a horizontal surface (being the water level). In this paper we present a novel approach of a DEM-based volume extraction through the creation of synthetic reference surfaces on which the volumetric measurements are homogenized to compensate the uncertainty of the reference system or to simulate a pristine topography. Two applications are shown, from the microscale of pitting and chemical weathering on carbonates to the macro-scale of measuring geomorphological features like lava tube collapses on planetary bodies like Earth and Mars.

8.1.1 Measuring the Recession Rates of Carbonate Rocks

Carbonate rocks (limestones, marbles) are among the most commonly used building materials (both as dimension and ornamental stones), and are highly vulnerable to weathering, especially in polluted areas. Present-day models of surface erosion for carbonate rocks are damage or dose-response functions based on factors including air pollution (gas and solid), atmospheric CO_2 concentration, rainfall, and temperature (Brimblecombe and Grossi,

3D Digital Geological Models: From Terrestrial Outcrops to Planetary Surfaces, First Edition.
Edited by Andrea Bistacchi, Matteo Massironi, and Sophie Viseur.

2008). Several of these models have been formulated using linear or multiple regression techniques of surface recession data, obtained by short-term measurements (a few weeks to a few years) of weight loss on limestone plates (e.g. Searle and Mitchell, 2006; Kucera et al., 2007) and specific environmental and climate parameters such as precipitation, rain pH, and SO_2 and HNO_3 concentrations in air and deposition rates (e.g. Lipfert, 1989; Tidblad et al., 2001; Kucera et al., 2007; MULTI-ASSESS Report, 2007). These models have been used to predict long-term (centuries) rates of limestone erosion (Grossi et al., 2008; Brimblecombe and Grossi, 2009; Sabbioni et al., 2010), and build limestone vulnerability recrudescence or mitigation maps according to projections of future climate changes as outlined in the IPCC (Intergovernmental Panel on Climate Change) Reports (Bonazza et al., 2009a, 2009b). Different methods have been used to measure recession in stones exposed outdoor for long periods (tens of years) in specific localities (Meierding, 1981; Feddema and Meierding, 1987; Trudgill et al., 1989; Dragovich, 1991; Roberts, 2005; Inkpen, 2012a; 2012b). Identification of surface recession patterns and estimate of material loss in heritage buildings have been also performed via a probabilistic Bayesian approach based on digital photogrammetry data (Jalón et al., 2021), an approach that allows rapid examination of large portions of a monument surface. Although in theory long-term measurements provide a better estimate of average damage response under given environmental conditions, they require measurement repetitions after decades (Inkpen, 2012a, 2012b), which is unfavourable to the calibration of general recession-rate equations to be applied to different materials and environments, and they may be affected by considerable uncertainties on the estimation of the original stone surface. On the other hand, short-term measurements easily provide a damage estimate of different carbonate rock types, under different environmental conditions, but the error on such micrometric recession values dramatically propagates when pertaining dose-response equations are used to assess long-term stone recession. In this paper, we apply a DEM-based volume extraction approach for surface analysis through confocal microscopy, allowing precise measurement of surface recession and volume loss in samples subjected to accelerated aging tests (see Section 8.2.1 and Salvini, 2018, for details).

8.1.2 Lava Tubes on Earth and Mars

Lava tubes are a type of cave found within several volcanic environments, but particularly in the presence of shield volcanoes and basaltic pahoehoe flows.

Lava tubes are mainly straight tubular conduits that generally form when the outer surfaces of lava channels cool more rapidly, forming a hardened crust; subsequently the remaining lava flows out of the tube, leaving a void space (Calvari and Pinkerton, 1999; Dragoni et al., 1995; Greeley, 1971a, 1971b). The mechanisms of lava tube formation can also be related to inflating lava emplacement underneath pre-existent lava sheets (Keszthelyi and Self, 1998; Greeley et al., 1998 Self et al., 1996).

These conduits can be found in several places on Earth: in Mt Etna (Italy), in Hawaii, such as Kazumura Cave in Mauna Loa, which is the longest lava tube on Earth (up to 65 km) (Léveillé and Datta, 2010; Allred and Allred, 1997), Undara in Queensland (Australia), in the Galapagos and in the Canary Islands (La Corona, Lanzarote, and Cueva del Viento, Tenerife).

From optical imaging remote sensing, the presence of a lava tube is inferred from alignment of collapses of the ceiling of lava tube sections. These collapses, called jameos or skylights, are most frequently elliptical, with the minor axis corresponding to the maximum width of the tube and the major axis highlighting the path of the tube, and rarely circular when the lava tube ceiling is still overhanging and not entirely collapsed (Kempe and Al-Malabeh, 2005; Haruyiama et al., 2012).

Images from the Viking Orbiters first highlighted the presence of lava tube candidates on Mars (Greeley and Spudis, 1981), where the largest shield volcanic province is located in the Solar System: the Tharsis region (Hodges and Moore, 1994). With the outcomes of the first high-resolution images from the Mars Global Surveyor, Mars Express, and Mars Reconnaissance Orbiter, more and more elongated collapse pit chains possibly related to lava tubes have been localized (Hodges and Moore, 1994; Wyrick et al., 2004; Cushing et al., 2007; Keszthelyi et al., 2007). In particular, tube-fed flows and ridges have been detected in Olympus Mons' flanks, Alba Patera (Bleacher et al., 2006), and Hadriaca Patera (Léveillé and Datta, 2010).

Although lava tube sections can develop in several levels within the same tube due to different pulses in the lava flowing rate, it is straightforward that the volumes of the collapsed sections can provide a first estimate of the minimum underground volume to be expected on Mars or unexplored tubes on Earth and compare these values to what we know on Earth explored tubes.

Here we propose a volume extraction and comparison between a Martian lava tube and one of the best terrestrial analogs in the Canary Islands. The Canary Islands are indeed located on a hot spot over a so-called "slow moving plate": here a long-lasting focused magmatic activity controls the type of volcanism, effusion rates, and related morphologies, along with geochemistry. Mars being a one-plate shell planet, volcanic processes and landforms in the Canary Islands can be readily compared to Martian equivalents (Meyzen et al., 2015). Among them, the Corona lava tube in Lanzarote is one of the largest in the world in terms of size and its volume is likely to be more similar to those on Mars and on the Moon (e.g. Sauro et al., 2019, 2020; Theinat et al., 2020).

The Martian lava tube chosen for comparison is located in the surroundings of Hadriaca Patera, a large volcanic edifice of ~450 km in diameter on the north-eastern side of the Hellas impact basin, more precisely centered at 37° 24' S, 89° 37' E. The lava tube is highlighted by a sinuous ~75 km-long alignment of elongated collapse pits that ends in a 6 km-wide lava channel. The dimension of the collapse pits suggests a lava tube width varying between 300 and 500 m.

On Earth, the Corona lava tube of Lanzarote Island in the Canary archipelago has been chosen as a comparison to the Martian example, applying the same DEM-based approach. This structure is ~8 km long and departs from the eastern flank of the Corona volcanic edifice, formed approximately 20 thousand years ago (Carracedo et al., 2003). The last ~2 km section of the tube is flooded by seawater and the conduit terminates 64 m under the sea level (Isler, 1986). It manifests its presence on the surface with a sinuous series of 25 collapses, presenting a subcircular/elongated shape in plan view, with a width between 25 and 50 meters roughly corresponding to the inner diameter of the tube.

8.2 Micro-elevation Maps and DEMs Production

8.2.1 Carbonate Samples Preparation and Confocal Microscopy Scan

A set of eleven carbonate rocks differing in their mineral composition (e.g. the presence and abundance of dolomite, clay minerals, hydroxides) and textural features (grain-size of calcite grains, porosity) underwent aging tests consisting of immersion cycles in two different artificial rainwaters corresponding to those of Bologna (Panettiere et al., 2000) and Stresa environment (Rogora et al., 2004) (northern Italy) respectively. The experiment was interrupted after 240 cycles because, from the damage function described by Lipfert (1989) and the measured recession on Carrara marble, the results were shown to correspond to about 1,000 mm of rainfall (see Salvini, 2018, for details). Samples were prepared in plates ($\approx 2 \times 2 \times 1$ cm) and the surface lapped on sandpaper with a grit of 1,200. For each plate, a stainless steel flat washer was cut in two halves and attached to the opposite sides of the plate. The freshly cut flat metal surface was then polished together with the stone surface, in order to provide a reference height for the evaluation of stone recession.

Surface topography was acquired using an Olympus Lext OLS4000 CLSM (confocal laser scanning microscope) on areas of approximately 2.5×2.5 mm. For each sample, the same area was scanned before the aging test, and after 54, 140, and 240 cycles. Each acquisition also included a narrow strip of the attached metal surface. Native CLSM topographic acquisitions resulting in a 2,550 × 2,550 matrix whose cell values represented the actual height (in μm) with respect to a horizontal instrumental surface (Ficker and Martišek, 2012). These matrices were converted to an *xyz* point cloud with a Matlab script and, by knowing the sampling steps (0.6 μm), subsequently sampled as micro-DEMs (or elevation maps) in the ArcGis environment (Figure 8.1).

8.2.2 Stereo DEM Extraction for Mars

The DEM used to extract the volumes in the Mars case was obtained from stereo overlapping images from CTX (Context Camera; Malin et al., 2007) onboard Mars Reconnaissance Orbiter, with a nominal resolution of 6 m/pixel of the southern part of the Hadriaca Patera lava tube. This particular section encompasses 15 collapsed pits aligned into a sinuous path and was selected because of the good quality of both images in terms of an absence of hard shadows within the collapses and overall low noise. In fact, the eventual presence of noise and especially hard shadows would have negatively affected the stereo correlation algorithm producing artifacts and no-data gaps in the final DEM, turning the results into unreliable volume estimates.

The images (B18_016733_1427_XN_37S270W, P19_008439_1403_XI_39S270W) were downloaded from the PDS (Planetary Data System) and ingested into the ISIS3 software suite (Integrated Software for Imagers and Spectrometers; Gaddis et al., 1997) to be radiometrically calibrated and destriped. This allows the removal of linescan artifacts and the vignetting effect, which lowers the brightness at the edges of the image. Then images were projected using the related SPICE kernels, which contain the pointing information of the satellite at the moment of a specific image acquisition. These data are used to locate the image on the surface of the planet and, together with reference system information (i.e. ellipsoid and projection system), are essential for georeferencing and map projection.

To perform the stereo matching we used the ASP (Ames Stereo Pipeline plugin for ISIS3; Moratto et al., 2010). The most important step in this phase, before stereo matching,

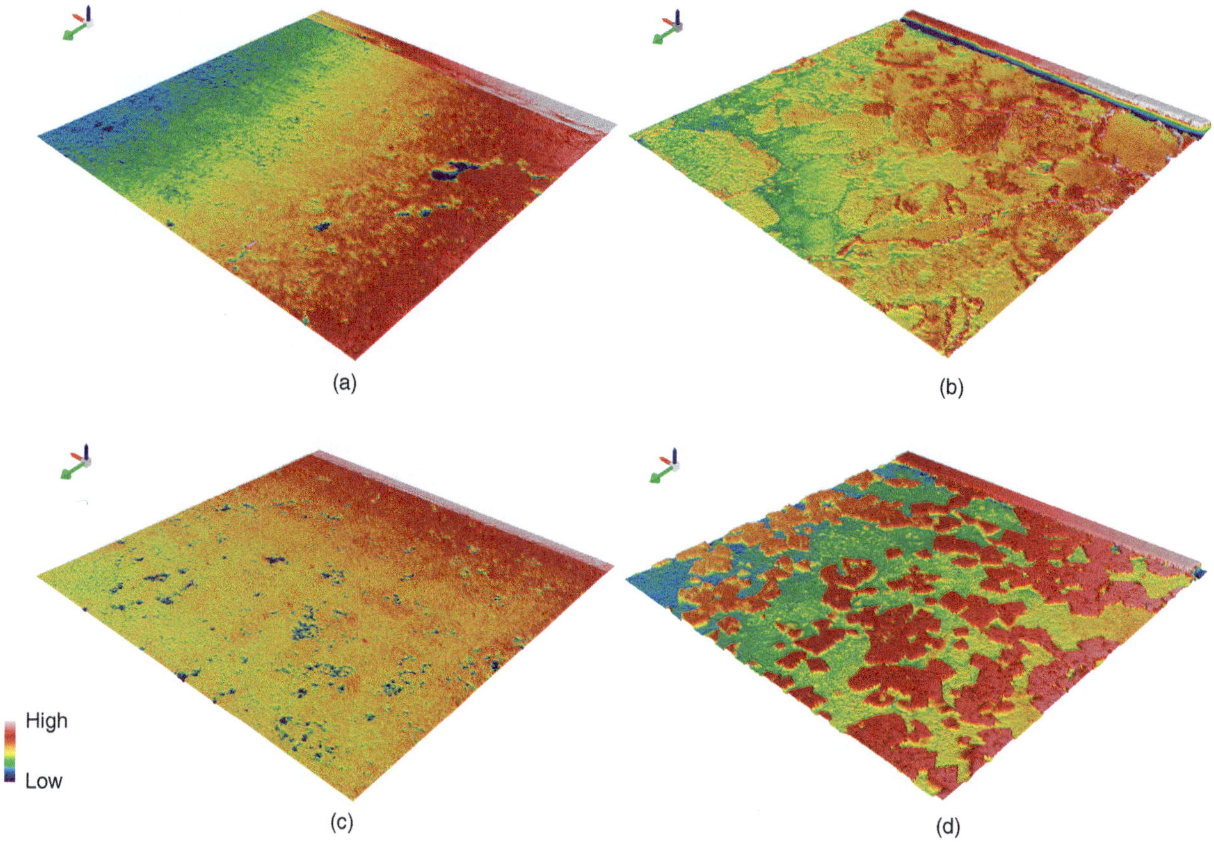

Figure 8.1 Examples of CLSM 3-D elevation maps (area is about 2.5 × 2.5 mm): Botticino Stone before aging (a) and after 240 aging cycles (b); Chiampo Paglierino Stone before aging (c) and after 240 aging cycles (d). In both cases, aging tests were performed using an artificial water corresponding to Stresa rainwater. Different colors correspond to different elevations (not in scale). Note the roughening of the surface after aging cycles, with differential regression among domains characterized by different textural and compositional features (e.g. bioclasts vs. surrounding micrite in "b"; dolomite euhedral crystals vs. surrounding calcite matrix in "d"). The lateral color shift in some of the maps reflects the inclination of the sample during acquisition. Note the flat metal strip along the far side of the map, used to anchor the reference surface.

is the bundle adjustment of images, in order to obtain a validated terrain model suitable for scientific analyses.

This procedure corrects the potential errors incoming from inaccurate pointing information of the images that could produce artifacts or altered slopes in the final DEM. Bundle adjustment can be performed in a completely automated way into ASP and takes care of iteratively adjusting the position of the camera until the eventual shift of the retro-projection of the pixels of the image is minimized (Moratto et al., 2010). The stereo-matching algorithm of ASP generates a point cloud that has to be aligned to a fixed reference surface in order to have absolute height values in the final product. To this aim, we aligned the point cloud to the MOLA (Mars Orbiter Laser Altimeter, onboard Mars Global Surveyor) PEDR (Precision Experiment Data Records) that are the actual laser shots on the Martian surface. The point cloud was then interpolated according to Moratto et al. (2010) using a 3× original ground pixel resolution of the images. The obtained DEM has 18 m post-spacing and a vertical estimated precision of 2 meters, obtained with the USGS Pilot tool (Figure 8.2b and c).

For the calculation of the volumes of the Corona lava tube collapses we used a raster DEM interpolated at 5 m resolution based on the 2.5 m LiDAR (light detection and ranging) acquisitions provided by the CNIG (Centro Nacional de Informaciòn Geogràfica; Figure 8.3a).

8.3 Volumes Extraction

8.3.1 Carbonate Rock Slabs

The analysis of the CLSM scans of the rock plates before aging showed that the height of the metal surface always resulted 10–20 microns higher than the stone surface, probably due to the hardness contrast of the two materials.

Moreover, since the height values in the scans are related to a fixed internal horizontal reference in the CLSM, the same sample, re-scanned after aging, suffers errors due

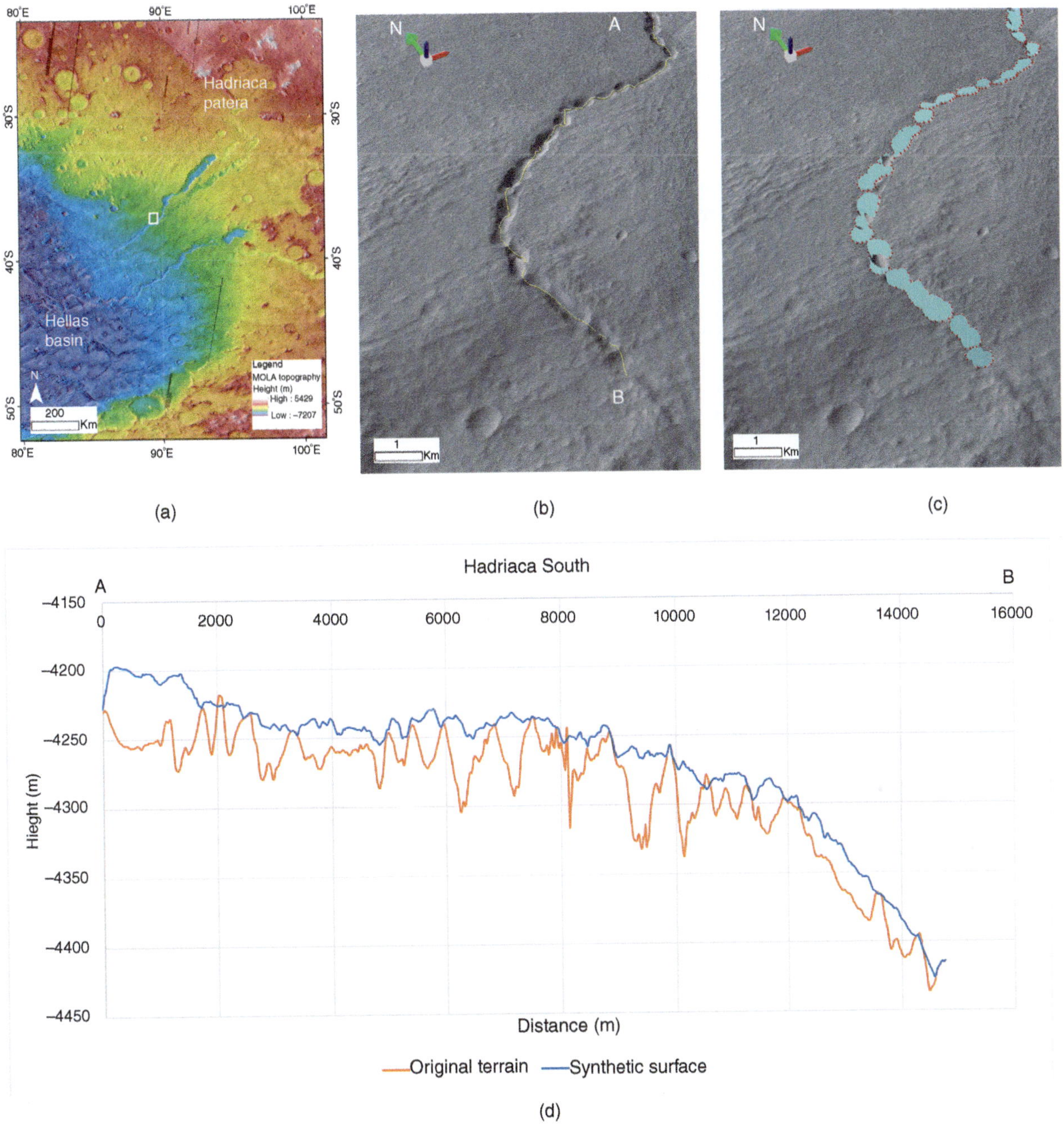

Figure 8.2 In a) the location of the studied Hadriaca lava tube section, on the borders of eastern Hellas Basin and located on the edge of a large lava channel departing from the north-eastern volcanic edifice. MOLA DTM and THEMIS daytime IR global mosaic in transparency. In b) a perspective view of the CTX orthoimage draped onto the DTM with 3x vertical exaggeration of the sinuously aligned collapses. The yellow line marks the topographic profile traced in d). In c) it is visible how we contoured the collapses and filled the voids by triangulating the nodes on the borders reconstructing the pristine topography marked in cyan. In d) the synthetic surface and the original terrain are shown. These are used in the calculations to create a finite volume taking into account the slope and roughness of the terrain.

to hand-positioning both in terms of distance from the reference plane and tilting in x, y, z, so to get absolute recession values we needed to build a reference synthetic surface attached to the sample itself and normalize each calculation to it. We chose to anchor the reference surface to the metal plate attached to the sample that is not affected by the acidic immersions and remains virtually unchanged through the aging tests.

In this way, the stone recession was calculated as the variation in the relative distance of the stone surface from

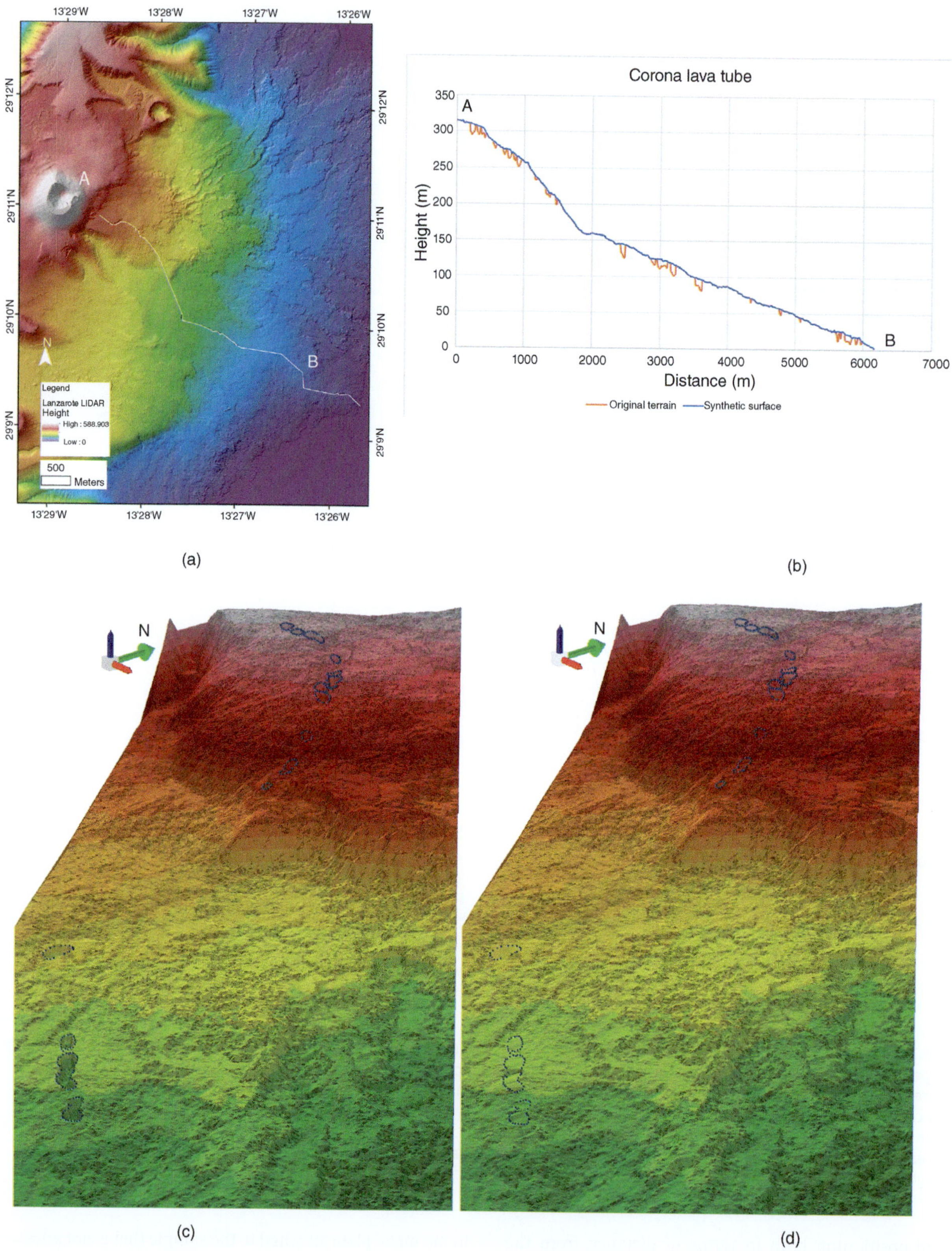

Figure 8.3 in a) the 5m LIDAR topography and highlighted the profile trace of the 8 km-long lava tube departing from the eastern Corona volcano. The white line traces the path of the lava tube and marks the topographic profile shown in d). In b) a perspective view of the LIDAR with 3x vertical exaggeration showing some of the collapses contoured by the nodes used for triangulating the synthetic surface. In c) the LIDAR with the voids filled by the synthetic surface, best approximating the actual topography. In d) the actual topographic profile in orange and in blue the topography that takes into account the synthetic surfaces filling the voids.

the reference surface. To compensate for the initial difference in height (depending on the sample, 10–20 μm from the metal plate to which we anchored the synthetic reference surface), we subtracted it from each recession measurement.

Two difficulties arose in placing the reference surfaces: (1) considering that CLSM 3-D elevation maps only occupied a fraction of the total stone surface, the reference surface could only be anchored on one of the four flat metal surfaces attached to the plate and (2) sample surfaces from different acquisitions (e.g. after different numbers of aging cycles) are not coplanar and are generally not perfectly horizontal. Therefore, a reference surface was constructed for each of the four CLSM 3D elevation maps (e.g. before the aging test and after 54, 140, and 240 cycles), each anchored to the same points of the flat metal surface, and extended over the sample, parallel to the stone surface, assuming that recession equally operated over the whole stone surface (see Figure 8.4). The original sample surface and the same surface after 240 cycles are shown in Figure 8.1. This was accomplished by fitting the absolute elevation values of the sample surface along two profiles parallel to the metal strip, one close to the metal and the other to the opposite side, and calculating the corresponding absolute elevation of the reference surface in such a way that this resulted in contact with the metal strip while remaining parallel to the stone surface. The average elevation difference between the sample surface and the reference surface was then calculated within ROIs (regions of interest) of aged samples, and the recession was obtained from this value by subtracting the elevation difference of the same ROI before aging. The average surface recession after a certain number of cycles was calculated by selecting an ROI as large as possible from the CLSM 3-D elevation maps. To investigate the effect of grain size on recession, different ROIs were selected on regions of the sample surface characterized by different grain sizes. Reliability of this approach was tested by observing whether ROIs located in different parts of the sample surface, but characterized by the same grain size of the calcite, provided the same values of recession. Stone recession values obtained can be expressed either as linear recession (perpendicular to the stone surface) or as volume loss, multiplying this value by a reference area (see Figure 8.7 later in Section 8.4).

8.3.2 Mars and Earth

Since our work is dealing with collapse pits and depressions that are very elongated and of huge size (up to ~1 km in length and 500 m in width for a single collapse in the case of Hadriaca Patera), we had to perform several preprocessing procedures on the DEM before getting to actual volume calculations.

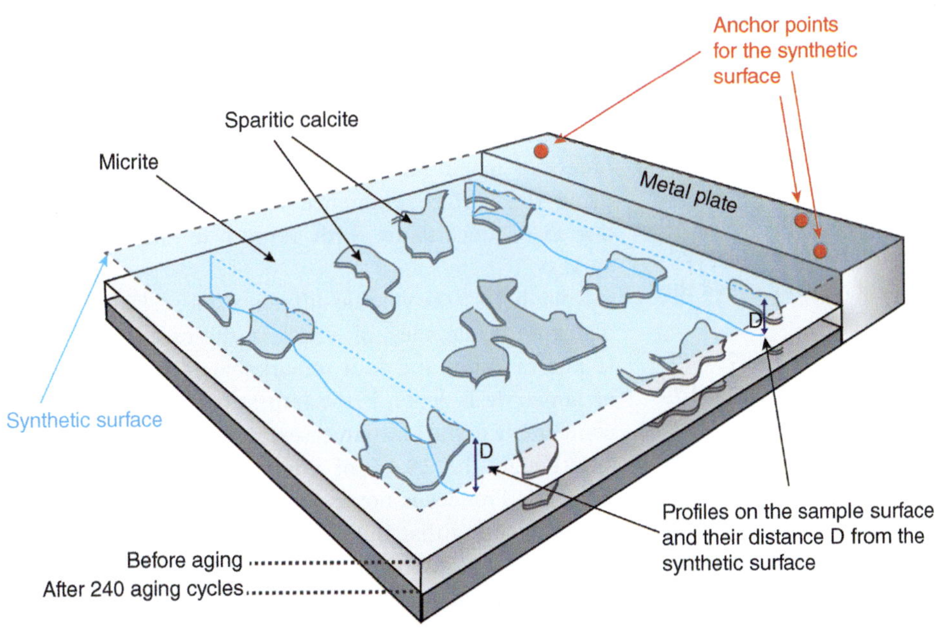

Figure 8.4 schematic representation of the process of creating a reference synthetic surface attached to the metal plate where the carbonate rock samples are attached to, and the carbonate rock recession after 240 aging cycles. Being the metal plate not affected by accelerated aging tests, this provides an internal constrain used to build a reference system to calculate the recession rate of the rock. The profiles extracted on the scan of the sample (presenting spiritic calcite, micrite) is projected on the reference synthetic surface and normalized taking into account its difference D from its height, considered as absolute. This is calculated taking into account its inclination in 3D space by using the three points of anchor to the metal plate.

Indeed, all these collapses are aligned and, as aforementioned, very elongated; thus, a simple volume calculation below an average height referred to the ground was not possible due to the ground slope steepness.

To overcome the problem of steepness and roughness of the overall topography, we made use of a synthetic surface attached to the borders of the collapse similar to the approach adopted for creating the reference surfaces for recession estimates on stone plates. The obtained uncollapsed DEM is the only way to approximate the pre-existing topography and to allow an estimation of the *true* volume of the depressions.

In order to obtain measurements of the uncollapsed surface, we had to locate precisely the edges of the collapses in order not to under- or overestimate the final volumes.

As a first step, we precisely located the edges of the collapses by calculating the maximum profile curvature on the DEM, highlighting the maximum change in slope (i.e. the highest gradient) along the slope direction (Wood, 1996). We contoured and digitized the edges and overlapped them onto the CTX ortho-image at 6 m/pixel for a visual check (Figure 8.2c). We then placed a node approximately every 50 meters, extracted the height of each one, and using them as nodes proceeded with the creation of the synthetic surface through triangulation. Along with that, we also extracted the TIN (Triangulated Irregular Network) from the original raster DEM by converting the height value of each cell and their spacing into a triangular mesh. We substituted the collapsed depressions on the original DEM with the synthetic surfaces filling the collapses, obtaining a mesh of a pre-collapse pristine surface.

In order to calculate the volume of the depressions we then compared the two meshes together by means of a dedicated tool in the 3D Analyst toolbox of ArcGIS, called the *surface difference*.

This algorithm is able to calculate the volume encompassed between two surfaces by means of a geometric comparison of the triangles of both of them. One of these surfaces is used as input (the original DEM) and another one as reference (the synthetic uncollapsed surface). The triangles of the input meshed surface (i.e. the original DEM) are compared with those of the reference surface and categorized as a function of their position, which can be above, below, or co-planar to the uncollapsed surface. Groups of contiguous triangles of the same class are polygonized and the volume of each polygon contributes to the total volume estimate. The resulting output contains a polygon feature class that contours what is above or below the reference surface providing the relative volumes in cubic meters. The areal sectors where the meshed surfaces (original DEM and uncollapsed surface) are coincident have output volumes equal to 0. The entire process is displayed in Figure 8.5.

All this information and the python code are usable via the *arcpy* module contained in https://pro.arcgis.com/en/pro-app/tool-reference/3d-analyst/surface-difference.htm#L_ and references therein.

We were therefore able to extract the volumes of every pit and collapse as well as the minimum bounding geometry (rectangular convex hull, using the *minimum bounding geometry* ArcGIS tool), deriving the length of the major and minor axes (corresponding to the width of the collapse). The results are shown in Table 8.1.

8.3.3 Validation of Volume Extraction

In order to validate these measurements, we compared those extracted in the Corona lava tube with high-resolution drone photogrammetry (flight height: 20 m; ground resolution: ~2 cm) and TLS (Terrestrial Laser Scanner) surveys of three of the accessible collapses (Santagata et al., 2018). The collapses range from small ones to the largest end-members: Jameo de Prendes of 36 × 28 m; Jameo de la Puerta Falsa of 116 × 44 m, and Jameo de la Jente of 178 × 50 m (the Jameo de la Jente collapse is shown in Figure 8.6).

In Table 8.2 we show how the performance of our method compares to these high-end survey techniques. Given that we are comparing volumes calculated with centimetric precision with those extracted from datasets with 5 m/pixel of DEM grid size (therefore of three orders of magnitude higher), all the volumes extracted using the surface difference technique have a certain degree of uncertainty with respect to the TLS+drone surveys.

Nonetheless, the volume differences of 18.4% in the case of Jameo de Prendes, of ~5.5% in the case of Jameo de la Puerta Falsa, and to a remarkable ~2.0% in the case of Jameo de la Jente, is the largest one and hence more comparable to the Martian cases.

Although the 18.4% of volume underestimation for Jameo de Prendes is not as low as the other two cases, it is worth noting that this is one of the smallest collapses where the measurement approximation due to the large grid size is greater.

Since, as shown in Table 8.1, all the minimum axis of the collapsed lava tubes sections on Hadriaca Patera are from 2 to 3 times larger than the largest on Corona, we are confident that the *surface difference* extraction method can provide a reliable estimate of the actual collapse volumes on Mars.

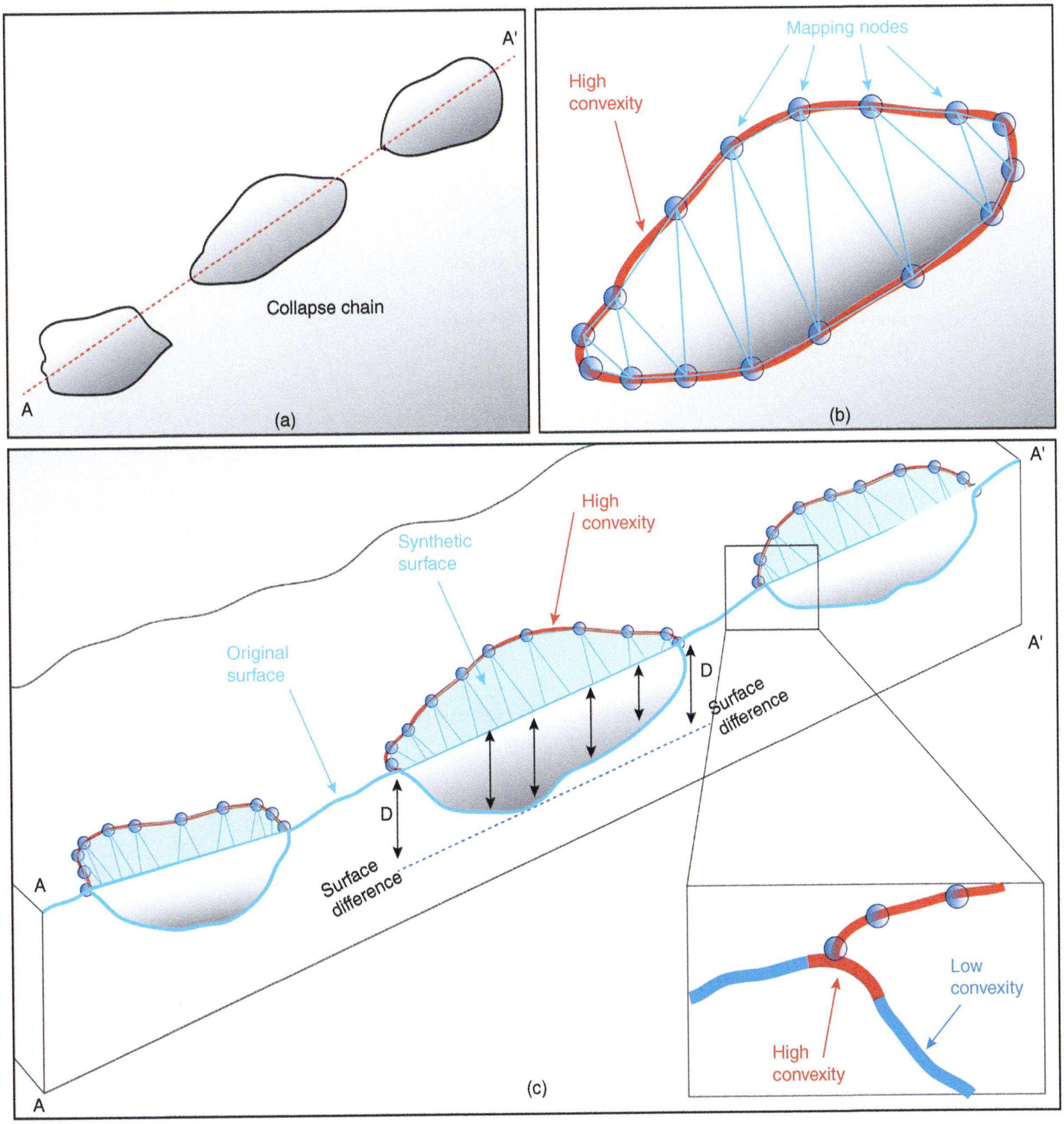

Figure 8.5 detail of the process used to create the synthetic surfaces that approximate the pristine topography before the collapse. In b) the collapses are digitized, then triangulated and merged with the topography. In c) along the cross section A-A' it is shown the difference between the synthetic surface and the actual topography calculated with the surface difference tool gives a volume estimation of the collapses. In the box in c) it is shown how we used the profile curvature, with the most convex part highlighted in red, in order to map more accurately the collapses.

8.4 Results and Discussion

In this work we tested the effectiveness of GIS-based techniques in retrieving volumetric information on both micro-DEMs from CLSM scans of laboratory rock samples and DEM from remote sensing on Earth and Mars. In both cases, we needed to constrain synthetic reference surfaces to the DEMs in order to obtain a fixed reference system constrained to the model itself and use that as a fixed reference for all the subsequent measurements.

In the case of the laboratory carbonate rock plates, this technique allowed coping with two main problems encountered: (1) to refer all the heights of the micro-DEM to the synthetic surface in order to eliminate the effect of different

Table 8.1 Width and volumes derived with the hole filling technique in each of the Corona lava tube and Hadriaca Patera collapses.

EARTH (Corona lava tube)		MARS (Hadriaca Patera lava tube)	
Width (m)	*Volume (cubic meters)*	*Width (m)*	*Volume (cubic meters)*
49	22019	488	17222086
40	8832	358	2960634
45	11928	431	3710842
36	4570	412	3842495
45	7718	310	803660
55	14005	366	1739869
25	247	411	4704175
46	6034	424	3469816
52	14921	447	5042341
37	2372	501	6509060
41	5693	407	5614545
29	2337	392	2603571
50	27316	378	5884632
33	7046	309	1920688
46	28873	296	635899
43	20041		
44	25038		
17	626		
29	2995		
20	515		
33	5913		
31	2579		
45	20394		
38	4030		
22	896		

distances from the fixed reference plane in the CLSM and (2) different orientations in x, y, z of the rock sample in subsequent CLSM scans, both caused by hand-positioning of the sample in the instrument.

More in detail, in the case of carbonate rocks subjected to 240 immersion cycles corresponding to 1,000 mm of rainfall, this morphometric approach allowed recognizing the control of grain size of calcite crystals (e.g. fine micrite, coarse micrite, dense calcitic bioclast structures, and sparite) on the recession rate. The different parts of the stone surface characterized by fine-grained calcite systematically registered greater recession after 240 aging cycles compared to those characterized by coarse-grained calcite. The same recession was measured in different samples on areas bearing the same grain size of calcite crystals (Figure 8.7). Furthermore, different recession values were measured in areas characterized by the same grain size of the calcite crystals after the same number of aging tests, but using either of the two types of water corresponding to rainwater of Bologna and Stresa respectively (Figure 8.7), congruently with their respective difference in pH (Salvini, 2018). Besides the information on the control of calcite grain size on the recession rate, these data confirm the potentiality of the approach described here to the analysis of topographic features with micrometric resolution.

On the planetary scale, synthetic surfaces constrained to DEMs were also used aiming to reconstruct the pristine topography before the collapse of lava tubes sections, in order to have a realistic estimation of the total volume of the voids. These synthetic surfaces were obtained by triangulating equally spaced nodes mapped on the edges of the collapses, generating a topography that accounted for the slopes and roughness of the surroundings. This served as a reference to calculate the volume encompassed between this and the underlying collapse on the DEMs.

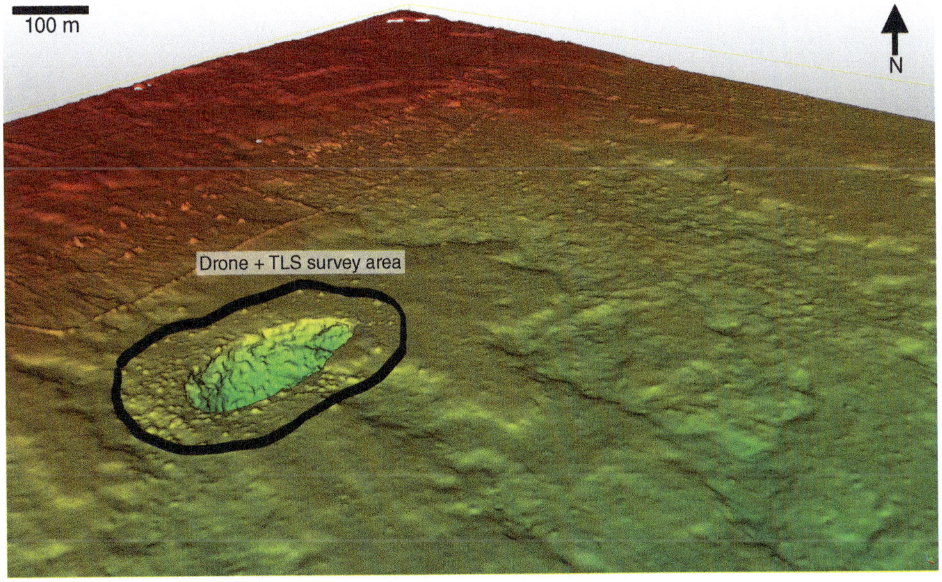

Figure 8.6 Example of dataset used for validation of the surface difference technique in the case of collapsed sections of lava tubes. The black circle highlights the area surveyed by TLS and drone at 2 cm resolution, that is merged with airborne LIDAR at gridded at 2.5 m. This dataset was compared with an airborne LIDAR at 5 m of grid size.

Table 8.2 Comparison of the volume values obtained by surface difference and those obtained by TLS+ drone surveys.The last column show the percentage of volume underestimation caused by the 5 m grid size of the DEM used to analyze all the 25 collapses with the 2 cm grid size of the high-resolution topographic surveys. It is straightforward that with the increase of the size of the collapse in size the volume difference decreases.

Name of the collapses (Jameos)	Surface Difference (m^3)	TLS+Drone photogrammetry (m^3)	Collapse size in plan view (m)	Volume difference (m^3)	Volume difference (%)
Prendes	2337	2865	36x28	528	18.4%
Puerta Falsa	25038	26493	116x44	1455	5.5%
Jente	27316	27894	178x50	578	2.0%

This extraction technique is validated for Earth (and therefore also for Mars, given the largest size of the collapses) by comparison with TLS and drone surveys on Corona collapses with centimetric resolution. The volumes of the collapsed lava tubes sections for both Earth and Mars are visible in Table 8.1 and the resulting datasets are plotted in log/log (Figure 8.8): the width throughout both datasets is almost constant, as expected, whereas the volumes are very variable, depending on the length of the collapsed sections. However, more interestingly, by this analysis we could infer that the collapses (and thus most likely the related subsurface voids) on Hadriaca Patera are at least one order of magnitude larger than those of the Corona lava tube, whose conduit size is one of the largest known on Earth. This numerically confirms what has been hypothesized by Cushing (2012) and is in agreement with the values calculated with morphometric methods in Sauro et al. (2020). Moreover, these results were indeed expected since in planetary bodies with lower gravity larger effusion rates can be even some orders of magnitude larger than on Earth (Gregg et al., 1996), thus facilitating the formation of larger conduits.

8.5 Conclusions

In this work we have shown that by applying a unified technique based on DEM and GIS analysis it is possible to derive volumes and use them to infer different processes at scales that differ in several orders of magnitude: from the microscale of laboratory samples dealing with cultural heritage to the macro scale for geomorphological

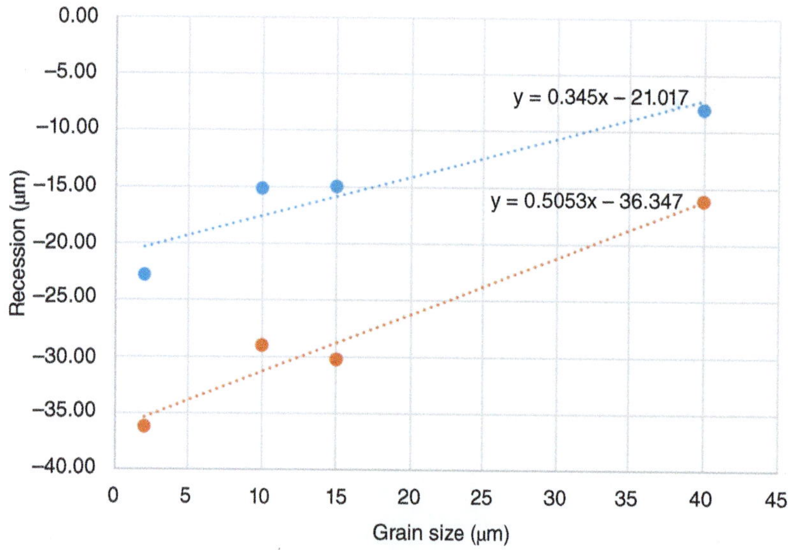

Figure 8.7 Correlation between the average grain size of the different textural elements (from lower to higher grain size: micrite, coarse-grained micrite, calcite in bioclasts, and sparite) of the considered 11 samples and the relative average recession value measured after 240 aging cycles under the Stresa and Bologna waters, respectively.

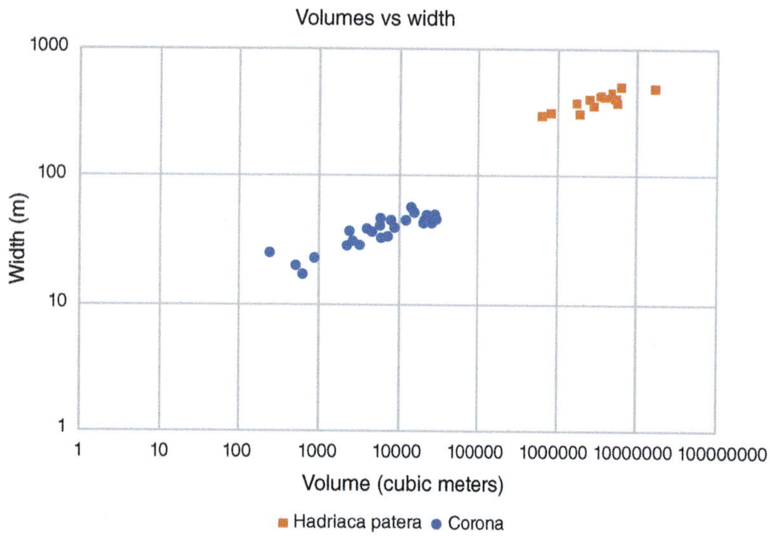

Figure 8.8 Plot showing in log/log the Volumes vs width of the Martian and Earth lava tubes with each point marks a single collapse. It is clearly visible that all the martian collapses are 1 order of magnitude larger than those on Earth, being the Corona lava tube the largest known in terms of volume of the subsurface void.

studies of planetary bodies. We were able to import in ArcGIS environment the CLSM scans from carbonate rocks subjected to cycles of aging and estimate the recession rate according to the calcite grain size, which appears consistent throughout different parts of the samples (see Figure 8.7).

In the case of planetary bodies, we dealt with the estimate of the volumes of collapsed sections of lava tubes on Hadriaca Patera on Mars and on the Corona lava tube as the Earth analog, also for data extraction validation. We reconstructed the best approximation of the pristine ground morphology before the collapse happened and we were able to take into account the slope and the expected roughness of the pre-collapse terrain and provide a realistic value of the volumes of the voids. By comparing the obtained results, it was possible to evaluate the difference in magnitude of the collapses on Mars and Earth, being one order of magnitude larger on the former than on the latter, both considering the volume and width. This confirms previous hypotheses (Sauro et al., 2020; Cushing, 2012) on the presence of very large subsurface conduits on the red planet.

References

Allred, K. and Allred, C. (1997). Development and morphology of Kazumura Cave, Hawaii. *Journal of Cave and Karst Studies* 59 (2): 67–80.

Bleacher, J.E., Greeley, R., Williams, D.A., and Neukum, G., and HRSC Co-investigator Team 411 (2006). Comparison of effusive volcanism at Olympus, Arsia, Pavonis, and Ascraeus Montes, 412 Mars from lava flow mapping using Mars Express HRSC data. In *37th Lunar and Planetary 413 Science Conference*.

Bonazza, A., Messina, P., Sabbioni, C. et al. (2009a). Mapping the impact of climate change on surface recession of carbonate buildings in Europe. *Science of the Total Environment* 407: 2039–2050. https://doi.org/10.1016/j.scitotenv.2008.10.067.

Bonazza, A., Sabbioni, C., Messina, P. et al. (2009b). Climate change impact: Mapping thermal stress on Carrara marble in Europe. *Science of the Total Environment* 407: 4506–4512. https://doi.org/10.1016/j.scitotenv.2009.04.008.

Brimblecombe, P. and Grossi, C.M. (2008). Millennium long recession of limestones facades in London. *Environmental Geology* 56: 463–471. https://doi.org/10.1007/s00254-008-1465-z.

Brimblecombe, P. and Grossi, C.M. (2009). Millennium long damage to building materials in London. *Science of the Total Environment* 407: 1354–1361. https://doi.org/10.1016/j.scitotenv.2008.09.037.

Calvari, S. and Pinkerton, H. (1999). Lava tube morphology on Etna and evidence for lava flow emplacement mechanisms. *Journal of Volcanology and Geothermal Research* 90 (3–4): 263–280. https://doi.org/10.1016/S0377-0273(99)00024-4.

Carracedo, J., Singer, B., Jicha, B. et al. (2003). La Erupcion y el Tubo Volcanico Del Volcan Corona. *Estud. Geológicos* 59: 277–302. https://doi.org/10.3989/egeol.03595-6104.

Cushing, G.E. (2012). Candidate cave entrances on Mars. *Journal of Cave and Karst Studies* 74 (1): 33–47. https://doi.org/10.4311/2010EX0167R.

Cushing, G.E., Titus, T.N., Wynne, J.J., and Christensen, P.R. (2007). THEMIS observes possible cave skylights on Mars. *Geophysical Research Letters* 34 (17): https://doi.org/10.1029/2007GL030709.

Dragoni, M., Piombo, A., and Tallarico, A. (1995). A model for the formation of lava tubes by roofing over a channel. *J. Geophys. Res.* 100: 8435–8447.

Dragovich, D. (1991). Marble weathering in an Industrial Environment, Eastern Australia. *Environmental Geology and Water Sciences* 17: 127–132. https://doi.org/10.1007/BF01701568.

Feddema, J. and Meierding, T. (1987). Marble weathering and air pollution in Philadelphia. *Atmospheric Environment* 21: 143–157. https://doi.org/10.1016/0004-6981(87)90279-4.

Ficker, T. and Martišek, D. (2012). Digital fracture surfaces and their roughness analysis: Applications to cement-based materials. *Cement and Concrete Research* 42 (6): 827–833. https://doi.org/10.1016/j.cemconres.2012.03.007.

Gaddis, L., Anderson, J., Becker, K. et al. (1997). An overview of the integrated software for imaging spectrometers (ISIS). In: *Lunar and Planetary Science Conference*, vol. 28, 387.

Greeley, R. (1971a). Observations of actively forming lava tubes and associated structures. *Hawaii. Modern Geology* 2: 207–223.

Greeley, R. (1971b). Lava tubes and channels in the Lunar Marius Hills. *Earth, Moon, and Planets* 3 (3): 289–314. https://doi.org/10.1007/BF00561842.

Greeley, R. and Spudis, P.D. (1981). Volcanism on Mars. *Reviews of Geophysics* 19 (1): 13–41. https://doi.org/10.1029/RG019i001p00013.

Greeley, R., Fagents, S.A., Harris, R.S. et al. (1998). Erosion by flowing lava: Field evidence. *Journal of Geophysical Research: Solid Earth* 103 (B11): 27325–27345. https://doi.org/10.1029/97JB03543.

Gregg, T.K.P. and Fink, J.H. (1996). Quantification of extraterrestrial lava flow effusion rates through laboratory simulations. *Journal of Geophysical Research* 101 (E7): 16891–16900. https://doi.org/10.1029/96JE01254.

Grohmann, C.H. and Sawakuchi, A.O. (2013). Influence of cell size on volume calculation using digital terrain models: A case of coastal dune fields. *Geomorphology* 180–181: 130–136. https://doi.org/10.1016/j.geomorph.2012.09.012.

Grossi, C.M., Bonazza, A., Brimblecombe, P. et al. (2008). Predicting twenty-first century recession of architectural limestone in European cities. *Environmental Geology* 56: 455–461. https://doi.org/10.1007/s00254-008-1442-6.

Haruyama, J., Morota, T., Kobayashi, S. et al. (2012). Lunar holes and lava tubes as resources for lunar science and exploration. In: *Moon: Prospective Energy and Material Resources* (ed. V. Badescu), 139–163. Berlin, Heidelberg: Springer https://doi.org/10.1007/978-3-642-27969-0_6.

Hodges, C.A. and Moore, H.J. (1994). *Atlas of Volcanic Landforms on Mars. US Geological Survey Professional Paper 1534*, 194. DC: Washington.

Inkpen, R.J., Viles, H., Moses, C., and Baily, B. (2012a). Modelling the impact of changing atmospheric pollution levels on limestone erosion rates in central London, 1980–2010. *Atmospheric Environment* 61: 476–481. https://doi.org/10.1016/j.atmosenv.2012.07.042.

Inkpen, R.J., Viles, H., Moses, C. et al. (2012b). Thirty years of erosion and declining atmospheric pollution at St Paul's Cathedral, London. *Atmospheric Environment* 62: 521–529. https://doi.org/10.1016/j.atmosenv.2012.08.055.

Isler, O. (1986). 1986 International expedition to the Tunel de la Atlantida_O. Isler.PDF. In: *Caves and Caving*, 45.

Jakobsson, M., Björck, S., Alm, G. et al. (2007). Reconstructing the Younger Dryas ice dammed lake in the Baltic Basin: Bathymetry, area and volume. *Global and Planetary Change* 57 (3–4): 355–370. https://doi.org/10.1016/j.gloplacha.2007.01.006.

Jalón, M.L., Chiachío, J., Gil-Martín, L.M., and Hernández-Montes, E. (2021). Probabilistic identification of surface recession patterns in heritage buildings based on digital photogrammetry. *Journal of Building Engineering* 34, 101922: https://doi.org/10.1016/j.jobe.2020.101922.

Kalmár, J., Papp, G., and Szabó, T. (1995). DTM-based surface and volume approximation. Geophysical applications. *Computers and Geosciences* 21 (2): 245–257. https://doi.org/10.1016/0098-3004(94)00069-7.

Kempe, S. and Al-Malabeh, A. (2005). Newly discovered lava tunnels of the Al-Shaam plateau basalts, Jordan. *Geophysical Research Abstracts* 7, 03204.

Kempe, S., Bauer, I., Bosted, P. et al. (2010). Inflationary versus crusted-over roofs of pyroducts (lava tunnels). In: *Proceedings of the 14th International Symposium on Vulcanospeleology*, 12–17 August 2010,, 93–101. Australia: Undara.

Keszthelyi, L. and Self, S. (1998). Some physical requirements for the emplacement of long basaltic lava flows. *Journal of Geophysical Research* 103464: 447–527. https://doi.org/10.1029/98JB00606.

Keszthelyi, L., Jaeger, W.L., Milazzo, M.P., McEwen, A.S., and HiRISE Team (2007). High resolution imaging science experiment (HiRISE) images of volcanic terrains from the first 6 months of the Mars reconnaissance orbiter primary science phase. Seventh International Conference on Mars, Abstract #3314.

Kucera, V., Tidblad, J., Kreislova, K. et al. (2007). UN/ECE ICP Materials dose-response functions for the multi pollutant situation. *Water, Air, and Soil Pollution: Focus* 7: 249–258. https://doi.org/10.1007/s11267-006-9080-z.

Léveillé, R.J. and Datta, S. (2010). Lava tubes and basaltic caves as astrobiological targets on Earth and Mars: A review. *Planetary and Space Science* 58 (4): 592–598. https://doi.org/10.1016/j.pss.2009.06.004.

Lipfert, F.W. (1989). Atmospheric damage to calcareous stones: Comparison and reconciliation of recent experimental findings. *Atmospheric Environment* 23: 415–429. https://doi.org/10.1016/0004-6981(89)90587-8.

Malin, M.C., Bell, J.F., Cantor, B.A. et al. (2007). Context camera investigation on board the Mars Reconnaissance Orbiter. *Journal of Geophysical Research* 112, E05S04.: https://doi.org/10.1029/2006JE002808.

Martha, T.R., Kerle, N., Jetten, V. et al. (2010). Landslide volumetric analysis using cartosat-1-derived DEMs. *IEEE Geoscience and Remote Sensing Letters* 7 (3): 582–586. https://doi.org/10.1109/LGRS.2010.2041895.

Meierding, T.C. (1981). Marble tombstone weathering rates: A transect of the United States. *Physical Geography* 2: 1–18. https://doi.org/10.1080/02723646.1981.10642201.

Meyzen, C.M., Massironi, M., Pozzobon, R., and Zilio, L.D. (2015). Are terrestrial plumes from motionless plates analogues to Martian plumes feeding the giant shield volcanoes? *Geological Society London, Special Publication* 401: 107–126. https://doi.org/10.1144/SP401.8.

Moratto, S.Z.M., Broxton, M.J., Beyer, R.A. et al. (2010). Ames Stereo Pipeline, NASA's Open Source Automated Stereogrammetry. In: *41st Lunar and Planetary Science Conference*. no.2364.

MULTI-ASSESS Report (2007). Model for multi-pollutant impact and assessment of threshold levels for Cultural Heritage. Deliverable 02.

Nichols, A.B. and Lange, D.A. (2006). 3D surface image analysis for fracture modeling of cement-based materials. *Cement and Concrete Research* 36 (6): 1098–1107. https://doi.org/10.1016/j.cemconres.2006.01.002.

Oberbeck, V.R., Quaide, W.L., and Greeley, R. (1969). On the origin of Lunar sinuous rilles. *Modern Geology* 1: 75–80.

Panettiere, P., Cortecci, G., Dinelli, E. et al. (2000). Chemistry and sulfur isotopic composition of precipitation at Bologna, Italy. *Applied Geochemistry* 15: 1455–1467. https://doi.org/10.1016/S0883-2927(00)00012-3.

Roberts, S.M. (2005). Surface-recession weathering of marble tombstones: New field data and constraints. *GSA Special Paper* 390: 27–37. https://doi.org/10.1130/0-8137-2390-6.27.

Rogora, M., Mosello, R., and Marchetto, A. (2004). Long-term trends in the chemistry of atmospheric deposition in Northwestern Italy: The role of increasing Saharan dust deposition. *Tellus* 56B: 426–434. https://doi.org/10.3402/tellusb.v56i5.16456.

Sabbioni, C., Brimblecombe, P., and Cassar, M. (ed.) (2010). *The Atlas of Climate Change Impact on European Cultural Heritage*. London, UK: Anthem Press.

Salvini, S. (2018). Deterioration of carbonate rocks and vulnerability of cultural heritage in a changing climate. PhD thesis, University of Padova.

Santagata, T., Sauro, F., Massironi, M. et al. (2018). Subsurface laser scanning and photogrammetry in the Corona Lava Tube System, Lanzarote, Spain, vol. 20, EGU2018- 5290. In: *EGU General Assembly 2018*.

Sauro, F., Pozzobon, R., Santagata, T. et al. (2019). Volcanic caves of Lanzarote: A Natural Laboratory for understanding Volcano-Speleogenetic processes and planetary caves. In: *Lanzarote and Chinijo Islands Geopark: From Earth to Space* (ed. E. Mateo, J. Martínez-Frías and

J. Vegas), 125–142. Cham: Springer International Publishing. https://doi.org/10.1007/978-3-030-13130-2_9.

Sauro, F., Pozzobon, R., Massironi, M. et al. (2020). Lava tubes on Earth, Moon and Mars: A review on their size and morphology revealed by comparative planetology. *Earth-Science Reviews* 209 (July): 103288. https://doi.org/10.1016/j.earscirev.2020.103288.

Searle, D.E. and Mitchell, D.J. (2006). The effect of coal and diesel particulates on the weathering loss of Portland Limestone in an urban environment. *Science of the Total Environment* 370: 207–223. https://doi.org/ https://doi.org/10.1016/j.scitotenv.2006.07.005.

Self, S., Thordarson, T., Keszthelyi, L. et al. (1996). Inflated pahoehoe lava flow fields: Frozen lava flow d I. *Geophysical Research Letters* 23 (19): 2689–2692. https://doi.org/10.1029/96GL02450.

Theinat, A.K., Modiriasari, A., Bobet, A. et al. (2020). Lunar lava tubes: Morphology to structural stability. *Icarus* 338: 113442. https://doi.org/10.1016/j.icarus.2019.113442.

Tidblad, J., Kucera, V., Mikhailov, A. et al. (2001). UN ECE ICP materials: Dose-response functions on dry and wet acid deposition effects over 8 years of exposure. *Water, Air, and Soil Pollution* 130: 1457–1462. https://doi.org/10.1007/978-1-4020-5885-1_27.

Trudgill, S.T., Viles, H.A., Inkpen, R.J., and Cooke, R.U. (1989). Remeasurement of weathering rates, St Paul's Cathedral, London. *Earth Surface Processes and Landforms* 14: 175–196. https://doi.org/10.1002/esp.3290140302.

Tseng, C.M., Lin, C.W., Stark, C.P. et al. (2013). Application of a multi-temporal, LiDAR-derived, digital terrain model in a landslide-volume estimation. *Earth Surface Processes and Landforms* 38 (13): 1587–1601. https://doi.org/10.1002/esp.3454.

Wood, J. (1996). The geomorphological characterization of digital elevation models. PhD Thesis, University of Leicester, UK.

Wyrick, D., Ferrill, D.A., Morris, A.P. et al. (2004). Distribution, morphology, and origins of Martian pit crater chains. *Journal of Geophysical Research E: Planets* 109: E06005. https://doi.org/10.1029/2004JE002240.

9

Robust Detection of Circular Shapes on 3D Meshes Based on Discrete Curvatures: Application to Impact Craters Recognition

Jean-Luc Mari[1], Sophie Viseur[2], Sylvain Bouley[3], Martin-Pierre Schmidt[1], Jennifer Muscato[1], Florian Beguet[1], Sarah Bali[1], and Laurent Jorda[4]

[1]*Aix Marseille Univ., CNRS, LIS, Marseille, France*
[2]*Aix Marseille Univ., CNRS, IRD, INRAE, Coll France, CEREGE, Aix-en-Provence, France*
[3]*GEOPS – Géosciences Paris Saclay, Univ. Paris-Sud, CNRS, Université Paris-Saclay, Orsay, France*
[4]*Aix Marseille Univ., CNRS, CNES, LAM, Marseille, France*

Abstract

The robust extraction of feature areas on surface objects embedded in 3D, like circular shapes, is a challenging problem. Classical approaches generally rely on image processing and template matching on a 2D flat projection of the 3D object (for instance a high-resolution photograph). In this chapter, we propose two full 3D methods using curvature operators to extract these features. In the first approach, a tagging stage of regions of interest is performed over a 3D model of the object in order to extract the features. A single circular connected line is constructed, exclusively composed of edges of the initial mesh around each detected area. The second approach starts with the detection of pits. Then, "best fit circles" are found according to the rims of the circular shapes. We apply these approaches to the detection of craters on celestial bodies observed by space missions, here the asteroids Vesta and Lutetia.

9.1 Introduction

In recent years, astrophysicists have frequently used the field of digital imaging to characterize the properties of different celestial objects. This led to the development of many 2D methods. Today, thanks to embedded systems, we are able to exploit 3D models at high resolutions containing all information required for the detection of geomorphological features.

Most celestial bodies display impacts of collisions with asteroids and meteoroids. These traces are called craters. The possibility of observing and identifying these craters and their characteristics (radius, depth, and morphology) is the only method available to measure the age of different units at the surface of the body, which in turn allows its conditions of formation to be constrained (see Marchi et al., 2010, 2012a, 2012b; Fassett, 2016).

Interplanetary space probes always carry at least one imaging instrument on board. The visible images of the target are used to reconstruct high-resolution 3D models of its surface as a cloud of points in the case of multi-image dense stereo (cf. Gaskell et al., 2008), or as a triangular mesh in the case of stereo and shape-from-shading (cf. Preusker et al., 2012).

The goal of this work is to develop a methodology to automatically detect the craters lying on these 3D models. Different types of craters can be found on planetary surfaces (Melosh, 2011), but in this work we focus only on the detection of the craters' rims.

We have developed two variants based on discrete curvature computation.

The first method uses curvature estimations and vertices labelling to detect specific concave depressions on the surface. Concentric rings are then built around these depressions to find a circular rim using the curvature values of the points on the ring. This method has been tested on a 3D model of the asteroid Lutetia observed by the ROSETTA (ESA) space probe.

The second method is based on the detection of closed circular areas of crater edges. The deepest point gives the contour of the area and then a circle is fitted.

This chapter is divided as follows: in Section 9.2, the existing approaches related to crater detection are briefly described. In Section 3, we relate some basic notions and then develop the core of our approaches in Sections 4 and

3D Digital Geological Models: From Terrestrial Outcrops to Planetary Surfaces, First Edition.
Edited by Andrea Bistacchi, Matteo Massironi, and Sophie Viseur.

5. These sections also present the results we obtain and the validation of the two methods.

9.2 Related Work

Multiple techniques exist to solve the problem of crater detection. The most accurate method consists in manual detection directly based on surface images using Geographic Information System (GIS) software for localization. This well-established method offers the best detection rate as well as a near-zero false positive rate (Figure 9.1). However, having to manually point each crater on the image makes this technique highly time-consuming, tedious, and subjective (cf. Robbins et al., 2014).

Many alternate and faster approaches have been developed. Most of them are based on automatic detection by pattern recognition.

The method described by Pedrosa et al. (2012), developed to detect impact craters on Mars, uses mathematical morphology functions directly onto high-definition pictures of the surface. Afterward, a step using a template matching via Fast Fourier Transform is applied. This approach achieves a high rate of 90% effective detection, but produces 25% false positive detections.

A different approach (Sawabe et al., 2006) aims at using Hilditch's thinning algorithm (Hilditch et al., 1969) on a binarization of the initial image followed by a fuzzy Hough transform (Duda and Hart, 1972) to detect curves. This method achieves a rate of 80% effective detection on lunar images (Figure 9.2).

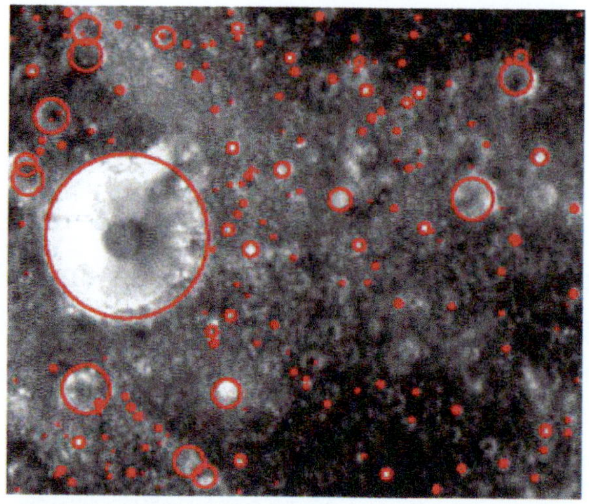

Figure 9.2 Results obtained on an image of the Moon with the method of Sawabe et al.(2006). *Source:* Sawabe Y., Matsunaga T., Rokugawa S. (2006), 37:1, 21–27 / With permission of Elsevier.

Many works have conceptualized and categorized the properties of *elementary forms* based on morphometric variables (Shary, 1995; Shary et al., 2002, 2005; Minár and Evans., 2008).

9.3 Basic Notions

9.3.1 Triangle Mesh

A triangle mesh is a discrete geometric structure to represent 3D objects by juxtapositions of triangular elements.

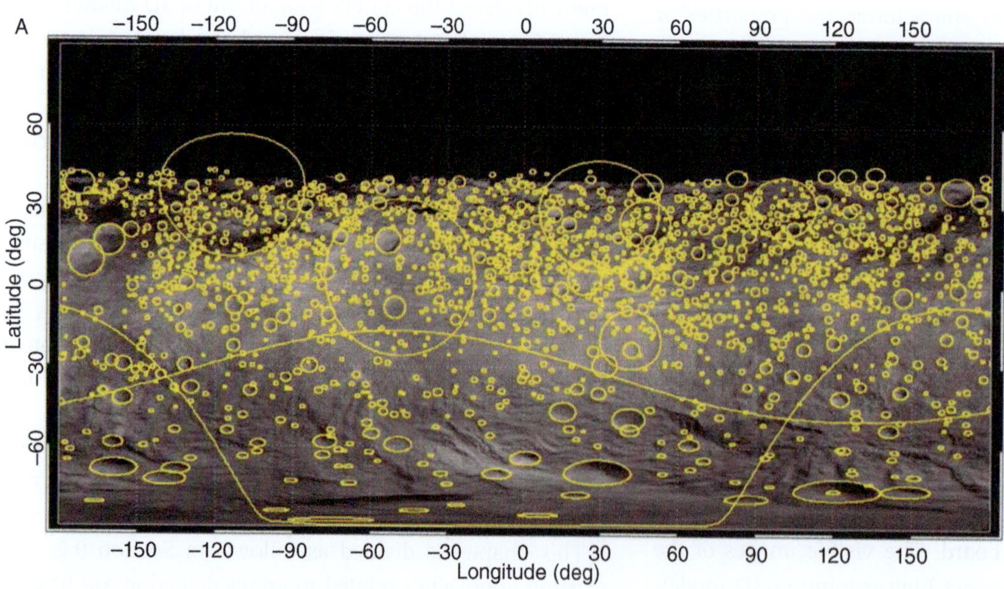

Figure 9.1 1872 craters with a diameter superior to 2 km were detected in the case of the asteroid Vesta. *Source:* Courtesy of Marchi S., et, al., (2012), 336:6082, 690–694/With Permission of AAAS.

Any polyhedron can be transformed into a triangle mesh without any loss of information simply by adding edges on the non-triangular faces. Each triangular face has a unique and easy-to-determine normal vector. Besides, the meshes we use present a homogeneous distribution of vertices onto the surface due to the constant sampling rate used during the reconstruction process.

9.3.2 Mesh Smoothing

It is possible to smooth a surface in order to minimize the noise induced by the data acquisition from the space probe. The Laplacian smooth (see Vollmer et al., 1999) consists in moving every vertex to the barycenter of its neighbors. This fast smoothing method tends to contract low definition objects, but it is very effective at removing noise on highly detailed models (hundreds of thousands of points).

9.3.3 Curvatures over a Surface

The curvature is a differential descriptor of the local appearance of a curve or a surface. This indicator is used extensively in mesh processing because it offers information on the concavity or convexity of a surface along its principal directions. This can be used in order to get a classification of the vertices lying on the discrete surface.

Let us denote K_1 and K_2 respectively as the maximal and minimal curvatures of a point P on a smooth surface along the two principal directions defined by T_1 and T_2 (Figure 9.3).

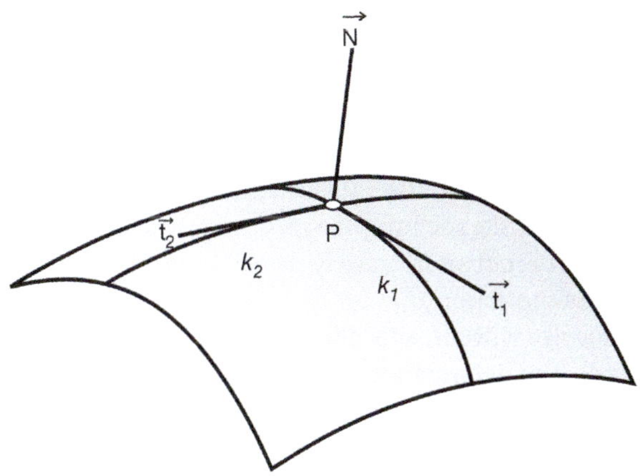

Figure 9.3 A representation of K_1, the maximal curvature, and K_2, the minimal curvature, on a point P lying on a surface.

We use the algebraic point set surfaces variant based on local fitting of algebraic spheres to estimate the two principal curvature values of each point on the surface (see Guennebaud and Gross, 2007; Guennebaud et al., 2008). We then define the mean curvature H and the Gaussian curvature K as the following:

$$H = \frac{K_1 + K_2}{2}$$

$$K = K_1 \cdot K_2$$

A crater, due to an impact that hits the surface at an angle higher than 15 degrees, is characterized by a pit surrounded by a circular rim. This is due to the isotropic diffusion of the kinetic energy of the impactor on the surface during the impact. Through the study of the mean and Gaussian curvatures of each vertex (cf. Haala et al., 2004), we can delimit pit areas as vertices, where $H > 0$ and $K > 0$, and rim areas as vertices, with $H < 0$ and $K \approx 0$ (Figure 9.4). However, this criterion does not take into account the possible overlapping of multiple craters.

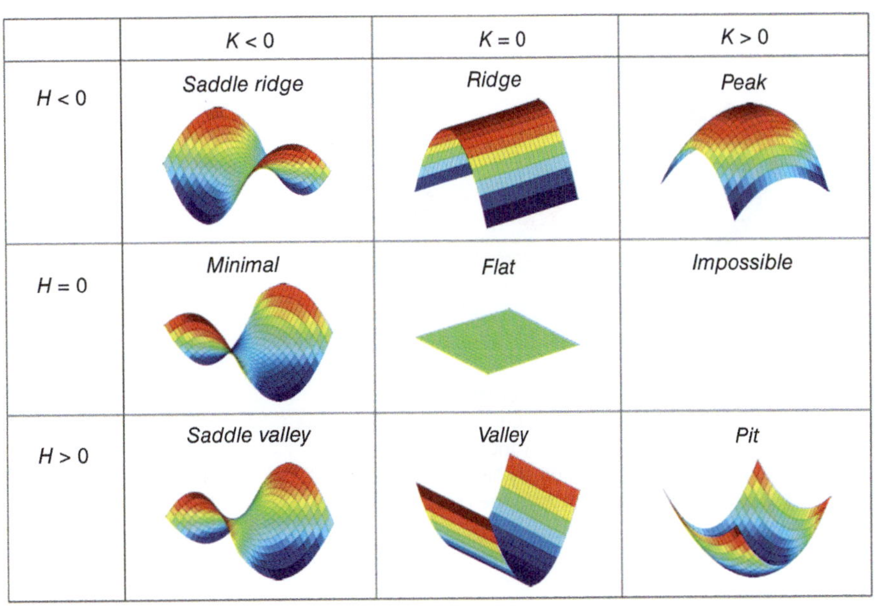

Figure 9.4 Using the signs of K and H, we can define eight types of fundamental shapes.

9.3.4 Levels of Detail

The estimation of the curvature on a discrete object can be obtained through approximation over a small portion of the surface. This can be made, for example, by local fitting of the osculating circle in the principal directions of this area. Therefore, the filter defining the portion of surface used for computing the curvature must be closely related to the level of detail of the object.

A filter size will correctly allow fitting of a limited range of shapes. If a deformation of the surface is smaller than the filter, it will be passed over and treated as noise. If a deformation of the surface is larger than the filter, it will appear flat.

On a high-resolution mesh, only the smaller craters will correctly depict the characteristics of our description of a crater (Figure 9.5). The idea is to run multiple passes of detection with different levels of detail. This can be done by heavily simplifying and smoothing the input mesh to a degree where the smaller details correspond to the concavity and convexity of the larger craters. Thus it allows the detection of these deformations (Figure 9.6).

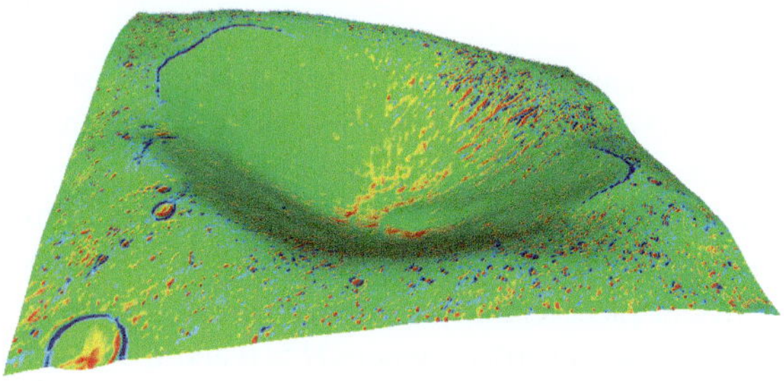

Figure 9.5 Visualization of the mean curvature on a high-resolution surface (6,000,000 faces). Convex areas are represented in blue, relatively flat areas are in green, and concave areas are in red. Small craters are visible on the periphery of the surface, but the curvature of the main crater in the center is not properly detected.

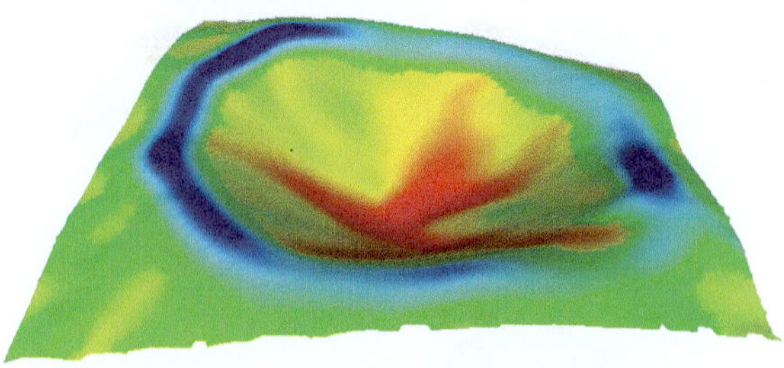

Figure 9.6 Visualization of the mean curvature on a low-resolution surface (10,000 faces). Convex areas are in blue, relatively flat areas are in green, and concave areas are in red. On this extremely simplified version of the surface, small craters are invisible, but the the main crater in the center is properly defined.

9.4 Approach Based on Ring Propagation

9.4.1 Overview

The proposed approach consists in two steps of data pre-processing:

- Adjustment of the level of detail,
- Computation of the curvatures,

followed by three steps of detection:

- Detection of concave areas,
- Selection of the probable centers,
- Expansion of multiple rings around each center,

resulting in a set of elementary cycles made of a subset of edges of the original mesh. We can then estimate the possibility of each ring lying on the periphery of a crater by the study of the curvature value, in order to consolidate their localization.

9.4.2 Seeds Search

The first step of our approach is to mark each vertex having positive Gaussian and mean curvatures, as well as being above a specific threshold (Figure 9.7). These concave parts of the mesh are thus transformed into a binary map. The thresholds vary according to multiple factors, including the level of detail of the mesh and the degree of erosion

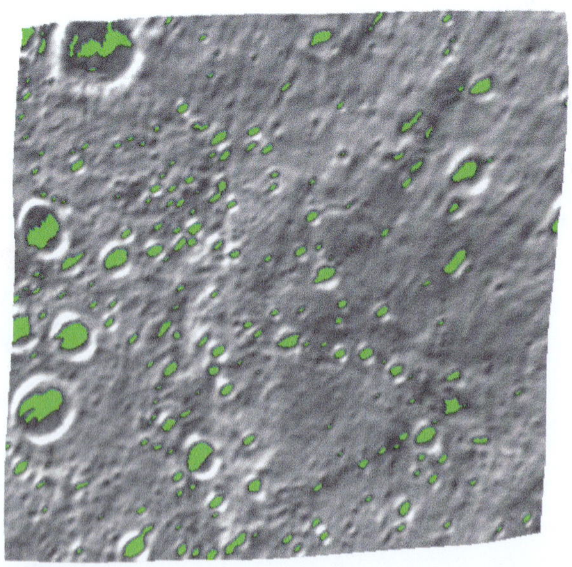

Figure 9.7 Visualization of the detection of concave areas. Each connected component displayed in green is considered as a candidate for the circular pit detection.

of the craters. Because it is difficult to change manually the threshold values for each mesh, a normalization of the curvature values of the whole mesh in the interval [0, 1] is performed.

In the second step, a process similar to the watershed algorithm (Beucher, 1982) expands these regions to *fill* the concave areas. This helps to improve the localization of the probable centers while lowering the false-positive count that would come out from a valley-shaped topography.

It is then necessary to give a distinct label to each connected component of this subset of vertices. This can be done by propagating labels through the edges of this subset.

For each connected component, we then compute the 3D coordinates of its barycenter. The points on the mesh whose coordinates are the closest to these barycenters are marked as the seeds for the next step of the detection.

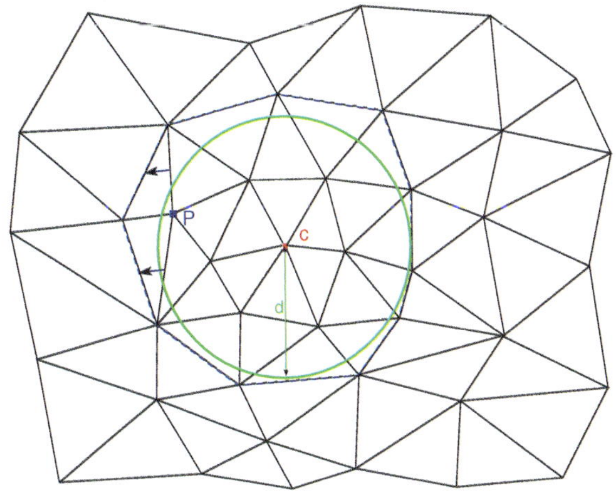

Figure 9.8 Illustration of the process of updating a point p with a distance to C less than d, resulting in a ring (blue dotted line) close to a circle centered on C with a radius of d composed exclusively of edges of the initial triangle mesh.

9.4.3 Ring Construction

Once the probable center of each crater is set, an initial ring is constructed with the immediate neighbors of these seed points (see Section 9.4.2). The next step is to stretch out the rings in all the directions at a constant rate. The objective is to maintain their circularity during the process.

If the center is properly set in the middle of a crater, then at one point the ring associated with this center matches the peripheral rim of this crater and therefore has the following characteristics:

- Low ring point to center distance variation.
- High mean curvature.
- Near-zero Gaussian curvature.

Because the curvature of each vertex of the mesh is pre-computed, the calculation of the mean and Gaussian curvatures of each ring is straightforward. Therefore, maintaining the circularity of the ring during its expansion is at the core of the developed algorithm (see Algorithm).

For that reason, we choose to represent the ring as a cyclic double-linked list of vertices on the mesh. This allows it to conveniently loop through each point of the ring while expanding it outward.

Let us denote d as an arbitrary distance to the probable center C of the ring and p as one of the points of the ring. If the distance between C and p is less than d, the ring is updated by replacing p with its neighbors exterior to the ring.

By repeating this process until no point remains to update, a circular ring is constructed, centered on C, with a radius equal to d (depending of the density of points on the mesh). This ring has the advantage of being exclusively composed of edges existing on the initial mesh (Figure 9.8).

By progressively increasing the distance d, we obtain multiple concentric rings around C. Through the study of the curvature values of the vertices attached to each ring, we can compute the likelihood of one of them being on the periphery of a crater (Figure 9.9).

It is important to note that, most of the time, only the first significant increase in the curvature value of a ring is the corroboration of a circular rim. For that reason, the algorithm developed uses the radius of the ring as a minimizing factor of the likelihood.

Let d_1 be the minimal search distance (i.e. the smallest crater radius allowed) of the current pass.

Let d_2 be the maximal search distance (i.e. the largest crater radius allowed) of the current pass.

Let n be the number of points of the current ring.

Let K_1^i be the maximal curvature of the point i of the ring.

Let d be the approximate radius of the current ring.

We define the *coefficient of similarity* of a ring with this definition of a crater:

$$V = \frac{d_2 - d_1}{n \cdot d} \sum_{i=1}^{n} K_1^i$$

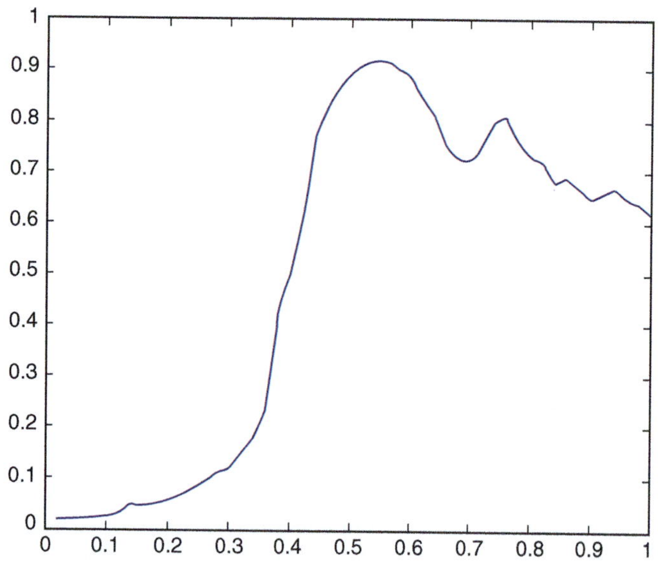

Figure 9.9 Graph of the average of the maximal curvature k_1 (on the Y-axis) of the ring built with 50 different distances d (on the X-axis). The first significant spike of the average k_1 value matches the rim around the crater.

Algorithm 9.1 ***Ring construction***

Input: List of vertices on the mesh
Output: List of rings on the mesh

Let n be the count of probable centers.
Let d_1 be the minimal research distance.
Let d_2 be the maximal research distance.
Let *nStep* be the desired count of concentric rings.
Let *centers*[] be the list of probable centers.
Let *rings*[] be an array of n rings.
Let *bestRings*[] be an array of n rings.
for $i = 0$ to n **do**
  Let *size* be the count of points in the ring.
  Let c be the count of up-to-date points.
  Let p be a point of the ring.
  rings[i]$\leftarrow$ ring formed by the direct neighbors of *centers*[i]
  for $d = d_1$ to d_2 in *nStep* **do**
    while $c < size$ **do**
      if *distance*(p, *centers*[i]) $< d$ **then**
        update(p)
        $p \leftarrow$ following point on the ring
        $c \leftarrow 0$
      end if
      $c \leftarrow c + 1$
    end while
    Let V be the likelihood of *ring*[i] being a correct detection.
    Let *bestV* be the likelihood of *bestRing*[i] being a correct detection.
    if $V >$ *bestV* **then**
      bestRing[i] $\leftarrow$ *ring*[i]
      bestV $\leftarrow V$
    end if
  end for
end for

9.4.4 Results and Validation

In order to evaluate the reliability of our approach, we ran the algorithm on a triangular mesh characterizing the surface of the asteroid Lutetia (stereophotoclinometry method; see Gaskell et al., 2008). The resolution is 160 m per element. The range of diameters is 0.8–5 km (about 5–30 elements of resolution). We then compared the result with a manual detection provided by a planetary scientist expert in the visual identification of craters. Because of the partial coverage of the surface by the space probe, we defined a working region with minimal data loss (Figure 9.10).

It is important to note the existence of an acquisition bias caused by the angle of incidence of the light. It induces a deformation of the surface perception. On the curvature map of the surface, the rims oriented North-West to South-East are clearly visible while the rims oriented North-East to South-West are almost flat (Figure 9.10). To compensate for this data flaw, the method considers the average curvature of each ring, even if it can alter the characterization of strongly deformed craters and valleys.

The proposed method offers an effective detection rate of 85% and a false positive rate of 10% within a single pass (Figure 9.11). Some of the larger rings are slightly off-centered as a result of an incorrect epicenter placement. Because of the difference in scale, the primary craters need a complementary pass at a lower level of resolution, in order to be detected.

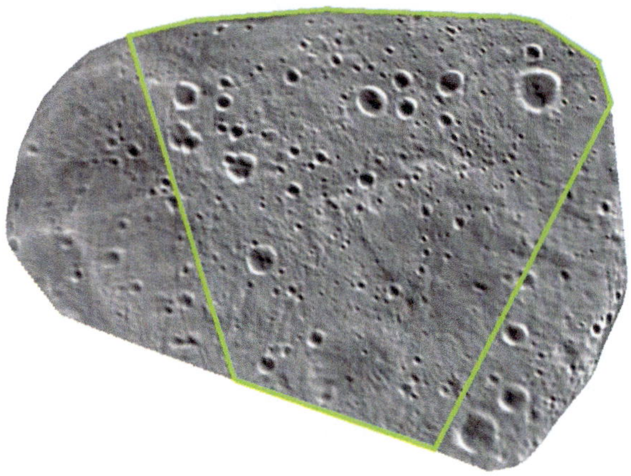

Figure 9.10 Visualization of the mean curvature map of the surface (300,000 faces). The mesh is displayed with false color (gray-scale with white for high convexity and black for high concavity). The working area is marked by the green border.

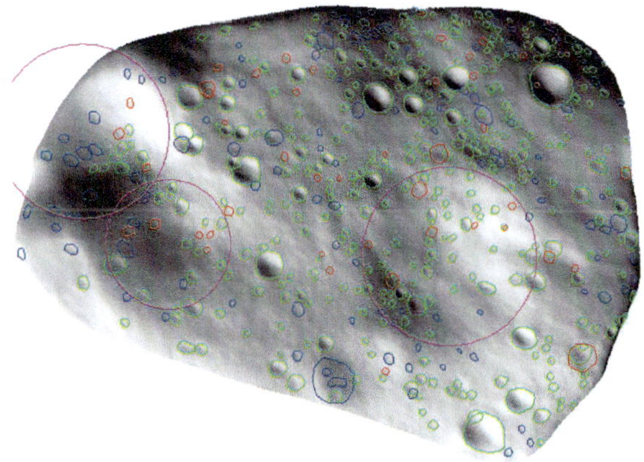

Figure 9.11 Visualization of the result of the automatic detection. Green circles are true positives, red are false positives, blue are missed detections, and pink are primary craters.

9.5 Approach Based on Circle Fitting

9.5.1 Description of the Approach

First, a skeleton (cf. Kudelski et al., 2013) is computed from an area of interest representing the regions of maximum and average positive curvature. The skeletonization is performed on surface regions of the original mesh, by homotopic thinning with a generic mesh adjacency (not a four- or eight-adjacency like in 2D images). We apply a pruning algorithm to reduce the noise perturbing the circular shapes we want to retrieve (see Figure 9.12). The next step is the determination of the centers of the circles from which we propagate an outline.

The final step is a fitting of the points at the edge of the pit by a circle. The fitting is performed after rotating these points in a 2D frame defined by the average plane.

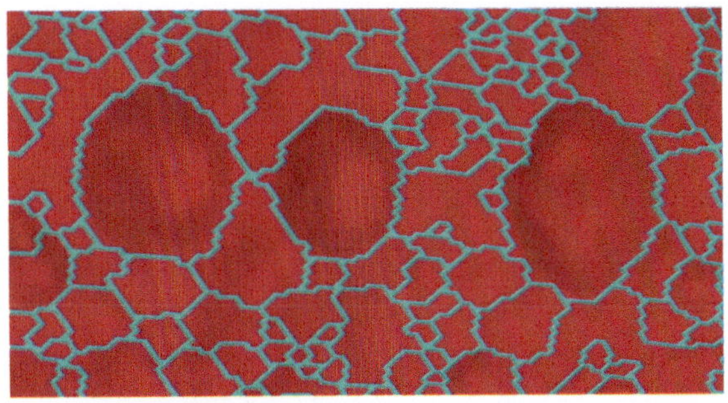

Figure 9.12 Pruning of the skeleton of an area of interest on the model of asteroid Vesta.

We obtain the contour circularity index by calculating the relative radial distance between the circle and the point.

9.5.1.1 Area of Interest and Skeletonization

To get an area of interest, we take all points of average or maximum positive curvature. The skeletonization method of Kudelski et al. (2013) is then applied to this area.

9.5.1.2 Circle Fitting

A switch from 3D to 2D is required. This is done by computing the approximating plan of the contour points. The computation of the eigenvalues of the following matrix gives the normal vector of the plane:

$$M = \begin{bmatrix} \Sigma x_i^2 & \Sigma x_i y_i & \Sigma x_i z_i \\ \Sigma x_i y_i & \Sigma y_i^2 & \Sigma y_i z_i \\ \Sigma x_i z_i & \Sigma y_i z_i & \Sigma z_i^2 \end{bmatrix}$$

We use the diagonal matrix M' to project the points on the plane in 2D. The 2D co-ordinates of these points are used to minimize the system of equation of the circular regression:

$$\begin{cases} (\Sigma x_i^2)\,\alpha + (\Sigma x_i y_i)\beta + (\Sigma x_i)\gamma = \Sigma x_i\,(x_i^2 + y_i^2) \\ (\Sigma x_i y_i)\alpha + (\Sigma y_i^2)\,\beta + (\Sigma y_i)\gamma = \Sigma y_i\,(x_i^2 + y_i^2) \\ (\Sigma x_i)\alpha + (\Sigma y_i)\beta + n\gamma = \Sigma x_i^2 + y_i^2 \end{cases}$$

The center corresponds to the 2D coordinates :

$$x = \alpha/2$$

$$y = \beta/2$$

They are converted back in 3D using the inverse matrix M'^{-1} (see Figure 9.13).

9.5.1.3 Circularity Criterion

Let R be the radius of the circle, R_i the Euclidean distance between the center and a point p_i, and n the number of points. An index of circularity Φ from the residuals between the circle and the points is

$$\Phi = \sum (R - R_i)^2 / n.R$$

This ratio gives the proximity of the extracted points to the fitted circle.

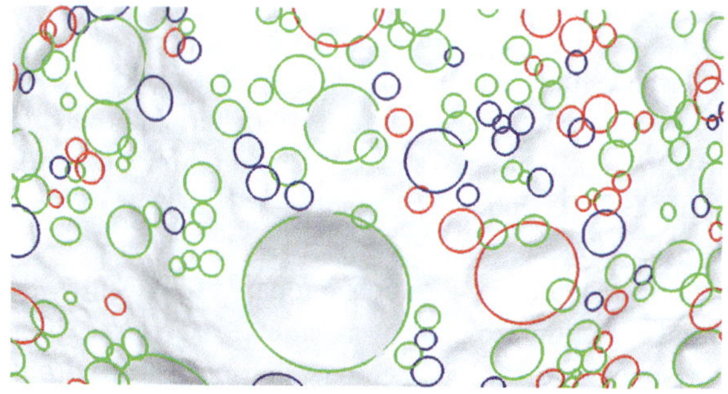

Figure 9.14 Detection obtained from the 3D model of the asteroid Vesta. The mesh consists of 1,572,866 vertices.

9.5.2 Results and Validation

After tests on the asteroid Vesta model (a triangular mesh with a resolution of 800 m per element, obtained using a stereophotoclinometry method; see Gaskell et al., 2008), we obtain a rate of 75% detection and 10% false positive with a selection threshold for centers at 0.20 (see Figure 9.14). The diameter of the smallest crater is equal to about 4 km (five elements of resolution). With a threshold of 0.15, we obtain a rate of 83% for detection and 15% for false positive. The threshold is the minimum of maximum curvature value to select probable pit centers. Note that the diameters of the "small" craters of our sample are often smaller than 10 elements of resolution of the 3D model. Note also that the manual identification of craters of up to 10 pixels (elements of resolutions of the images) in diameter are also known to be difficult to identify in images, even among experts (Robbins et al., 2014).

9.6 Conclusion

The approaches to automatically extract craters developed in this chapter rely on an appropriate usage of discrete curvature mapping of the vertices of a surface mesh.

In the first approach, we process mesh smoothing and mesh decimation algorithms in order to set the level of resolution of the surface. This level of detail conforms to the size of the concave shapes or circular depressions we want to

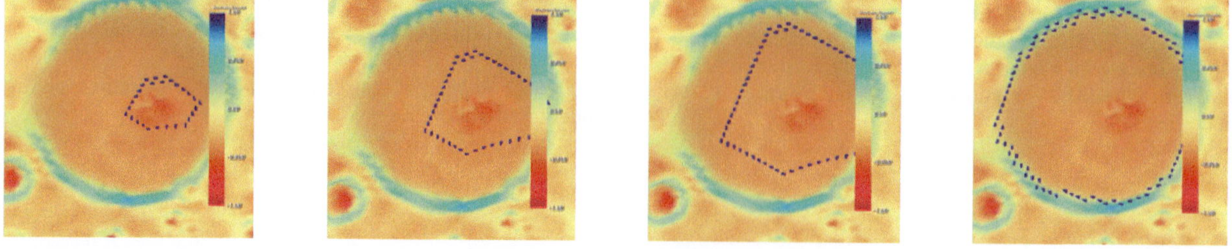

Figure 9.13 Determination of the center of a circle and outline propagation.

detect. Ideally, most of these circular depressions or craters show a high maximal curvature K_1 on their periphery and a low mean curvature H in their center. We then detect the probable centers using the mean curvature of the area as an estimator of its concavity. Finally, we expand multiple circular concentric rings around each center and through the analysis of the curvature values we determine which of these concentric rings is the most likely to be on a circular rim.

Based on 3D triangle meshes, the proposed method offers a quick extraction of a specific type of circular deformations on a surface. One of the advantages of this approach is its speed because the only time-consuming step is the calculation of the curvature, which however needs to be executed only once. Afterward, the determination of each ring is fast and only dependent on the search distance.

On a 2.6 GHz processor, with a complete mesh of the asteroid Vesta (3,000,000 faces), the curvature estimation takes 2 minutes, and the the detection of 3,500 pits covering between 1/100 and 1/30 of the size of the mesh takes 5 seconds.

The second approach is a variant of the first one, but instead of propagating rings until they reach the rims of craters, we compute "best fit circles" using a simple optimization technique.

The results are comparable to those of 2D detection algorithms (Pedrosa et al., 2013) with the advantage of having fewer false positives. The performances are the same as the first approach.

As future work and for the first approach, we plan to validate the method using reconstructed models of rocky planets and satellites instead of asteroids. Indeed, the resolution of models grows fast and could potentially allow the detection of a huge number of new craters. This would in turn allow a better understanding of planetary formation and evolution.

For the second approach, future improvements should be considered as well. During the phases of skeletonization and pruning, an over-segmentation can split a crater. Removing branches can delete some unclosed craters. To overcome this problem, a multiresolution approach could allow the removal of small-size craters and could prevent over-segmentation of the mesh. Moreover, a reconstruction method of the rim of the circles prior to the pruning step could be associated.

Moreover, we plan to integrate these two approaches in the near future within a larger testbed integrating the very same datasets for both methods. Despite the fact that the two approaches have not been applied on the same areas of both asteroids, these areas have the same geometric properties and the characteristics of the craters are very close.

Acknowledgments

We would like to thank Dimitri Kudelski for providing his mesh structure library, Pierre-Marie Victor Zanetta for his work on the manual validation of our results, and François Colas and David Baratoux for their precious help and support.

References

Beucher, S. (1982). Watersheds of functions and picture segmentation. In: *IEEE International Conference on Acoustics, Speech, and Signal Processing*, 1928–1931. Paris, France.

Duda, R.O. and Hart, P.E. (1972). Use of the Hough transformation to detect lines and curves in pictures. *ACM* 15: 11–15.

Fassett, C.I. (2016). Analysis of impact crater populations and the geochronology of planetary surfaces in the inner solar system. *Journal of Geophys. Res. Planets* 121: 1900–1926.

Gaskell, R., Barnouin-Jha, O., Scheeres, D. et al. (2008). Characterizing and navigating small bodies with imaging data. *Meteoritics and Planetary Science* 43 (6): 1049–1061.

Guennebaud, G. and Gross, M. (2007). Algebraic point set surfaces. *SIGGRAPH* 2007: 26. (23).

Guennebaud, G., Germann, M., and Gross, M. (2008). Dynamic sampling and rendering of algebraic point set surfaces. *Eurographics* 27 (3).

Haala, N., Reulke, R., Thies, M., and Ashoff, T. (2004). Combination of terrestrial laser scanning with high resolution panoramic images for investigations in forest applications and tree species recognition. *IAPRS and SIS* XXXIV: part 5/W16, Dresden.

Hilditch, C.J., Meltzer, B., and Michie, D. (1969). Linear skeletons from square cupboards. *Machine Intelligence* 4: 403–420.

Kudelski, D., Viseur, S., and Mari, J.-L. (2013). Skeleton extraction of vertex sets lying on arbitrary triangulated 3D meshes. In: *17th IAPR International Conference on Discrete Geometry for Computer Imagery, DGCI 2013*, Lecture Notes in Computer Science, LNCS 7749, 203–214. Seville, Spain: Springer.

Marchi, S., Barbieri, C., Kueppers, M. et al. (2010). The cratering history of asteroid (2867) Steins. *Planetary and Space Science* 58: 1116.

Marchi, S., McSween, H.Y., O'Brien, D.P. et al. (2012a). The violent collisional history of Asteroid 4 Vesta. *Science* 336 (6082): 690–694.

Marchi, S., Massironi, M., Vincent, J.-B. et al. (2012b). The cratering history of Asteroid (21) Lutetia. *Planetary and Space Science* 66: 87.

Melosh, H. (2011). *Planetary Surface Processes (Cambridge Planetary Science)*. Cambridge University Press.

Minár, J. and Evans, I.S. (2008). Elementary forms for land surface segmentation: The theoretical basis of terrain analysis and geomorphological mapping. *Geomorphology* 95 (3–4) Elsevier: 236–259.

Pedrosa, M., Silva, E.A., and Nogueira, J.R. (2012). Impact crater detection on Mars from digital image. *Lunar and Planetary Science Conference* 43: 2004.

Pedrosa, M., Pina, P., Machado, M. et al. (2013). Automated crater detection in the surface of Mercury in mdis-nac-imagery. In: *Lunar and Planetary Science*, 2472.

Preusker, F., Scholten, F., Matz, K. et al. (2012). Topography of Vesta from Dawn FC stereo images. In: *Lunar and Planetary Science Conference*. 43rd edition.

Robbins, S.J., Antonenko, I., Kirchoff, M.R. et al. (2014). The variability of crater identification among expert and community crater analysts. *Icarus* 234: 109–131.

Sawabe, Y., Matsunaga, T., and Rokugawa, S. (2006). Automated detection and classification of lunar craters using multiple approaches. *Advances in Space Research* 37 (1): 21–27.

Shary, P.A. (1995). Land surface in gravity points classification by a complete system of curvatures. *Mathematical Geology* 27 (3): 373–390.

Shary, P.A., Sharaya, L.S., and Mitusov, A.V. (2002). Fundamental quantitative methods of land surfaceanalysis. *Geoderma* 107 (1–2): 1–32.

Shary, P.A., Sharaya, L.S., and Mitusov, A.V. (2005). The problem of scale-specific and scale-free approaches in geomorphometry. *Geografia Fisica e Dinamica Quaternaria* 28 (1): 81–101.

Vollmer, J., Mencl, R., and Muller, H. (1999). Improved Laplacian smoothing of noisy surface meshes. *Eurographics* 18: 3.

Part III

3D Modelling of the Subsurface from Surface Data

10

Remote Sensing and Field Data Based Structural 3D Modelling (Haslital, Switzerland) in Combination with Uncertainty Estimation and Verification by Underground Data

Roland Baumberger[1,2], Marco Herwegh[1], and Edi Kissling[3]

[1] *Institute of Geological Sciences, University of Bern, Baltzerstrasse 1+3, CH-3012 Bern*
[2] *Federal Office of Topography, Swiss Geological Survey, Seftigenstrasse 264, CH-3084 Wabern*
[3] *ETH Zürich, Institute of Geophysics, Sonneggstrasse 5, CH-8092 Zürich*

Abstract

Ideally, the three-dimensional (3D) representation of the subsurface relies on both underground and surface data. Commonly, the former is scarcely available or entirely lacking and, therefore, the latter has to be used when projecting surface information to depth. We present an approach to investigate the large-scale 3D deformation pattern in the Central Alps (Switzerland) from surface data only, while assessing uncertainties. Up to now, little attention has been paid to the assessment of uncertainties related to input data, the extrapolation of surface data, and the correlation of surface and underground information. Our uncertainty estimations concentrate on aspects related to both 2D and 3D input data. We introduce the concepts of the Central Extrapolation Surface (CES) and the Extrapolation Uncertainty Area/Volume (2D: EUA; 3D: EUV). With this study, we document how (i) geometrical correlations between surface and underground data within their respective EUA/EUV may be used to construct a 3D model and (ii) we demonstrate that this approach delivers geologically relevant results. The suggested workflow allows the generation of reliable structural 3D models. The workflow will help to predict geology at depth in various fields (e.g. underground constructions and improved geodynamic understanding of mountain building processes).

10.1 Introduction

The use of geologic surface information for the prediction of subsurface continuation of rocks and structures represents not only an important task for science but also applied projects (e.g. underground constructions, deep geothermal energy, fossil energy and ore resources, CO_2 sequestration, and nuclear waste disposal), to be used, for example, for planning purposes, resource estimation, visualization, and risk assessment (e.g. Wellmann et al., 2018; Xiang et al., 2018; Schneeberger et al., 2018). Different powerful software tools (e.g. Move, GoCAD, 3DGeomodeller) have been developed during past years to facilitate extrapolation of surface data to depth, benefitting from advances in computer capacities. Despite this progress in construction capabilities, development of three-dimensional (3D) geological models still depends on the amount and quality of surface information available and assumptions regarding projection to depth. Ideally, geological models are built based on combined surface and underground information (e.g. Turner et al., 2021; Yang et al., 2021; Xiong et al., 2018; Wellmann et al., 2018; Schneeberger et al., 2017; Sala et al., 2014; Bistacchi, 2010; Gallerini and De Donatis, 2009; Zanchi et al., 2009; Kaufmann and Martin, 2008; Lemon and Jones, 2003). Reliable underground data (e.g. borehole data or subcrop information), however, is commonly sparse or lacking altogether (e.g. Sala et al., 2014). Consequently, such 3D representations of the subsurface often must rely on available surface data only (e.g. Turner et al., 2021; MacCormack et al., 2019; Schober and Exner, 2011; Dhont et al., 2005; Sprague and de Kemp, 2005). In contrast, Yang et al. (2021) describe implicit modelling with thousands of boreholes. In cases of subplanar structures like brittle faults or ductile shear zones, as the topic of this study, 3D geometry can be obtained by the 3D intersection line of the plane of interest with surface topography. Given the enhanced erodibility and, hence, visibility of these tectonic structures, the resulting lineaments represent the ideal type of input data for regional-scale geological 3D models.

Two methodologies exist to find and assess geological structures at the surface: field observations/measurements and remote sensing information. The strength of field

3D Digital Geological Models: From Terrestrial Outcrops to Planetary Surfaces, First Edition.
Edited by Andrea Bistacchi, Matteo Massironi, and Sophie Viseur.

observation datasets is their large number of local data points, which provide a variety of relatively precise information (e.g. strike, dip, lineations, and outcrop descriptions). Often, such datasets are punctually complete. In contrast, their weaknesses are characterized by (i) incomplete coverage due to non-continuous field work, large study areas, or bad terrain accessibility, (ii) unknown consistency due to non-uniform mapping or measuring, and (iii) unknown representativeness due to inconstant measuring of lineaments. The power of remote sensing bases on the possibility of fast and long-range detection of 2D lineaments (e.g. Dar et al., 2010; Rahiman and Pettinga, 2008; Morelli and Piana, 2006; Raharimahefa and Kusky, 2006). From a 2D point of view, such datasets are complete, continuous, and representative. They feature a steady and unvarying regional-scale data acquisition as well as a long-ranging digitization methodology. On the contrary, weak points are (i) the limitation to apparent strike information only (no direct dip information without coupling to a DEM), (ii) the difficulty of distinguishing different fault zones at or close to the spatial resolution of the applied imaging technique, and (iii) their lack of local field information regarding characterization of host rock and associated tectonites. As outlined in detail below, we believe that a combination of the two aforementioned types of surface information allows their individual limitations to be overcome (e.g. Baumberger, 2015; Banerjee and Mitra, 2004). The techniques have complementing strengths allowing for a better control on limitations (i.e. cross-check remote sensing data with field measurements) and subsequent derivation of uncertainties based on verified remote sensing data. Thus, more reliable 3D geologic models can be developed. Currently there exist two fundamentally different approaches in the generation of 3D structural models (e.g. Turner et al., 2021; MacCormack et al., 2019; Wellmann and Caumon, 2018). Both approaches include all information (surface, underground) available to make predictions at depth. (a) The explicit approach is mainly based on the knowledge and the experience of the modeller, besides the available input data. This "expert's guess" approach has the goal to derive a single 3D model, which fulfills the available structural information best (e.g. Cowan et al., 2003; Vollgger and Cruden, 2013). Ideally, such 3D models contain estimates on uncertainties, which unfortunately are often missing. (b) In contrast, implicit modelling or probabilistic 3D approaches also use the available structural information but calculate thousands of potentially possible 3D models evaluating the likelihood of the derived models to generate a likelihood space for the most "realistic" model (e.g. Caumon et al., 2009; Wellmann et al., 2010, 2018; Wellmann and Regenauer-Lieb, 2011; Schneeberger et al., 2017). In the current study, we follow approach (a), with the major aim to shed new light on the estimation of the uncertainties.

The overall goal of this study is to predict the large-scale 3D geometry of the main tectonic structures in the central Aar massif region (Figure 10.1), which are modelled from surface data and evaluated with available underground references. Caumon et al. (2009) provide a general overview of key concepts, modelling rules, guidelines, and workflows. The development of 3D geological models from surface data only is reported by, for example, Tonini et al. (2009), Dhont et al. (2005), and Sprague and de Kemp (2005). Approaches describing the correlation of surface data with underground information are presented by various authors. Sala et al. (2014) find their 3D geological model partly confirmed by integrating speleological surveys of four caves. Dag et al. (2013) present a methodology for ore-body determination based on borehole data, whereas Bistacchi et al. (2008) integrate uncertainty considerations for depth-projected structures in their preliminary study related to the Brenner Basis Tunnel in the Eastern Alps. Tonini et al. (2008) recreated cross-sections out of projected structural surface measurements (dip, dip azimuth) to depth in order to predict the geological situation at tunnel level. Bond et al. (2007, 2015) focus on the human bias introduced to geological models, based on the interpreter's experience, the techniques applied, and the availability of sparse data.

Little attention, however, has been paid so far to the assessment of uncertainties related to different input data (overview in Turner et al., 2021), the projection of surface to depth, and correlation methodologies, although precision and reliability of 3D geologic models has long been a key topic in geological 3D modelling (e.g. Wellmann and Caumon, 2018; Wellmann et al., 2010, 2018; Wellman and Regenauer-Lieb, 2011; Wycisk et al, 2009; Tacher et al., 2006; Davis and Keller, 1997; Pang et al., 1997). The estimation of intrinsic uncertainties of individual data types (e.g. resolution of Digital Elevation Models (DEM), influence of digitization accuracy, etc.) is essential for the interpretation of 3D models. So far, most uncertainty considerations focus on the three-dimensional results (e.g. Wellmann et al., 2011; Zhu and Zhuang, 2010; Tacher et al., 2006) or on modelling methodologies (e.g. Bistacchi, 2009; Bistacchi et al., 2008; Kaufmann et al., 2008; Tacher et al., 2006).

Unlike our own study, uncertainties related to input data and extrapolation methodologies are not discussed in previous works, or if they are, to a little extent only (e.g. Bistacchi et al., 2008). Recently, Rauch (2021) focuses explicitly on cartographic evidence, bedrock complexity and outcrop information, while Turner et al. (2021) specifically point out the uncertainty of geological data. In this study, we present a new and integrated approach

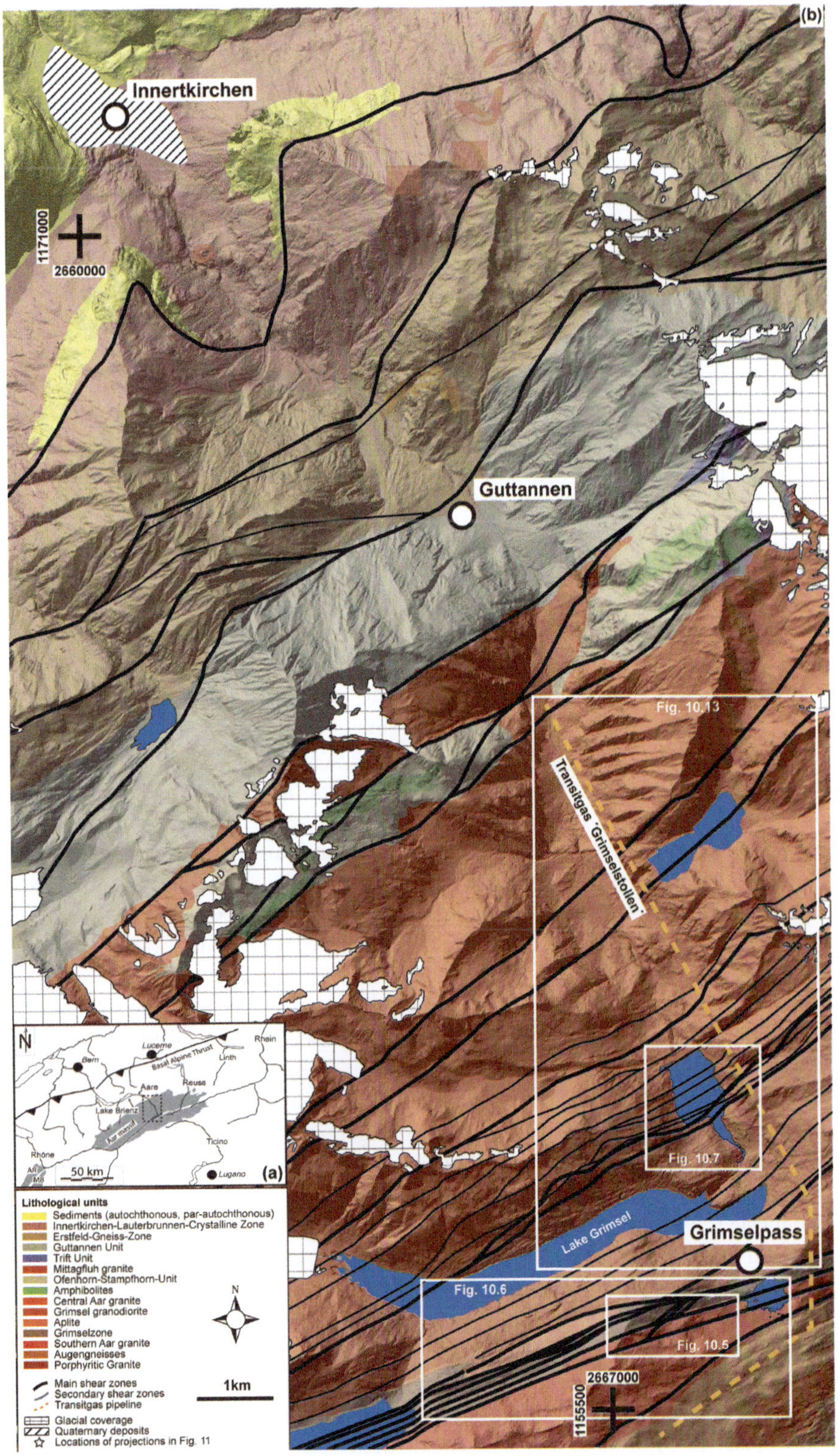

Figure 10.1 Geological maps of the study area (a, b). (a) Tectonic sketch shows the position of the study area within the Aar massif. (b) The Aar massif is subdivided into (i) its (par-)autochthonous sedimentary cover in the N, (ii) its pre-Variscan units (Innertkirchen-Lauterbrunnen-Crystalline Zone, Erstfeld gneiss Zone, and Guttannen Zone) in the central part, and (iii) the post-Variscan plutonites (Mittagfluh granite, Central Aar granite, Grimsel granodiorite, Southern Aar granite) in the southern part. It becomes obvious that all lithological boundaries are marked with subparallel striking shear zone traces. Additional shear zones occur within the different lithological units. Squares denote positions of other figures. Dashed orange line shows the trace of the Transitgas tunnel. Overall dimensions are 10 × 20 km. Map courtesy of DIMAAR Project (University of Bern).

for the development of a 3D structural geologic model from surface data only and its comparison with selected subsurface (tunnel) information. We account for the characteristic uncertainties related to the 2D remote sensing data as well as to data obtained by field measurements and integrate these findings into the extrapolation of geological structures to depth.

The first part of the study is devoted to a careful analysis of various kinds of uncertainties of the different types of input data. In a second step, we establish and discuss the procedure to extrapolate surface information to depth to define the regional 3D geometries of fault/shear zones. We complete the study with an application and discussion on the uncertainties of the chosen approach, by controlling projected surface data by hard fact underground information.

10.2 Geological Setting

The Aar massif (Haslital area, Central Alps, Switzerland) is our study area. With 550 km^2 of surface exposition, it belongs to a chain of exhumed basement rocks of the former European passive continental margin, the so-called External Crystalline Massifs (e.g. Herwegh et al., 2022). Abundant lithologies in the Haslital area comprise pre-Variscan ortho- and paragneisses (Altkristallin), syn- to post-Variscan volcano-clastic suites and post-Variscan granitic intrusions (Grimsel Granodiorite (GrGr), Central Aar granite (CAGr), Mittagfluh granite (MiGr) and Southwestern Aar granite (SAGr) (see Abrecht, 1994; Niggli, 1965; Stalder, 1964; Berger et al., 2017a; and Figure 10.1a). The plutonites are of Early Permian age (Olson et al., 2000; Schaltegger, 1994; Wüthrich, 1963). Stalder (1964) reports an important difference in their biotite content (CAGr; GrGr), which controls style and intensity of solid-state deformation (Baumberger, 2015; Wehrens et al., 2016, 2017). Deformation history comprises Proterozoic, Variscan, and Alpine deformation events (Schaltegger, 1989; Labhart, 1977; Steck, 1968; Stalder, 1964). In the post-Variscan CAGr and GrGr, the two key lithologies of our study, however, any solid-state deformation must either be related to Mesozoic rifting or to the intense deformation of the Alpine orogeny (Stampfli et al., 2002). Several authors (e.g. Herwegh et al., 2020; Challandes et al., 2008; Bambauer et al., 2005; Choukroune and Gapais, 1983) propose a N–S oriented Alpine deformation gradient, pointing to Alpine metamorphism and deformation being highest at Grimselpass in the southern part of our study area. According to Goncalves et al. (2012) and Challandes et al. (2008), peak metamorphic conditions reached upper greenschist facies in the Grimselpass area (max. ~450 °C and ~6.5 kbar). During Alpine orogeny, strain localized and accommodated along mechanically weak zones (e.g. Herwegh et al., 2017, 2020; Wehrens et al., 2016, 2017; Oliot et al., 2010; Vouillamoz, 2009; Challandes et al., 2008; Baumberger, 2008; Marquer et al., 1985; Choukroune and Gapais, 1983; Steck, 1968, and Figure 10.1a). Following the nomenclature of Schmid and Handy (1991) we refer to these high strain domains as fault zones. They are several km long and cm to dm wide and developed deformation patterns at different scales, depending on deformation conditions and the kinematic framework (e.g. Herwegh et al., 2017, 2020; Wehrens et al., 2016, 2017; Belgrano et al., 2016; Mancktelow and Pennacchioni, 2005; Arboyela and Engelder, 1995; Gapais et al., 1987, 1989; Choukroune et al., 1987; Engelder, 1987, 1989; Aydin and Johnson, 1978). For fault zone initiation, progressive deformation either (i) accentuated or reactivated already existing mechanical anisotropies like pre-existing fault zones, compositional variations in the host rocks, and lithological contacts (Figure 10.2a–d; Wehrens et al., 2017; Vouillamoz, 2009; Baumberger, 2008; Keusen et al., 1989; Labhart, 1977; Steck, 1968) or generated new fractures as nuclei for ductile shearing (Wehrens et al., 2016; Mancktelow and Pennacchioni, 2005).

Wehrens et al. (2017) describe the evolution of distinct 3D deformation patterns and correlate them to an Alpine two-phase deformation of the Aar massif: a first compressional stage with NW-vergent reverse faulting (Handegg phase) followed by a younger transpressional stage with dextral strike-slip along ENE–WSW trending faults (Oberaar phase). In the North of the Aar Massif, Wehrens (2015), Berger et al. (2017b), Mair et al. (2018) and Herwegh et al. (2020) demonstrated that shortening along SE dipping thrust planes occurred (Pfaffenchopf phase), which is synchronous to the Oberaar phase in the South. Note that vertical displacement components from Handegg reverse faulting and NW-directed Pfaffenchopf thrusting substantially contributed to the overall exhumation of the Aar Massif (Herwegh et al., 2017, 2020, 2022). Both the Handegg as well as the Oberaar phase are in line with findings from Rolland et al. (2009), Challandes et al. (2008), and Burkhard (1988) and can be correlated with ductile deformation stages at 20–22 Ma and 14–12 Ma respectively. Ductile deformation ceased at about 11 Ma (Bergemann et al., 2014) and was followed by subsequent brittle deformation combined with hydrothermal activity (e.g. Wanner et al., 2019; Diamond et al., 2018; Egli et al., 2018; Belgrano et al., 2016; Hofmann et al., 2004). Damage zones of a few meters in width and fault gouges of a few millimeters to centimeters thickness result from the brittle deformation stages (Figure 10.2; see also Egli et al., 2018). Another consequence of multistage deformation are

Figure 10.2 Characterization of mechanical discontinuities in host rocks that may cause morphological incisions. (a) Ductile shear zones: two ultramylonitic bands in a coarse-grained granite at Gelmersee. *Source:* © R. Schneeberger. (b) Brittle fault zones: fault gouge in a coarse-grained granitic host rock at Grimselsee. *Source:* © M. Herwegh. (c) Intrusion contacts: metabasic dike in a granite at the Nagra Grimsel Test Site. *Source:* © R. Baumberger. (d) Arrays of joints: array of closely spaced joints dissecting a former mylonite at the intrusion contact between Mittagfluh granite to Altkristallin close to Mount Stampfhoren. *Source:* © M. Herwegh.

morphological incisions at the surface due to enhanced erodibility of the fault rocks. Particularly, brittle overprinting, in conjunction with the occurrence of fault gouges, results in severe surface incisions (Baumberger, 2015), which provide the base for fault zone mapping by remote sensing and the subsequent generation of a 3D structural model of the Haslital area in this study. In this way, a correlation between km-scale surface incisions, detailed outcrop information, and underground tunnel mapping becomes possible, as shown for two examples in Figure 10.3. Here outcrop-scale tectonites can be linked with both large-scale incisions and underground information. It becomes clear that shear zones prominently expressed on the large scale (e.g. orthophotos, aerial photographs) often rely on small structures only, which are only several cm wide.

10.3 Methodology

The major goal of our study is the prediction of subsurface continuation of the large-scale 3D geometries of major fault zones relying on surface data only. Remote sensing-based 2D lineament detection/quantification (Baumberger, 2015) combined with surface topography and structural field geology provide the basis for 3D structural modelling of the fault zones. To combine the characteristic strengths of the two methods, quantitative estimates of uncertainties of all surface data are prerequisites, since uncertainties in 2D propagate into the 3D projection. Therefore, special emphasis is paid to the following points:

- Identification and assessment of uncertainties of the different types of input data (2D), based on the generation of a 2D lineament map (see Appendix B, Figure B.1).
- Development of 3D representations from 2D surface data of fault zones and their extrapolation to depth.
- Verification and assessment of the extrapolated 3D geometries of fault zones by field observations at underground facilities.

Here we follow an approach that consecutively comprises the above-mentioned topics and guides through the

Figure 10.3 Link between km-scale surface incisions and outcrop-scale mechanical discontinuities for two representative examples. Orange line indicates the trace of fault cores. (a) Alplistock shear zone at Lake Gelmer: (a1) Morphological incision on the km scale *Source:* © swisstopo; (a2) Incision at the scale of hundreds of meters *Source:* © R. Schneeberger; (a3) Strongly schistose mylonitic granite indicating a ductile shear zone at the outcrop scale *Source:* © R. Schneeberger; (a4) Ultramylonitic bands in the core of the shear zone at the dm scale *Source:* © R. Schneeberger; (a5) Mapping shows the occurrence of mylonitic zones in the tunnel *Source:* redrawn after Schneider (1974). (b) Geisschälen shear zone at Lake Räterichsboden: Morphological incisions at the km scale (b1) (*Source:* © swisstopo) and the scale of hundreds of meters (b2) (*Source:* © Google). (b3) Granodiorite with strain gradient from weakly schistose to strongly schistose mylonite at the outcrop-scale *Source:* © M. Herwegh; (b4) Brittle fault gouge in the center of the shear zone at the m-scale *Source:* © M. Herwegh; (b5) tunnel mapping indicates occurrence of a strongly schistose zone *Source:* redrawn after Schneider (1974). In both examples the original mechanical discontinuity of the shear zone is always considerably smaller than the detectable large-scale morphological incision.

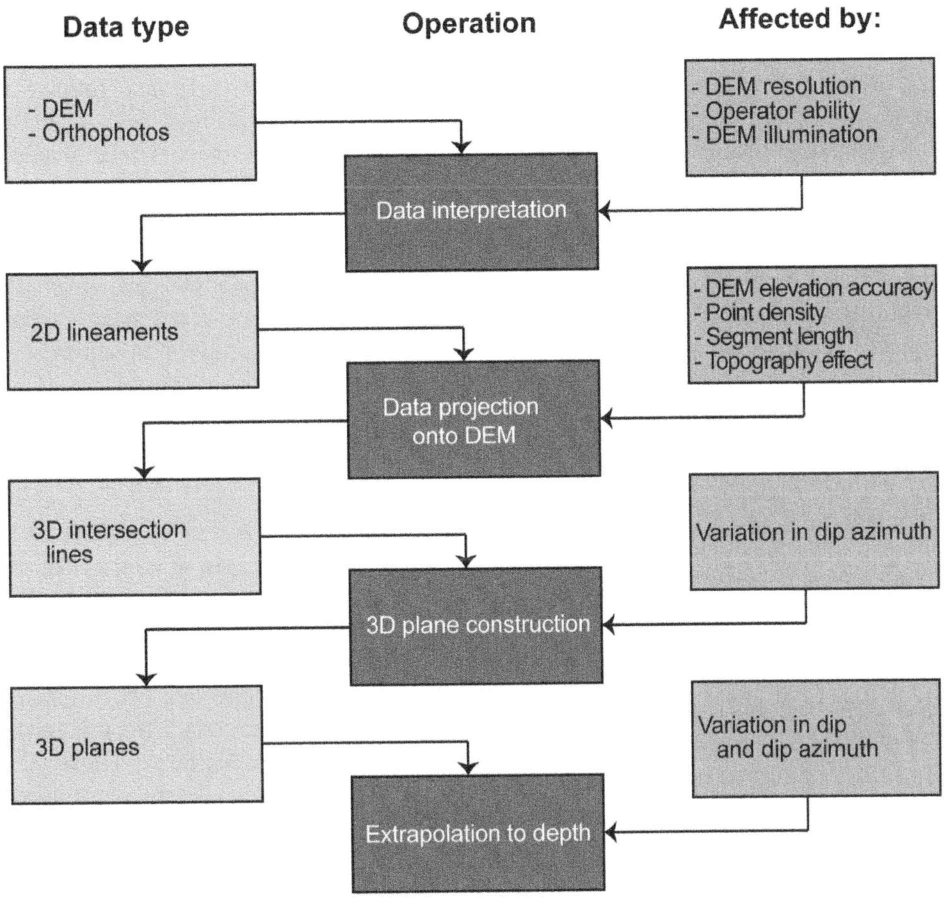

Figure 10.4 Four-step workflow applied to generate 3D planes from 2D lineaments. (1) Data interpretation comprised of the digitization of 2D lineaments based on DEM and orthorectified aerial photographs. This step was influenced by the DEM resolution, DEM illumination, and the operator's ability. (2) Data projection to DEM was used to generate 3D intersection lines from 2D lineaments as input data, which was influenced by DEM elevation accuracy, digitization point density, and segment length as well as the topography effect. (3) Using the 3D intersection lines as input data, the 3D planes were constructed. The most important influencing factor is the variation in strike. (4) Extrapolation to depth based on the generated 3D planes, while additionally taking the variation in dip into account.

transition from 2D surface information to 3D subsurface geometries (Figure 10.4).

10.3.1 Data Section

10.3.1.1 Definition of Terms

Lineament: We use the definition presented by O'Leary (1976) to describe linear topographic features mapped in 2D, obtained by remote sensing. Lineaments are linear or curvilinear features that represent the trace of an intersection between a planar or subplanar structural anisotropy (such as a fault) and the surface of the Earth. They can be recognized on aerial photos, satellite imagery, or topographic maps.

3D intersection lines: In our study, the intersection between two surfaces, e.g. topography (represented by a DEM) and a fault plane in 3D space, is defined as 3D intersection lines.

m-value: This term is used by Fernández (2005) to describe the variability of the dip orientation of the local relief of a 3D plane compared to its average dip orientation. For best results, the value should not be larger than 4.

k-value: Fernández (2005) uses this term to describe the variability in strike of the local relief in comparison to the overall mean strike of the 3D plane. The smaller the value, the lower the variability. Most accurate results are obtained with a value <1.

10.3.1.2 Input Data

Different types of input data were used, interpreted, and combined with each other to derive 3D fault zones from remote sensing-based 2D data acquisition.

DEM: A high-resolution DEM served as one of the most important input data sets for this study (Baumberger, 2015). The spatial resolution of the swissALTI3D LiDAR-based DEM (Source: © swisstopo) is 2 m/pixel and the accuracy in the vertical direction of the DEM is +/−1–3 m.

Aerial photographs: Orthorectified aerial photographs (swissIMAGE, spatial resolution: 0.5 m/pixel, vertical accuracy +/−0.5 m, Source: © swisstopo) supplemented the DEM as the basic input data set for this study.

Geological maps and cross-sections: Digital and analog geological maps and cross-sections of different quality, vintages, purposes, and scales were used to quality check the 2D remote sensing data acquisition as well as the final 3D structural model. Maps of hydraulic galleries, road tunnels, gas pipelines, and caverns provided valuable details to correlate projected surface information with information from the subsurface.

Remote sensing lineament map: High-resolution imagery (DEM and aerial photographs) was used to identify the lineaments and lithological boundaries by remote sensing to derive the remote sensing lineament map shown in Appendix B, Figure B.1. See Baumberger (2015) for details.

Field measurement and observations: Most of the data obtained by remote sensing was verified and complemented by field data. Thus, we amended structural orientation measurements (dip, dip azimuth) as well as geological characterizations (rock type, deformation, fabric, etc.) as additional attributes to the 2D information of remotely sensed fault zones.

10.3.2 Identification and Assessment of Uncertainties of Input Data Types

Turner et al. (2021) and Wellmann & Caumon (2018) describe uncertainties of the most used input data for 3D geological modelling, either based on implicit or explicit approaches. In line with these studies, our approach comprises a series of steps to derive from initial 2D input data to final 3D constructions of planes representing the geological structures. In this context, four successive types of datasets and their related uncertainties have to be evaluated in order to estimate the quality of the final 3D projection. These datasets are (i) original 2D remote sensing pixel data (i.e. orthophotos, DEM), (ii) their subsequent processing into 2D vector data, i.e. fault traces, (iii) the 3D planes being built by correlation of surface geometries of structures with topography, and (iv) the generated 3D projections of the formerly generated 3D planes. In this context four categories being subject of specific uncertainties can be considered, which are relevant for further extrapolation of surface data to depth:

1. Remote sensing data (2D) and related uncertainties: spatial resolution of the imaging technique and illumination direction during fault trace digitization.
2. Vector data (2D): point spacing along a fault trace line and coherence of data between 2D intersections of fault traces with topography and the orientation of future 3D fault planes.
3. Geometric properties of 3D planes derived from 2D data: dip extraction at the surface from 3D plane construction and variability of dip and dip azimuth along the surface trace of the fault zone.
4. Projection uncertainty of previously generated 3D planes: evaluation of projection dependency (strike vs. dip) and correlation of projection with subsurface information originated from maps.

10.3.3 Data Interpretation: From Remote Sensing to 2D Vector Data

High-resolution raster data: The remote sensing part of our approach allows the mapping of morphological incisions based on DEM and orthorectified aerial photographs. Along a vertical cross-section, the smallest still detectable incisions on a raster image are theoretically characterized by a minimum width of 3 pixels, where based on a change in gray value (depth information), the topographic depression in the central pixel separates the two elevated rims (Figure 10.5a). For the directional length, 2 pixels principally would be sufficient. In terms of later 3D extrapolation, however, a minimum of 3 pixels is required (see below). Consequently, a minimum area of 3×3 pixels is necessary for a proper identification of the location of an incision on a raster image (Figure 10.5a). The traceability of a fault zone depends on the spatial image resolution (e.g. Hopkinson, 2009; Gajski, 2004; Wolock and McGabe, 2000; Moore et al., 1991). To evaluate the related uncertainty, three different resolutions (2, 10, and 25 m) were chosen to digitize one single shear zone (Figure 10.5b–e).

Operator ability: During explicit geological modelling, the modelers are exposed to varying doctrines, changing geological concepts and data quality with time (e.g. Cowan et al. 2003, Vollgger and Cruden, 2013, Wellmann et al., 2018; Turner et al., 2021). This human bias may result in different evaluations or interpretations of input data by several of the modelers. Consequently, the uncertainty of explicit 3D geological models cannot be quantified or only with difficulties (e.g. Bond, 2007; Gunnink et al. 2010).

DEM illumination: A DEM can be illuminated from different directions to enlighten regions lying in the shadows and to trace a fault zone over long distances (Scheiber et al., 2015, and references therein). Given the illumination direction and orientation of topography shadings, the contrast gradient between incision and rim

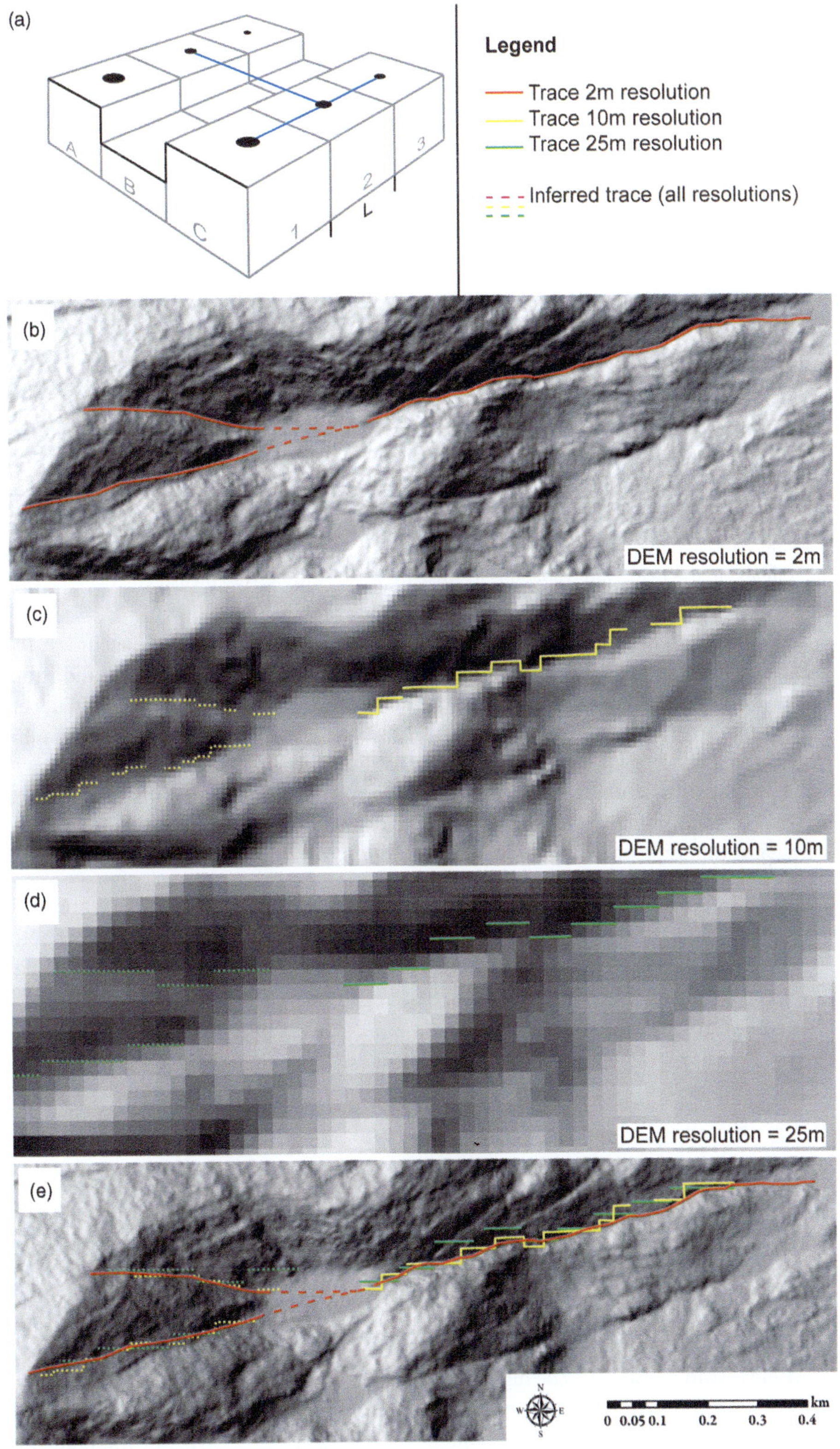

Figure 10.5 Influence of DEM resolution on digitization accuracy and the representation of lineaments. (a) A minimum of a 3×3 pixel matrix is required at the minimum to describe a lineament in 2D, regarding its topographic incision (a–c) and length (1–3). (b, c, d) DEM resolutions of 2×2 m, 10×10 m, and 25×25 m respectively require minimum interpolation distances of 6, 30, and 75 m (see text for explanation). Digitizable segments of shear zone trace are given in colors. (e) Integration of all digitized shear zone segments shows the resolution-derived deviations. The field domain is indicated in Figures 10.1 and 10.5.

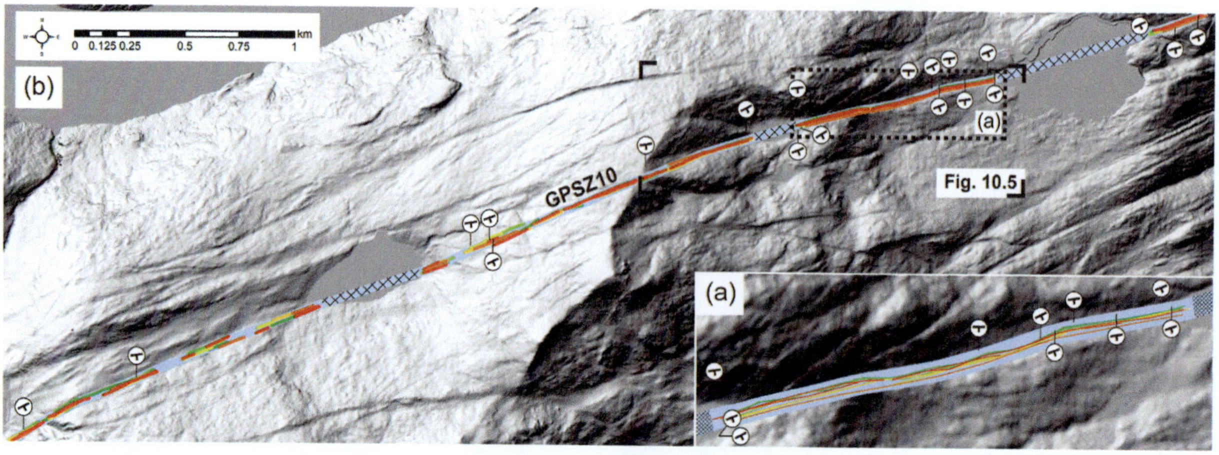

Figure 10.6 Variation of strike along the shear zone trace due to bandwidth error. Digitized trace of GPSZ 10 using eight different illumination directions of a DEM with 2 × 2 m resolution. (a) On the detailed scale, variations in shading directions on the DEM results in uncertainties of exact locations of shear zone traces (different colors generated from different shading directions). An uncertainty range results, which can be defined by an average bandwidth (bright blue stripe). (b) At the large scale, general variations in strike result, due to topographic effects (see text for explanation). Hatched blue domains indicate the uncertainty bandwidth from (A) at locations where no topographic information is available for shear zone tracing. Dip markers indicate 3D orientations measured with the compass in the field. Note variations in strike between field-derived measurements and the remote-sensing approach.

may shift spatially, yielding uncertainty in the location of the detectable fault trace. Resulting uncertainties are investigated for a fault zone using eight different illumination directions (multiples of 45°, with 315° as the standard direction; Figure 10.6). Figure 10.6a illustrates a digitized shear zone trace (GPSZ) on the local scale. GPSZ 10 is not represented by one continuous line, but by eight different non-continuous line subsets (shown by different colors). Variations in shading of the DEM cause uncertainties regarding the exact location of the shear zone trace, which is estimated by the bandwidth uncertainty (bright blue band in Figure 10.6). It is defined by the envelope surrounding the line segments, which were derived from the different illumination settings.

10.3.4 Data Projection onto to DEM: From 2D to 3D Data

Four important topics influence the comparison between 2D intersections with topography and their resulting 3D fault planes: (i) the DEM elevation accuracy, (ii) the point density along the line, (iii) the length of analyzed segments, and (iv) the topography effect.

DEM elevation accuracy: The higher the grid resolution of a DEM, the better the topographical representation (Warren et al., 2004). The precision of the digitization and projection of 2D vector data directly depends on the grid resolution (e.g. Hopkinson, 2009; Gajski, 2004; Wolock and McGabe, 2000; Moore et al., 1991). However, it is not only the horizontal but also the vertical accuracy that accounts for the precision of digitization and projection (Kenward et al., 2000). The latter generally depends on the availability (e.g. derivation from topographic maps, satellite imagery, LiDAR, etc.), the processing of the input data (e.g. stereo correlation, direct processing, generalizations, etc.), and the interpolation methods used for DEM creation (e.g. Wenzhong et al., 2014).

Point density along line: In contrast to the pixel-based DEM and orthophotos, digitized 2D lineaments consist of vectors connecting the digitization points. In general, the accuracy of the geometric representation of a 2D vector-based line in 3D depends on the point density along its trace (e.g. Turner et al., 2021), which is important for the presentation of both topography (vertical sections) and traces of fault zones (2D, 3D). The smaller the distance between points, the better the line mimics the topography. To analyze the influence of the point spacing and associated uncertainties of our fault traces, a single shear zone was resampled to different point intervals (Figure 10.7).

Length of analyzed segments: In terms of vectors used to describe a fault, two requirements need to be followed: a maximum spatial resolution for a precise location of the fault trace vectors and a maximum angular resolution of strike of the fault trace vectors. Unfortunately, the two requirements counteract each other, requiring a compromise by an optimization procedure. The minimum length of a fault vector defining a directional spatial extent of a fault zone is based on a 3 × 3 pixel matrix

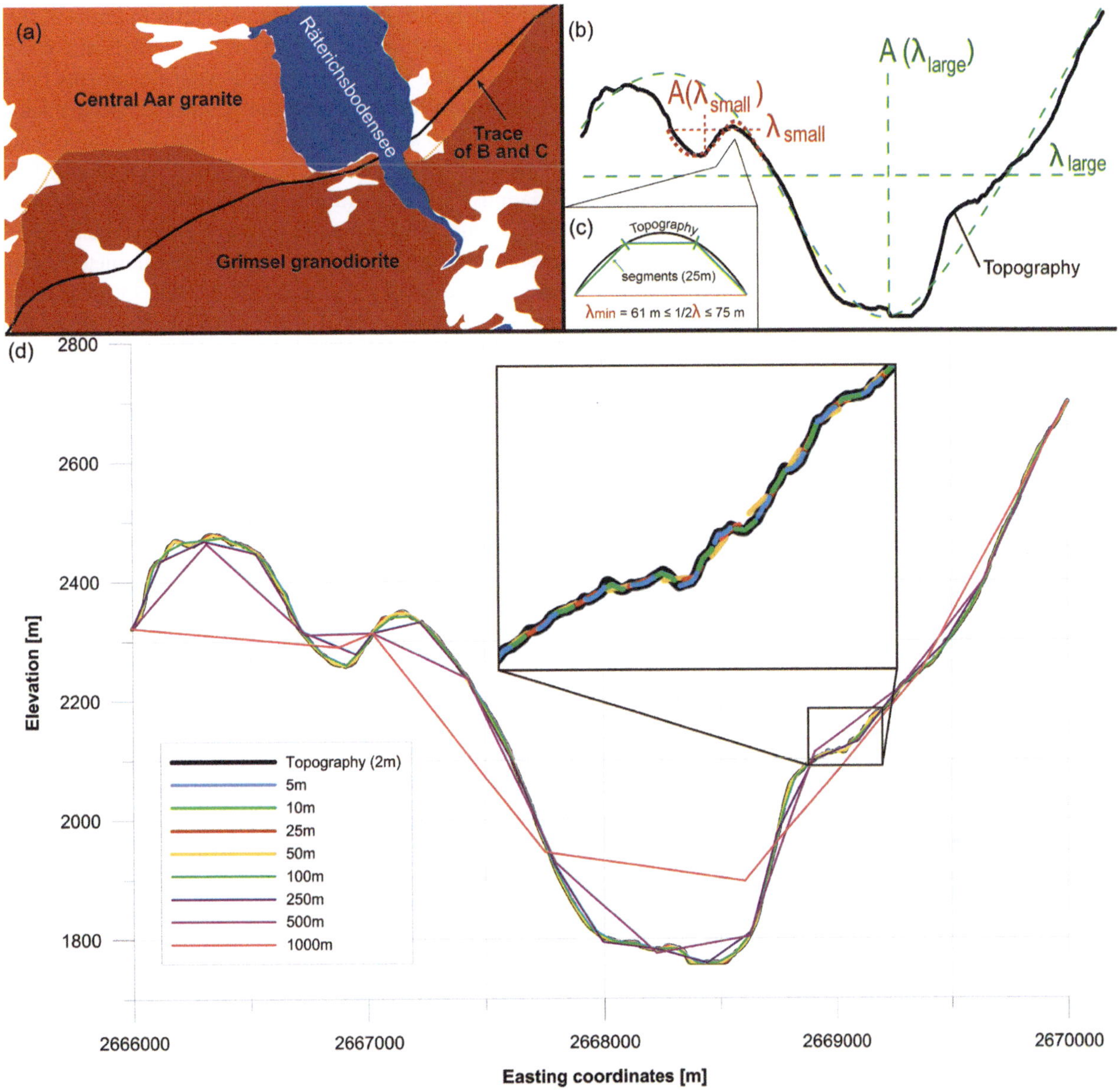

Figure 10.7 Representation of topography based on point density along a fault zone trace. (a) Geological map showing the location of the trace of Figures b and c; (b) The morphology of the landscape can be expressed by amplitudes (a) and wavelengths (l) of different orders (A_{large}, l_{large}, ...; A_{small}, l_{small}); (c) A minimum of three segments is needed to represent a simple topographic peak/valley; (d) The chosen DEM resolution therefore defines the minimum size of topographic variations still resolvable, as shown by the variations in grid resolution from 2 to 100 m. In the current framework a resolution of 10 m is suitable (see text).

(see above) defining the maximum resolution possible to define a fault vector. For this matrix, however, only strike directions (vectors) of a fault zone of multiples of 45° can be obtained (e.g. 0°, 45°, 90°, ..., 360°). To reduce this restriction in angular resolution, image analysis approaches usually build up on 10 × 10 or larger pixel matrices (e.g. Heilbronner and Barrett, 2014; Blaschke, 2010; Pirotti et al., 2010). Given the spatial resolution R [m] of the DEM, a matrix of $10R \times 10R$ results. Here the shortest (V_{min}: matrix edges) and longest (V_{max}: matrix diagonal) vector lengths are $V_{min} = 10\text{R}$ and $V_{max} = \sqrt{10R^2 + 10R^2}$, respectively. The use of V_{max} long fault segments allows for a minimum angular resolution A [°] of $A = \text{arcTAN}(R/V_{min})$, which we consider to be sufficient for the representation of the strike orientation of the fault segments. For that reason, we digitized the fault segments in a first step for maximum resolution. In a second step, obtained digitized fault traces are recalculated into segments optimized for the angular resolution in strike. For this purpose, a uniform vector length of fault zones of 25 m was chosen for the entire project. In the light of statistical analyses, fault traces of several 100 m to km length were subdivided into fault segments (vectors) uniform of length V_{max}.

Topography effect: Remote sensing data only provides information about the orientations of the fault segments of 3D fault zones in 2D map space. The apparent 2D strike of the 3D intersection line between a true 3D plane and topography can deviate from a true 3D strike direction of the plane. For example, for vertical 3D planes, the apparent strike of a fault segment in 2D of length V_{max} is identical to the true strike. For inclined 3D planes, however, the lower the plane dip (closest to horizontal and, therefore, the more crooked the 2D lineament trace), the larger the resulting deviation in 2D (see Appendix A). We call this inclination dependent deviation the topography effect.

To estimate uncertainties resulting from the length of analyzed segments and topography effects, the orientation of local strike was measured for each fault segment along a 4 km long fault zone and compared to variations in strike measurements obtained from field investigations (Figure 10.9).

10.3.5 3D Plane Construction: From 3D Intersection Lines to 3D Planes

10.3.5.1 3D Best-Fit Plane from 2D Lineaments

Lineament traces mapped in 2D serve as input data for the extrapolation to depth of surface information (Baumberger, 2015; Appendix B). The 3D intersection line of a fault zone with topography roughly defines the 3D spatial orientation of the respective fault plane. The 2D lineament traces are projected to a high-resolution DEM (2 m/pixel) and their intersection with the topography provides a general impression of the respective 3D geometry of the lineament. The results of the automated geological map deconstruction described by Jessell et al. (2021) directly serve as input data for 3D model construction, whereas Xiong et al. (2018) present an implicit function to derive a 3D plane from outcrop data directly. Ferńandez (2005) and Ferńandez et al. (2009) present an approach to develop so-called 3D best-fit planes from surface information only, which allows for the extraction of dip and dip azimuth values from initial 2D lines. After projection, the trace of example shear zone GSPZ10 is analyzed to identify the uncertainties related to the best-fit plane construction according to Ferńandez (2005). The influence of the local relief and the relevant input parameters (point distance, size of search window), which define the overall quality of the resulting best-fit plane, is measured by roughness of the individual 3-point planes relative to the average plane (m-value) and the variation in strike of the local individual 3-point planes (*k*-value). Roughness of the local relief and local strike variations of lineaments both are important components of the 3D intersection line and, therefore, also affect the overall roughness (*m*-value) and strike variation (*k*-value) of the derived best-fit plane. The variations of these parameters are evaluated by applying the best-fit plane methodology on the same shear zone several times using different windowing parameters each time (Figure 10.8). A constant interval of 250 m is used to move a 500 m wide search window along the trace of GPSZ10. Figure 10.8a illustrates the local variation in orientation from individual 25 m line segments compared to the deviation from the mean strike value per search window. Between profile distances 250–500 m, 2,000–3,000 m, and 4,000–6,000 m, the local orientation deviates remarkably from the search window mean values. The variations of *k*-values and *m*-values are displayed in Figure 10.8b. The k-values generally scatter between 0.5 and 1.5, reaching a minimum only slightly above zero at the water table of Lake Totensee in the E of GPSZ 10. For the *m*-value, values scatter between 4 and 12, but in general vary between 6 and 8. In line with their position along the strike of GPSZ 10, the outlier values correspond to a change in general orientation of GPSZ 10 (due to union with another structure) and a long distance with no change in direction (due to absence of measurement points while GPSZ 10 crosses a lake), respectively. Since the *k*-value and *m*-value essentially depend on the 3D intersection of the trace line with topography, no *k*-values and *m*-values can be detected at locations with a low relief (e.g. water surfaces, planes). By definition, the calculation algorithm sets the measurement values to zero at such areas.

The deviation in dip (*m*-value) and strike (*k*-value) per single search window (Figure 10.8c) vary between +/−2° and +/−5°, respectively, from the corresponding mean values of the entire GPSZ 10 shear zone (dip: 87°; strike: 71°). Note that the boundary effects (missing data points on both sides for mean strike and mean dip values per search window), caused by the principle of the moving window, are accounted for in Figure 10.8c.

Based on the size of the search window, *k*-values and *m*-values can be calculated for every point in the defined interval along the fault zone trace. Subsequently, the complete data set is filtered for values of $0.5 \leq k \leq 1.5$ and $4 \leq m$ per point, which are used to construct the best-fit plane for the corresponding fault zone. In our approach, we project the remote sensing based 2D fault traces onto the DEM to obtain the 3D intersection line between the fault plane and topography. In a second step, the points (coordinates in 3D space) along the 3D intersection line

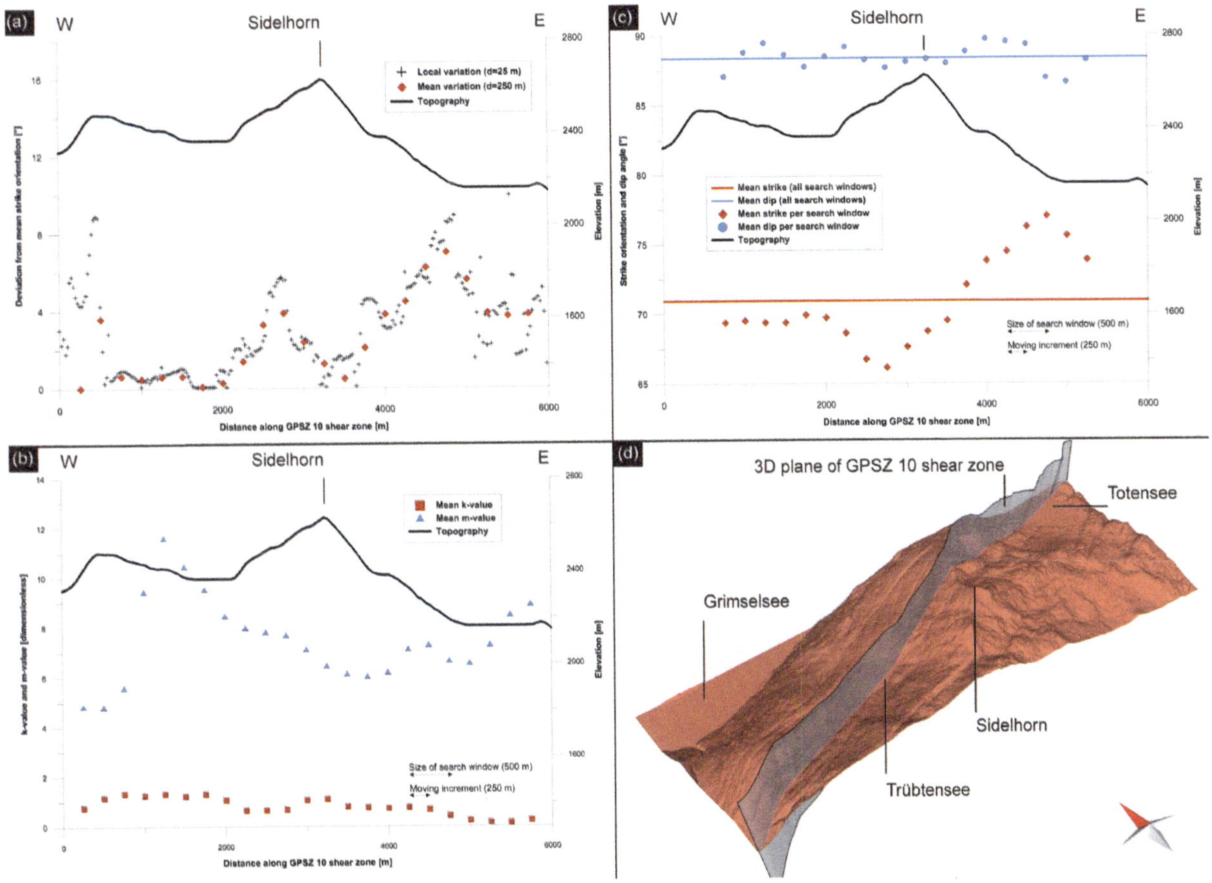

Figure 10.8 Calculation of the 3D orientation of a plane based on surface information. The quality (*k*-value and *m*-value, see Appendix B for details) of a best-fit plane through a given outcrop trace is investigated as a function of the variability in strike orientation (*k*-value) and in dip (*m*-value) within a search window size of 500 m and a constant moving distance of 250 m. Example analysis was performed along topography (black line) of the GPSZ 10 shear zone crossing Mount Sidelhorn (for location see Figure 10.1). (a) Values for orientation differ significantly between individual measurements and the averaged values per search window size. (b) Mean values for *k* and *m* vary within a small range along the strike, indicating a consistent planarity along the entire shear zone trace. (c) Averaged values of dip and strike per search windows deviate only within small bounds from mean values for the entire shear zone, pointing to smooth regional variations in strike and dip. (d) Visualization of the near-surface 3D geometry of GPSZ 10, nicely displaying the regional smooth variation along its trace.

are used to construct best-fit planes. At least three points (which comprise two fault segments of 25 m length) are used for this construction defining a minimum length of 50 m for a fault plane in the light of 3D extrapolations. Such best-fit planes further enable the extraction of dip and dip azimuth for each data point (25 m point distance) along the 3D intersection. For a local evaluation, the remote sensing-based fault plane orientations are compared with local field measurements (dip, azimuth; see Figure 10.9).

10.3.5.2 Dip Calculation for Surface Points Along the Lineament

Applying the best-fit plane approach for remote sensing data provides locally averaged azimuth and dip values at all scales $>1/2\lambda_{small}$. To estimate the quality of remote sensing derived azimuth and dip values, GPSZ10 is used as an example (sampling distance = 100 m), its orientation data are compared with field measurements generated along the surface trace of GPSZ10 (Figure 10.9). Using stereographic projections, we analyze the discrepancy between these two data acquisition methodologies and the ability of the field data to validate remote sensing based structural data (Figure 10.10a). A total of three data sets are investigated: (i) complete best-fit plane data, (ii) filtered best-fit plane values, and (iii) field measurements. (i) The complete, non-filtered dataset ($n = 62$, all *k*-values) is used to construct the general 3D orientation of GPSZ10; (ii) selected points only (*k*-value ≤ 1; see Ferńandez, 2005) contribute to the best-fit plane construction of GPSZ10 ($n = 30$); and (iii) field data ($n = 14$) located along the trace of GPSZ10 serve as a cross-check.

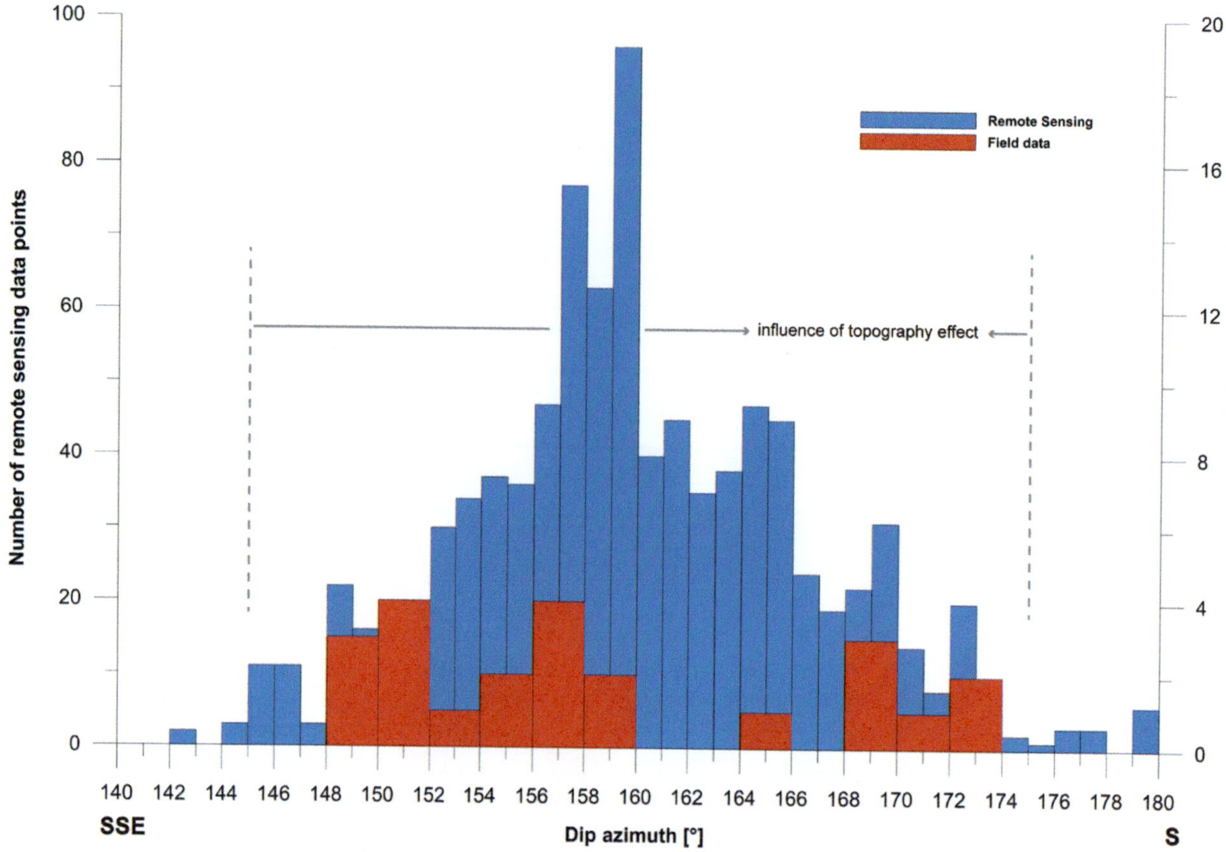

Figure 10.9 Variation of the strike along the outcrop trace of GPSZ 10. Histograms are of remote sensing data (blue: strike orientation of segments of 25 m length) and field observations (red: single data points). Mean value is 160°, 95% confidence interval equals 160 +/−15°, covering completely the data range of field measurements.

In general, remote data and field measurements correlate very well (Figure 10.10a). In agreement with Ferńandez (2005), the geometry of the best-fit-plane is most accurate if fault sticks with *k*-values ≤ 1 are used for plane construction only. The enlarged view in Figure 10.10b displays the variation in azimuth/dip between remote sensing data ($k \leq 1$: blue squares; all *k*-values: black crosses) and field measurement (red triangles). Single orientation values of best-fit plane data show an elongated, ellipsoidal pattern with steeply oriented planes. They are homogeneously distributed and the poles deviate from the mean orientation of the plane (160/84) by ~10° in dip and dip azimuth on average. In contrast, field measurements (red triangles) scatter more. Aberration is on average 20° for dip azimuth and 10° in dip angle from the respective mean value (156/78). The correlation between spatially correlated (search radius of 25 m) remote sensing data and field measurements (orange lines) reveal maximum aberrations of 20° in azimuth and 15° in dip, whereas the smallest difference is smaller than 5° for either value.

10.3.6 Extrapolation of Surface Data to Depth

Digitized 2D lines as well as constructed 3D planes show lateral variability in orientation. In 2D, this issue only comprises the variation in strike, whereas in the third dimension the dip angle must additionally be considered. To assess the correct 3D spatial location of fault zones, maximum variations of dip and dip azimuth of these geological structures must be defined. These 95% confidence intervals were obtained by resampling the values obtained from a best fit plane calculation (Ferńandez, 2005). The calculation interval is 200 m, in order to prevent influences from local relief to the calculation. Obviously, the real-world distribution of these values (e.g. shorter distances than 200 m) is not known. In such cases, bootstrapping helps to calculate, for example, the confidence intervals mentioned, based on the empirical distribution (Chernick, 2008). We used the Jackknife approach, a subset of the bootstrap resampling methods. For each shear zone under consideration in this study, we replaced the empirical distribution by generating 1,000 sets (a so-called bootstrap sample), each

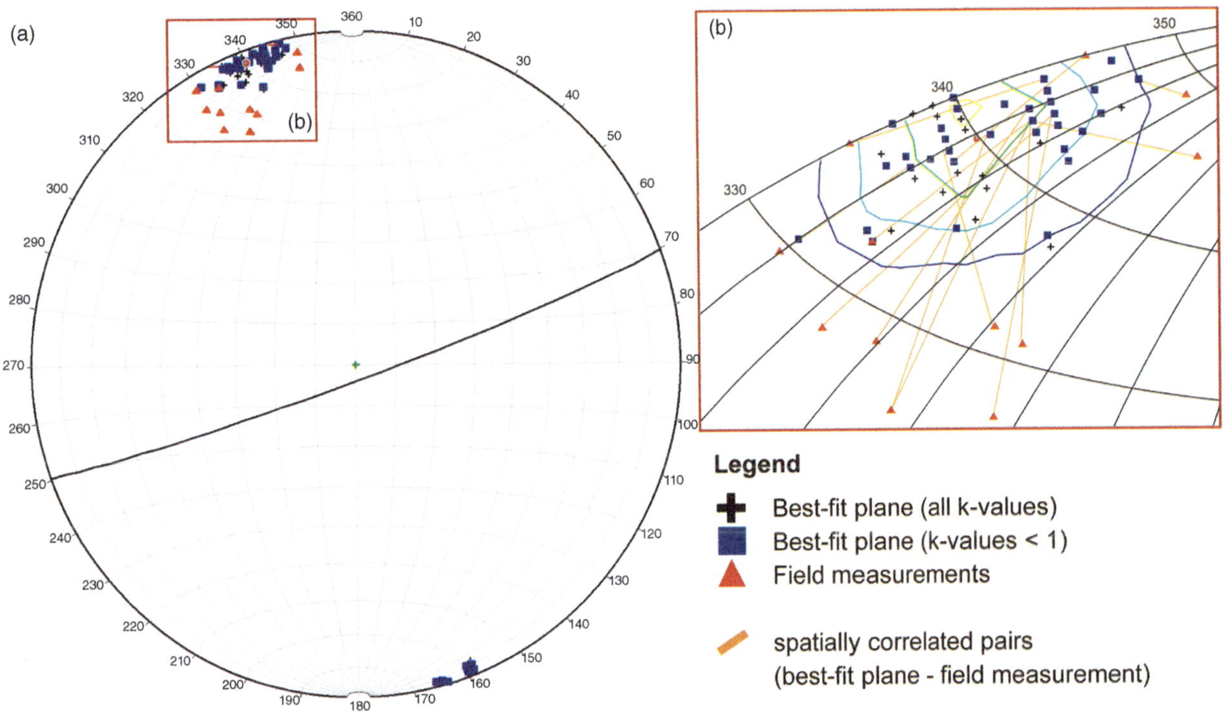

Figure 10.10 (a) Stereographic projections showing variability in dip and dip azimuth of data obtained by remote sensing approach (blue squares: best-fit plane approach; $n = 30$; black crosses: all measurements with $n = 135$) and measured in the field (red triangles). (b) Enlarged view of (A) showing the correlation of 14 remote sensing – field data pairs obtained at the same locations.

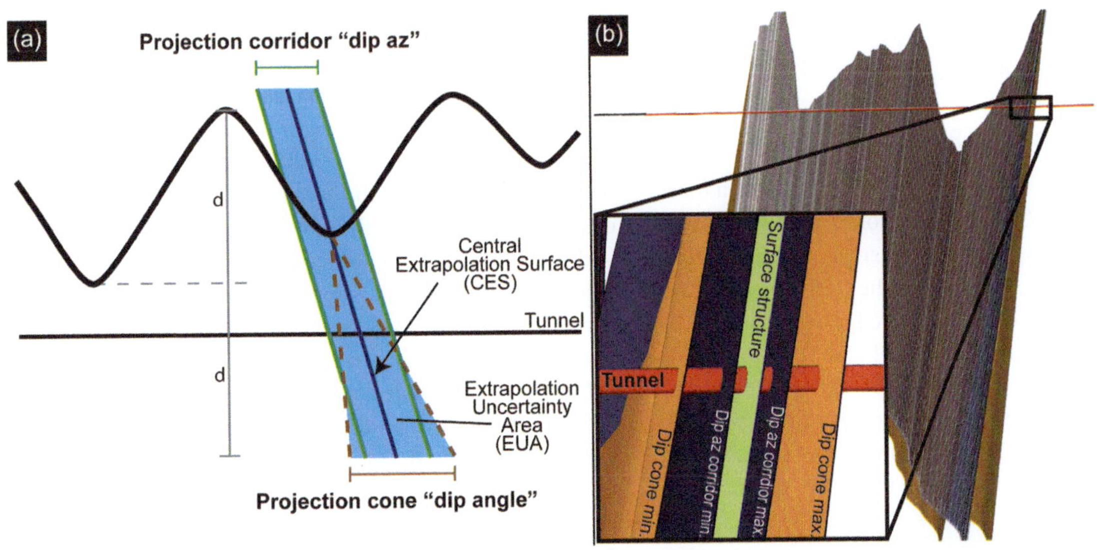

Figure 10.11 Uncertainty estimates for the extrapolation of surface structures to depth in 2D (a) and 3D (b). (a) 2D section showing central projection (blue line), projection corridor (based on bandwidth of error and associated dip azimuth; see Figure 10.5) and projection cone based on variability in the dip angle. Note that near and far surface sections, respectively, are mainly affected by uncertainty related to the bandwidth error and the dip angle.

of them consisting of four empirical samples. Next, 100 sets were selected randomly out of the bootstrap sample, which served as the basis for the calculation of both the mean dip and dip azimuth values as well as the dip and dip azimuth 95% confidential intervals. For more details see Chernick (2008).

Uncertainty considerations, which include the influence of the dip angle, are described by, for example, Bistacchi et al. (2008). These authors remain at general uncertainty considerations only and do not deduce their projection uncertainties based on their data. In contrast, we use our own data to derive the so-called "Extrapolation Uncertainty Volume/Area" (EUV: 3D, EUA: 2D section). Our approach comprises four steps:

1. Definition of minimum and maximum discrepancies in variations in strike and dip angle of our fault segments using the bootstrapping methodology (Jackknife resampling: Efron, 1979; Chernick, 2008).
2. Development of a central extrapolation surface (CES), which is defined by the mean dip and dip azimuth values obtained in step 1.
3. Allocation of the previously defined bandwidth uncertainty (projection corridor) to the CES.
4. Assignment of the projection wedge uncertainty, which defines the maximum variety of dip angles allowed for extrapolation.

To estimate the extrapolation uncertainty (projection uncertainty (PU) in Bistacchi et al., 2008), the projection corridor (i.e. bandwidth uncertainty) and projection wedge are combined to form the EUV/EUA (Figure 10.11a).

10.3.7 Assessment of 3D Plane Constructions

The final step of our approach comprises the evaluation of EUV/EUA and CES with fault zones at tunnel level obtained from mappings of the Transitgas Grimselstollen (scale: 1:500; Schneider, 1974). After the CES is projected to depth and the EUV/EUA is constructed, extrapolated fault zones are intersected with the tunnel axis (Figure 10.12). To identify the local relevance of the extrapolation (i.e. whether a single fault zone or a series of fault zones is enclosed by the EUV/EUA), an extended area of interest is defined, which in total spans three times the width of the EUA in a cross-section.

10.4 Results and Discussion

10.4.1 Remote Sensing and 2D Lineament Data

10.4.1.1 Uncertainties in 2D Lineament Data

Spatial Resolution of Image Data To evaluate the maximum image resolution necessary for a proper definition of the spatial continuation of fault traces, the Grimselpass shear zone 10 (GPSZ10) is used as a representative example for the study area. This fault trace was digitized three times on a DEM using resolutions of 2 m, 5 m, 10 m respectively (Figure 10.5b–d). The minimum digitization area of 3×3 pixels (Figure 10.5a) affects the minimum width of the structure and length of a line segment and defines the maximum interpolation distance between adjacent, but non-continuous digitized lines for lineament interpolation. In Figure 10.5b, we note that continuous mapping in full detail is possible when using high-resolution DEM (2 m/pixel). In this case, the minimum segment length to digitize and the maximum interpolation distance is 6 m. Figure 10.5c illustrates that a fully detailed digitization is not possible any more when using a DEM resolution of 10 m/pixel. The minimum line length and maximum interpolation distance equal 30 m, resulting in a discontinuous line where geometrical details become fuzzy. As Figure 10.5d documents, a very poor resolution of 25 m/pixel for topography is insufficient for mapping lineaments.

Note that in Figure 10.5b-d, due to the resolution being too low, no linking of the trace of GPSZ10 is made across the alluvial plane. An inferred trace can be drawn in Figure 10.5b only, whereas a continuous line would contravene the rule of maximum interpolation distance. Figure 10.5e illustrates the high-quality results when using high-resolution 2m raster data for remote sensing lineament digitization.

Influence of Varying Illumination Directions The 2D trace of GPSZ10 is represented by eight different non-continuous line subsets (represented by different colors), due to the influence of varying illumination directions. In the current example, the bandwidth uncertainty shows a constant width of 22 m. Traces of large-scale faults (Figure 10.6b) exhibit the effect of topography by a smooth and systematic variation in strike. As described in Appendix A, steeper oriented planes show a narrower strike confidence interval than gently dipping surfaces. For the steeply dipping GSPZ10 shear zone, all digitized 2D line segments of each subset are oriented subparallel to the general strike of GPSZ10.

Spatial Location of Remote Sensing and Field Data Strike orientations of lineaments are measured by means of remote sensing for 25 m long fault segments along the 4 km long trace of the GPSZ10 fault zone (Figure 10.6). Dip markers in Figure 10.6 represent the locations of additional field measurements. Hatched areas in Figure 10.6 denote regions within the bandwidth uncertainty, where no correlation can be done, due to exceeding the maximum interpolation

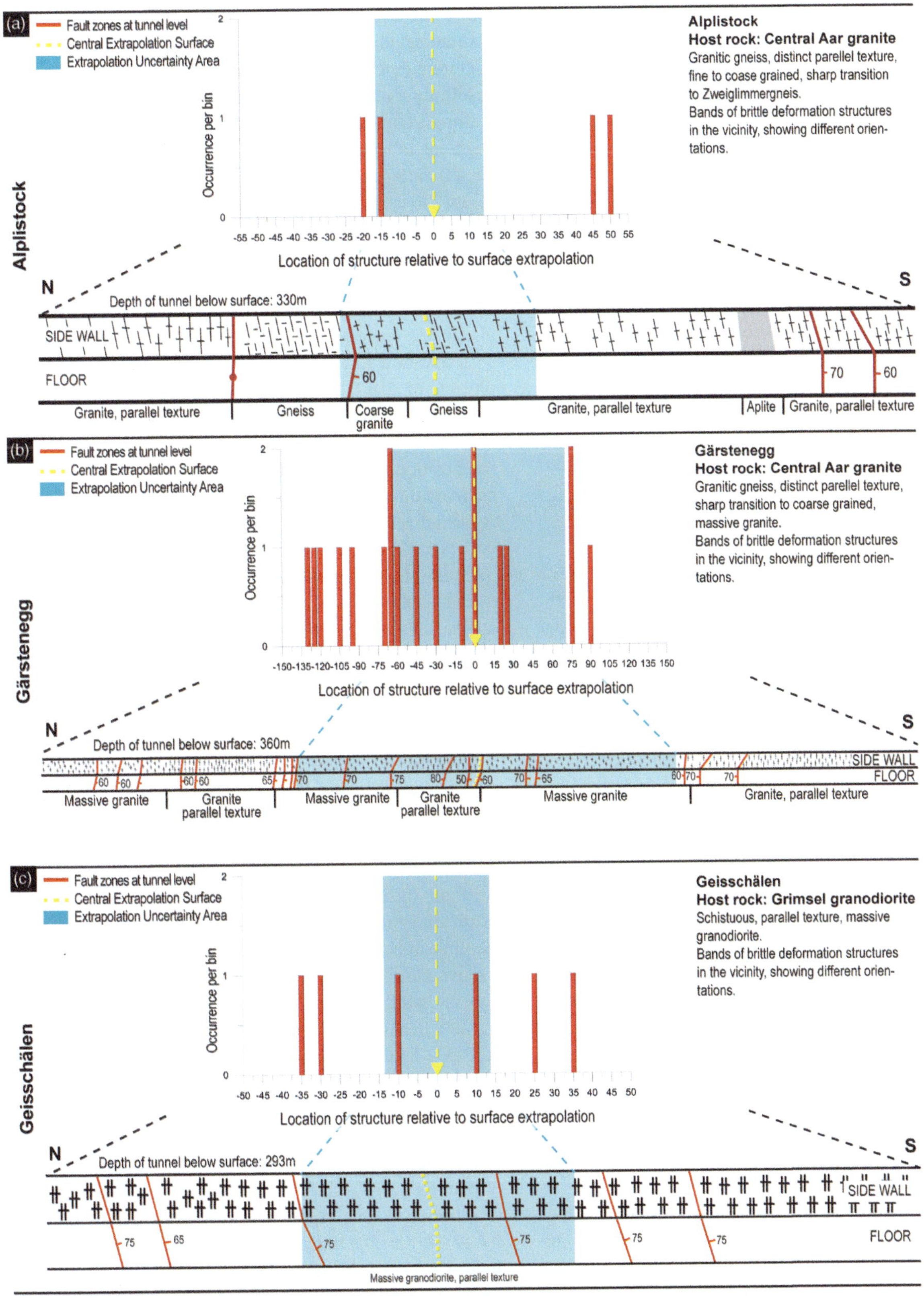

Figure 10.12 Projection of surface structures to depth and correlation with structures at tunnel level. (a) Alplistock-SZ: several meters wide mylonite zone granite just touches the rim of the Extrapolation Uncertainty Area (EUA). (b) Gärstenegg-SZ: projection cone uncertainty comprises in the CAGr a tunnel interval of 130 m showing 10 mylonitic shear zones and a gneissic granite domain. (c) Geisschälen-SZ: projection cone uncertainty defines in the schistose GrGr a 28 m wide tunnel interval, in which two mylonitic structures occur. See Appendix C for histograms for every major structure under investigation.

distance between non-continuous line segments. Note the differences in strike between remote sensing-based data and field observations. The histogram in Figure 10.9 illustrates the frequencies of the different segment orientations (blue) in comparison to data obtained by field measurements (red). Whereas the remote sensing data set covers almost the entire length of the fault trace (exceptions are lakes), the field data are restricted to a small number of specifically selected locations, as a result of limited accessibility or bad outcrop conditions. Remote sensing provides a regionally representative 2D dataset with a statistically well-based Gaussian normal distribution around a mean value of the azimuth of 160°. Field data scatter within +/−15° around 160° and plot within the 95% confidence interval of the remote sensing data set. It is important to note that owing to the topography effect, remote sensing-based strike values scatter between +/−15° of the main orientation of ~70°.

10.4.1.2 Discussion of Uncertainties Related to 2D Lineaments

The remote sensing detection of 2D lineaments based on DEM or orthophotos depends on the resolution of the underlying imagery (Figure 10.5). In this kind of image analysis, lineaments are represented as vectors, superimposed on a regular and rectangular gridded imagery, i.e. the pixel system. Kocal et al. (2007) denote length, location, and orientation of a digitized line segment (vector) as the most critical parameters for uncertainty estimation.

The minimum pixel matrix size required to detect a lineament is 3 × 3 pixels (Figure 10.5a). For the three chosen image resolutions of 2, 10, and 25 m respectively, matrices of 6 × 6, 30 × 30, and 75 × 75 m result. Figure 10.5b–d illustrates that the spatial resolution of digital imagery is a key factor in remote sensing (e.g. Kienzle, 2004; Hopkinson, 2009; Gajski, 2004). In our case, digitization of lineaments in full detail can only be performed using the highest resolution (2 × 2 m/pixel, Figure 10.5b and e). All other resolutions (10 × 10 m/pixel and 25 × 25 m/pixel) are too low and do not provide sufficient precision for an accurate digitization of the trace of a fault zone (Figure 10.5c and d). Based on the 3 × 3 pixel matrix argumentation, the minimum length of a digitized lineament is 3 pixels, which therefore defines the maximal spatial resolution possible for digitization.

Since *x*, *y*-coordinates of starting and end points of digitized vectors directly rely on the *x*, y coordinates of the image, the pixel spacing of the image may induce a loss of spatial accuracy of vector data in the case of low pixel resolutions, as shown in Figure 10.5c–d (e.g. Heilbronner and Barrett, 2014; Wellmann et al., 2010; Crosetto, 2000). This restriction of course influences the representation of a lineament in 2D map space, as proposed by, for example, Caumon et al. (2009).

Owing to aforementioned requirements on pixel-based resolution (10 × 10 pixel matrix, angular and spatial resolution), a two-step approach is necessary to generate a fault segment as accurate as possible. In a first digitization step, the 2D fault trace is defined by a maximum pixel resolution of 2 × 2 m using the 3 × 3 pixel matrix for local fault detection, followed then in a second step by a GIS-based automated intersection of the 2D fault trace into fault segments of 25 m length. This procedure is a prerequisite not only for a statistical orientation analysis of fault orientations but also for the correct further extraction of 3D directional data.

In addition to the pixel-based spatial and angular resolution of 2D fault segments (Kocal et al., 2007), the bandwidth uncertainty also reflects an additional parameter in the chosen remote sensing approach, affecting the positional accuracy of a fault segment due to the link between image resolution, topography, and DEM shadings. As visualized by Figure 10.6, the 2D variability in the *x*- and *y*-directions of the detected fault segments along a fault trace are defined by an envelope band, which is parallel to the mean orientation of the large-scale structure and whose width represents the bandwidth uncertainty. In the case of the chosen example, the bandwidth uncertainty is 22 m, a value that additionally affects the topography effect. Re-calculating the bandwidth uncertainties for other steeply dipping fault zones in the working area delivers similar values for the bandwidth uncertainty. Therefore, the uncertainty value of 22 m is representative throughout the entire area. In general, the bandwidth uncertainty becomes smaller the higher the resolution of the input data under use and the steeper the dip of the digitized fault zone (straight vectors). In general, higher resolution imagery results in better digitization accuracy and, thus, in smaller location uncertainty. As a great advantage, knowledge of the mean orientation of a fault trace (from fault segments) and the bandwidth uncertainty allows for the prediction of fault traces in regions where no surface data is available (e.g. lakes or alluvial plains; see the hatched areas in Figure 10.6).

In the case of 2D orientation statistics of fault segments, the topography effect needs to be considered. Since most of the fault zones in the study area are steeply oriented (dip angle >70°), the impact of the topography effects is minimized (see Appendix A, Figures A.1 and A.2). Based on the theoretical consideration (see Appendix A, Figure A.1), a topography-induced variability in orientations of an apparent 2D strike of up to 125° is possible. This is a very important aspect for 2D lineament analyses, since it apparently widens the topography-dependent orientation range in strike (e.g. rose diagrams). However, this topography effect is no disadvantage at all in the case of later 3D extrapolation of surface data to depth. On the contrary, topography effects deliver crucial information for the dip direction and dip values of 3D planes, particularly when fault structures are not steeply oriented.

To evaluate the quality of our remote sensing-based approach (Figure 10.9), the obtained orientations (blue, $n = 135$) of the 2D trace of the fault zone GPSZ 10 are compared with field measurements (red, $n = 14$). The obtained fault segments of GPSZ10 are normally distributed around a mean dip azimuth (~160°). Values towards the extremes of the distribution are owed to the topography effect. As illustrated in Figure A.2 in Appendix A, the topography effect for GPSZ 10 (dip angle of 65°) comprises a possible azimuthal deviation from the true strike of ~30°. The valid dip azimuth domain for GPSZ 10 is therefore restricted to 145° to 175° (160°+/−15°). In contrast, field measurements range from 148° to 174° (160°+/−14°), which is within the valid dip azimuth domain of GPSZ 10 as it is within the 95% confidence interval of the complete remote sensing dataset. Therefore, a constant strike along the entire fault zone trace can be assumed. The remote sensing and field data provide a very good estimation regarding the orientation of the fault zone trace in 2D map space, which is supported by Baumberger (2015).

10.4.2 Dip Extraction for Remote Sensing 2D Lineament Data

10.4.2.1 Uncertainties in Calculated Dip Values

Representation of Topography For the remote sensing-based projection of a fault plane to depth, azimuth and dip have to be calculated, a step for which the topographic variation along the fault trace is crucial. Figure 10.7 displays such a variation along a 3D trace of a fault zone (Figure 10.7a) in a vertical 2D section (Figure 10.7b). The morphological variation of the terrain can be considered in terms of scale-dependent amplitudes (A) and wavelengths (λ). At the regional scale, the landscape is expressed by large A and λ along section trace (A_{large}, λ_{large}). In contrast, relief is defined by small values of A and λ (A_{small}, λ_{small}) at the local scale (Figure 10.7b). In order to derive azimuth and dip of a fault plane, topographic variations of A over $1/2\lambda$ are sufficient (Figure 10.7c). To sufficiently describe the shape of a minimum topographic variation (hill or valley, i.e. $1/2A$ and $1/2\lambda$, Figure 10.7c), a minimum of three 3D fault segments is necessary for the calculation of the orientation of a local 3D plane. Given the image resolution dependent fault segment length (V), $1/2\lambda_{\text{small}}$ has to be at the minimum 3 times V (e.g. for 25 m long fault segments, $1/2\lambda$ should be a minimum 75 m). In terms of the minimum vertical variation necessary for the calculation of a local 3D plane, A_{small} can be defined by geological uncertainties derived from field experience and technical restrictions. The parameter A describes the variation in elevation needed along the lineament trace to identify the orientation of the 3D plane in the field, whereas the parameter λ represents the vertical accuracy of the underlying DEM (here 0.5 m). We propose A_{small} to be 10% of $1/2\lambda$ as a minimum to construct a valid 3D plane, which is about 8 m in our case.

Whereas scale-dependent values of A and λ describe the morphological variation of the terrain, the accuracy of its representation depends on the point spacing chosen for digitization. Figure 10.7d visualizes the XZ-plot of the fault zone trace shown in Figure 10.7a, which is resampled at different point spacings (2 m to 1,000 m). The minimum length of relief variability, which still can be identified by means of remote sensing, is defined by the resolution of the underlying imagery (in our case: 2 m) and the minimum 3×3 pixel matrix. While increasing the original spacing of 2 m (topography/DEM), the trace line becomes buckled and shows simplifications at inflexion points. This effect is clearly visible for point distances of 100 m, 250 m, 500 m and 1,000 m. Smaller point spacing (5 m, 10 m, 25 m, 50 m) represents the underlying topography more accurately (see inset in Figure 10.7d). As illustrated there, the 50 m point distance (orange line) still shows simplifications, which become prominent in regions with high variability in local relief. Point distances of 5 m (blue), 10 m (light green), and 25 m (red) represent the effective topography in a reasonable manner, implying spatial resolutions for local calculations of 3D orientation of a fault plane of 3×5 m, 3×10 m, and 3×25 m respectively.

3D Best-Fit Plane from Lineaments Lineament traces mapped in 2D serve as input data for the extrapolation to depth of surface information (Baumberger, 2015; Appendix C). They are projected to a high-resolution DEM (2 m/pixel) and their intersection with the topography provides a general impression of the respective 3D geometry of the lineament. Fernández (2005) presents an approach to develop so-called 3D best-fit planes from these 3D intersection lines only, which allows for the extraction of dip and dip azimuth values from initial 2D lines. After projection onto the DEM, the trace of example shear zone GSPZ10 is analyzed to identify the uncertainties related to the best-fit plane construction according to Fernández (2005). The influence of the local relief and the relevant input parameters (point distance, size of search window), which define the overall quality of the resulting best-fit plane, is measured by roughness of the individual three-point planes relative to the average plane (m-value) and the variation in strike of the local individual three-point planes

(k-value). A constant interval of 250 m is used to move a 500 m wide search window along the trace of GPSZ10. Figure 10.8a illustrates the local variation in orientation from individual 25 m line segments compared to the deviation from the mean strike value per search window. Between profile distances 250–500 m, 2,000–3,000 m, and 4,000–6,000 m, the local orientation deviates remarkably from the search window mean values. The variations of k-values and m-values are displayed in Figure 10.8b. The k-values generally scatter between 0.5 and 1.5, reaching a minimum only slightly above zero at the water table of Lake Totensee in the E of GPSZ 10. For the m-value, values scatter between 4 and 12, but in general vary between 6 and 8. In line with their position along the strike of GPSZ 10, the outlier values correspond to a change in general orientation of GPSZ 10 (due to union with another structure) and a long distance with no change in direction (due to absence of measurement points while GPSZ 10 crosses a lake), respectively. Since the k-value and m-value essentially depend on the 3D intersection of the trace line with topography, no k-values and m-values can be detected at locations with a low relief (e.g. water surfaces, planes). By definition, the calculation algorithm sets the measurement values to zero at such areas.

The deviation in dip (m-value) and strike (k-value) per single search window (Figure 10.8c) vary between +/−2° and +/−5° respectively, from the corresponding mean values of the entire GPSZ 10 shear zone (dip: 87°; strike: 71°). Note that the boundary effects (missing data points on both sides for mean strike and mean dip values per search window), caused by the principle of the moving window, are accounted for in Figure 10.8c.

Based on the size of the search window, k-values and m-values can be calculated for every point in the defined interval along the fault zone trace. Subsequently, the complete dataset is filtered for values of $0.5 \leq k \leq 1.5$ and $4 \leq m$ per point, which are used to construct the best-fit plane for the corresponding fault zone.

Dip Calculation for Surface Points Along the Lineament Applying the best-fit plane approach for initially 2D remote sensing data provides locally averaged azimuth and dip values at all scales $>1/2\lambda_{\text{small}}$. In order to estimate the quality of remote sensing-derived azimuth and dip values, GPSZ10 is used as an example (sampling distance = 100 m) and its orientation data are compared with field measurements generated along the surface trace of GPSZ10 (Figure 10.9). Using stereographic projections, we analyze the discrepancy between these two data acquisition methodologies and the ability of the field data to validate remote sensing-based structural data (Figure 10.10a). A total of three data sets are investigated: (i) complete best-fit plane data, (ii) filtered best-fit plane values, and (iii) field measurements. (i) The complete, non-filtered dataset ($n = 62$, all k-values) is used to construct the general 3D orientation of GPSZ10; (ii) selected points only (k-value ≤ 1; Ferńandez, 2005) contribute to the best-fit plane construction of GPSZ10 ($n = 30$); (iii) field data ($n = 14$) located along the trace of GPSZ10 serve as a cross-check.

In general, remote sensing data and field measurements correlate very well (Figure 10.10a). In agreement with Ferńandez (2005), the geometry of the best-fit plane is most accurate, if fault sticks with k-values ≤ 1 are used for plane construction only. The enlarged view in Figure 10.10b displays the variation in azimuth/dip between remote sensing data ($k \leq 1$: blue squares; all k-values: black crosses) and field measurement (red triangles). Single orientation values of best-fit plane data show an elongated, ellipsoidal pattern with steeply oriented planes. They are homogeneously distributed and the poles deviate from the mean orientation of the plane (160/84) by ~10° in dip and dip azimuth on average. In contrast, field measurements (red triangles) scatter more. Aberration is on average 20° for dip azimuth and 10° in dip angle from the respective mean value (156/78). The correlation between spatially correlated (search radius of 25 m) remote sensing data and field measurements (orange lines) reveal maximum aberrations of 20° in azimuth and 15° in dip, whereas the smallest difference is smaller than 5° for either value.

10.4.2.2 Discussion of Uncertainties Related to 2D Dip Extraction

The best-fit plane approach proposed by Ferńandez (2005) enables the transition from 2D mapping of geological structures to their 3D geometrical representations (e.g. Sala et al., 2014; Ferńandez et al., 2009; Banerjee and Mitra, 2004). The quality of remote sensing-based 3D orientations of local plane segments can be discussed on the basis of comparison with field measurements (Figure 10.10). The limited spread of +/−10° of individual poles of the best-fit plane data (blue squares) around the mean orientation of 160/84 points to both a precise 3D estimation and a homogenous and steep long-range orientation of GPSZ. In contrast, field data (red triangles) are more scattered. This discrepancy visualizes the strengths and weaknesses of either methodology very well (e.g. Baumberger, 2015; Bond et al., 2007; Kocal et al., 2007).

(A) Remote sensing data provides homogeneously distributed results in terms of location and orientation but does not account for information on the outcrop scale (e.g. Baumberger, 2015; Dar et al., 2010; Rahiman and Pettinga, 2008; Morelli and Piana, 2006). The influence of the topography effect while digitization and subsequent uncertainty propagation during further data processing accounts for the variation in this data set. In these terms, the location of remote sensing data and the precision of measurements still remains the most important aspects for its uncertainty evaluation (e.g. Baumberger, 2015; Wellmann et al. 2010; Kocal et al., 2007; Gahegan and Ehlers, 2000).

(B) Field data supplies detailed information for single locations but does not necessarily provide regional scale correlations. In cases where the accessibility of tectonites directly at the fault trace is limited (e.g. steep terrain or bad outcrop conditions due to coverage by debris, weathering products or soil), many of the measurements in the field have to be derived at the boundaries of the morphological incisions (Baumberger, 2015). Hence, the structural data are either generated at the rim of the fault core or even in its damage zone. An assumed parallelism of the measured structures to the main fault can therefore not be fully guaranteed, given the heterogeneous nature of deformation at the outcrop scale and the occurrence of secondary structures such as synthetic Riedel structures, splays, or bridges (Faulkner et al., 2010, and references therein). Owing to the restricted scale of maximum resolution and the larger spatial correlation of local fault planes over considerable distances (search window), the remote sensing approach is much less sensitive to local effects as is the case for field measurements. Consequently, a more consistent dataset results from remote sensing, which provides outcomes with better 3D precision.

10.4.3 3D Extrapolation to Depth

10.4.3.1 Results

Extrapolation Uncertainty Structural orientations extracted from 3D best-fit planes are used to extrapolate near-surface geometries to depth. The minimum and maximum lateral deviations in dip and dip azimuth of 3D fault zones define the projection wedge (dip dependency) and the projection corridor (strike dependency), respectively (Figure 10.11a). The values needed are calculated using the bootstrap resampling (Jackknife method; Efron, 1979; Chernick, 2008). The main surface is extrapolated to depth starting from the 3D intersection line of the fault zone, by applying the mean values for dip and dip azimuth obtained from Jackknife resampling (Figure 10.11a). The previously defined bandwidth uncertainty, which is 22 m in total width for all 3D fault zones (11 m on either side of the 3D intersection line), defines the corridor around the projected main surface (green lines in Figure 10.11a). Minimum and maximum strike values define the variation in orientation of the projection corridor along the 3D intersection line. Minimum and maximum dip angles along the 3D intersection line define the projection wedge (dashed brown lines in Figure 10.11a). Their respective combination, the so-called Extrapolation Uncertainty Volume/Area (3D, EUV; 2D cross-section, EUA), defines the maximum range of uncertainty to be expected at the tunnel level after extrapolation. Figure 10.11b illustrates an extrapolation of a trace of a 3D fault zone to depth and the respective evaluation of the projection type. As is clearly visible in the inset of Figure 10.11b, the 3D planes representing the projection cone envelope the three-dimensional projection corridor. The elevation of the tunnel determines which of the two projection uncertainties represents the worst case. In the chosen example, the EUV/EUA is defined by the dip angle, which has to be considered when dealing with the maximum uncertainty.

Projection of Surface Data In order to evaluate whether fault structures are indeed found within the predicted projection and associated EUA, 3D intersection lines are projected down to the level of the Transitgas Grimselstollen (Figure 10.1 for location details). The turquoise EUAs visualize the resulting extrapolation uncertainty areas at tunnel level in Figure 10.12a for three fault zones (for locations see Figure 10.13a): (A) the Alplistock shear zone, which is located in the northern part of the Central Aar granite (CAGr); (B) the Gärstenegg shear zone occurring in the southern part of the CAGr; and (C) the Geisschälen shear zone, which is located in the transition area between the CAGr and the Grimsel Granodiorite (GrGr). Yellow lines represent the central extrapolation surface (CES) and red lines document the occurrence of fault zones digitized from the original tunnel mapping by Schneider (1974). To evaluate the wide-range occurrence of fault zone arrays outside the EUA, an area of interest is defined for each extrapolation, which is three times larger than the EUA. (A) The CES intersects the tunnel level in the middle of an 8 m wide gneiss zone (original mapping by Schneider, 1974), terminated to the N and S by CAGr host rock of different deformation degrees (coarse-grained and parallel textured). Two singular fault zones occur at the left outermost rim (northbound) of the EUA. They denote the boundary between coarse-grained CAGr and a neighboring gneissic zone and a gneissic zone and parallel textured CAGr respectively. This transition can be correlated with the outcrop of the Alplistock shear zone. The remaining mappings towards the S are located far outside the EUA (width: 30 m). Based on the surface mapping and their location outside the EUA, they are not considered to be related to the Alplistock shear zone. Therefore, they are excluded from any further treatment.

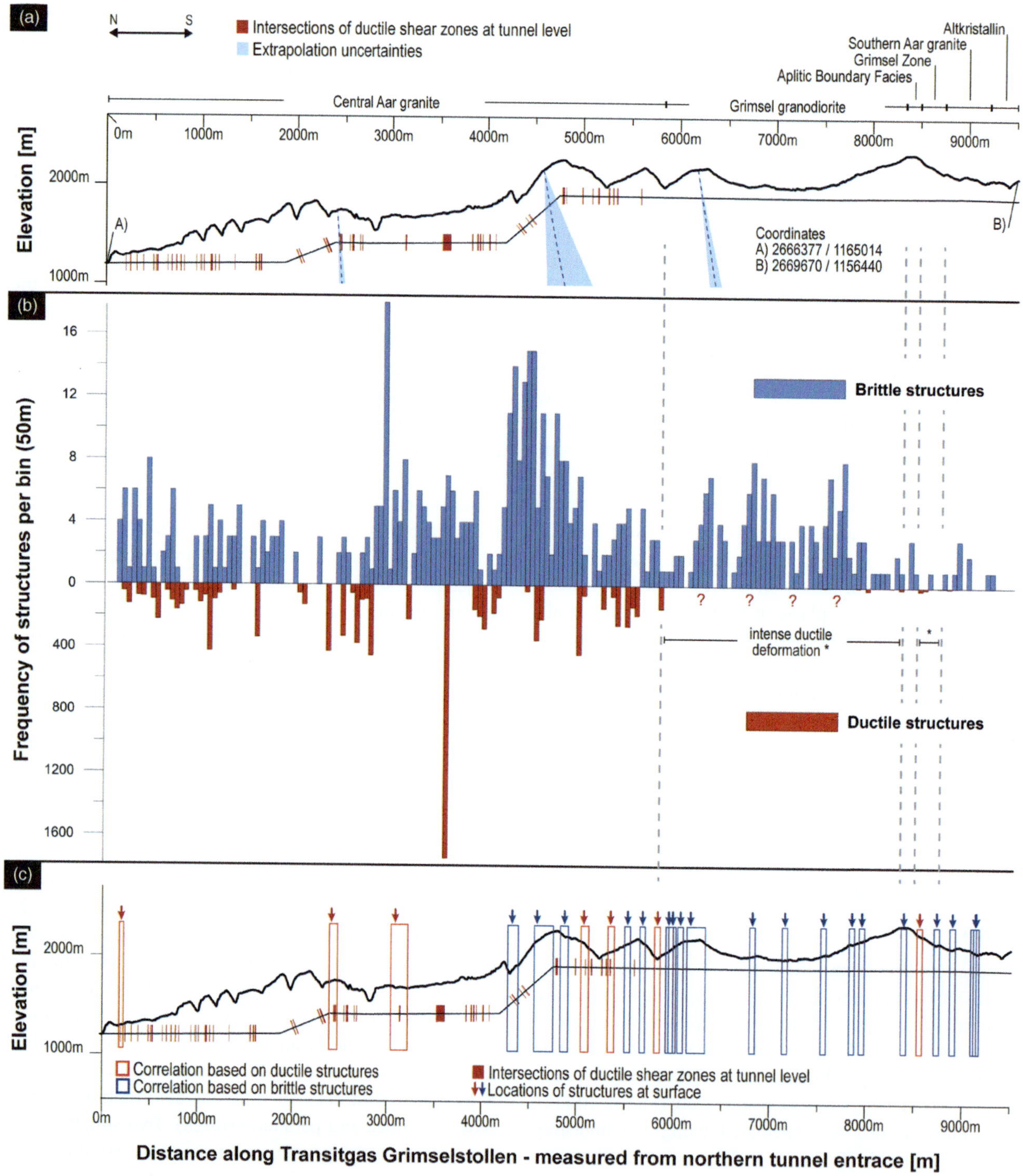

Figure 10.13 Correlation of NE–SW striking surface incisions with ductile shear zones and brittle faults at the tunnel level. (a) Vertical section showing topography, trace of the tunnel, locations of mylonitic domains, and projection cones (blue) of the three shear zones in Figures 10.10 and 10.11. (b) Histogram showing the frequency of brittle (blue) and ductile (red) structures along the tunnel trace. Frequencies are based on multiples of minimum widths of 5 cm per shear/fault zone. (c) Correlation of NE–SW striking surface incisions (arrows) with occurrence of brittle (blue) or ductile (red) faults within the uncertainty range (rectangles) at tunnel levels.

In contrast to (A), the CES of the Gärstenegg domain (B) hits an area of densely spaced fault zones at tunnel level. Thirteen occurrences are located within or in closest distance outside the EUA (width: 130 m). The density of fault zones is higher towards the left (northbound) of the CES than to the right. The correlation with the original tunnel mapping by Schneider (1974) unravels the fact that the high-density zone coincides with a parallel textured zone within surrounding massive CAGr. The CES is located at the southern sharp transition between these zones. (C) The EUA of the Geisschälen domain is entirely located within the commonly strongly foliated GrGr. Within the 25 m wide EUA, only two fault zones occur, which are symmetrically located around the CES. Evaluations of the ductile strain localization phenomena on the base of the old mapping are hardly possible owing to a lack of discrimination of different intensities of foliation. We are therefore resigned to further correlate surface with subsurface structures in the GrGr (Figure 10.13). Intersection analysis for all large-scale fault zones of the CAGr investigated in this study are presented in Appendix C.

Structural 3D Model and Shear Zone Map Fault zones mapped at the surface and in the tunnel differentiate in their frequency of occurrence, spatial location as well as geometrical appearance (e.g. orientation, width). To evaluate the relationship between both these datasets, criteria for the correlation of surface information with underground data need to be defined and the relevant fault zones must be selected. First, only subgroups of either data set are compared, which are oriented subparallel to the strike of the Aar massif (055°+/−15°). Second, a small number of prominent fault zones morphologically clearly expressed at the surface (width >1 m; $n = 30$ along the section length) is accompanied by a large number of subsurface fault zones at tunnel level (width >0.1 m; $n = 450$ along the section length). Figure 10.13a illustrates the 2D section of the Transitgas tunnel and the occurrence of CAGr and GrGr. Red areas in the CAGr denote intersections of ductile shear zones with the tunnel axis, whereas blue cones display the three example projections shown in Figure 10.12a–c. The frequency of fault zones along the trace of the tunnel is shown in Figure 10.13b. Brittle fault zones (blue) continuously occur along the trace in high frequencies of up to 17 fault zones per bin (bin size = 50 m). On the contrary, ductile shear zones are discretely located preferentially in the CAGr (artifact). For normalization purposes the frequencies are based on multiples of minimum widths of 5 cm per fault zone, e.g. a 50 cm wide fault zone appears as 10 entries in the histogram. Figure 10.13c shows possible correlations between surface data and underground mappings, based on the histogram in Figure 10.13b. Arrows indicate brittle (blue) or ductile (red) fault zones at the surface. The location of the arrows is connected with the peaks of either fault zone type in the histogram. Where arrows and fault zones of the same color occur at the same location, a correlation between surface exposed and underground mapped fault zones is possible. Owing to aforementioned strong ductile background deformation in the GrGr, no correlation for these structures is possible in this rock type (i.e. between tunnel meter 5,800 and 8,400). Additionally, no correlation is possible between tunnel meters 300 and 2,300 and 3,300 and 4,200, due to the absence of relevant fault zones for this study. The derivation and verification of a 3D structural model from surface data and field data only denotes the main goal of this study. Therefore, the complete set of 3D fault zones is combined with the 3D geometries of the lithological boundaries, which are constructed by following the workflow outlined before. Using this dataset, which is partially verified by field measurements, a shear zone map of the study area can be derived (Figure 10.14a). The NE–SW striking shear zones (red lines) are oriented subparallel to both the lithological boundaries and the orientation of the Aar massif. It is clearly visible that their density increases towards the S and that they change from a moderately south dipping orientation in the N to subvertical positions in the central and southern part of the Aar massif. Two additional sets of shear zones are visible, which are oriented E–W (yellow) and NW–SE (green). The findings related to the NE–SW striking fault zones can be cross-checked by the underlying 3D structural model (Figure 10.14b). Figure 10.14b comprises lithological boundaries and the most prominent shear zones only, whereas Figure 10.14c displays the complete structural model, containing all types of major fault zones and orientations. Note that in addition to these major fault zones of widths >1 m, a large number of large-scale fault zones with smaller widths exist (see Baumberger, 2015; Wehrens, 2015; Wehrens et al. 2017), which are not considered at all at the current stage.

10.4.3.2 Discussion of Uncertainties Related to Depth Projection

Projection Uncertainty Three-dimensional representations of traces of regional-scale fault zones are geometrically not uniform, but deviate laterally in dip and dip azimuth (Figure 10.11a). Bistacchi et al. (2008) introduce the concept of the Projection Uncertainty (PU), which incorporates the variation in orientation of fault zones along their trace. The PU solely depends on variations in dip angle, whereas the aberration in strike is not taken into account. Since their considerations were not verified by own field measurements, their concept remains theoretical and its relevance for long-range extrapolations is

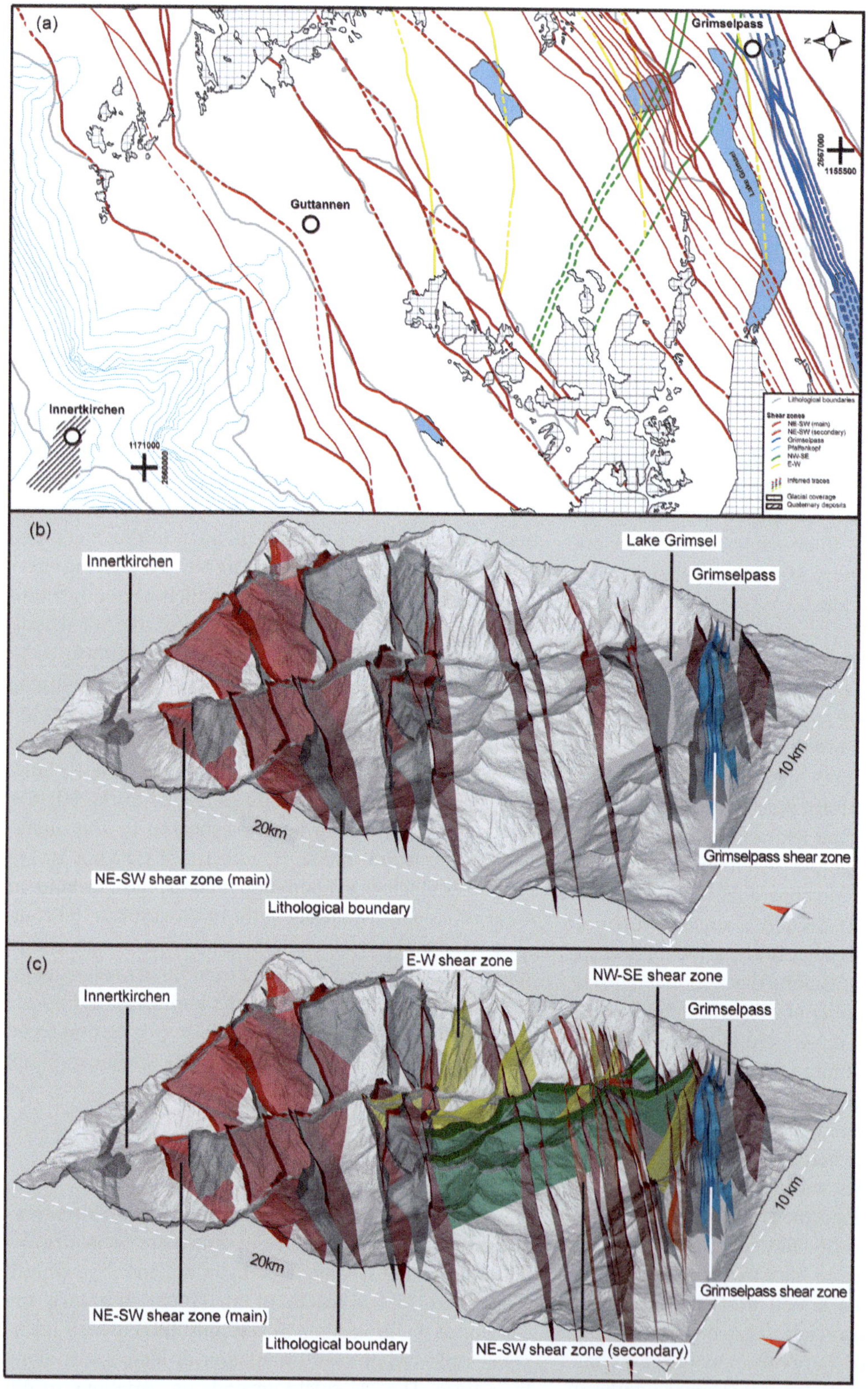

Figure 10.14 Shear zone map and structural 3D model of the Haslital. (a) Five different set of shear zones (width >1 m) are abundant: NE–SW (main), NE–SW (secondary), Grimselpass shear zone (GPSZ), NW–SW and E–W, which (i) often show a correlation between NE–SW striking shear zones and lithological contacts and (ii) a general increase of NE–SW striking shear zones in the S. (b) Structural 3D model containing the lithological boundaries, the main NE–SW trending SZ and the GPSZ only. Note the decreasing dip of the shear zones from S to N. (c) Complete structural 3D model, containing all sets of structures. Model dimensions are 20 × 10 km.

unclear. In contrast, the EUV/EUA of our study is based on our own surface data considering variations in both dip (projection wedge) and dip azimuth (projection buffer). Respective minimum and maximum mean values for the projection wedge are obtained from bootstrap resampling calculations (Jackknife method; Chernick, 2008; Efron, 1979), whereas the projection buffer, which denotes the 3D representation of the bandwidth uncertainty, remains constant in width (former bandwidth uncertainty). It follows from Figure 10.11a that the distance between the surface and underground target (e.g. tunnel, cavern) along the CES determines whether the uncertainty is related to the projection wedge or corridor. If the target area is located above the intersection point between the projection corridor and projection wedge, the maximum uncertainty of the projection depends on the surface variation in strike (i.e. projection corridor); otherwise it depends on the variations in dip (i.e. projection wedge). Minimum and maximum wedge angles in our study are 3.5° and 22°, respectively. Given the wedge-shaped geometry of the latter, the EUV/EUA widening leads to an increasing uncertainty with depth, if the area defined by the wedge angle intersects the projection buffer (Figure 10.11a). The differences in the definition of the EUV/EUA and in the spatial dependence denote the main discrepancy between our own approach and the PU introduced by Bistacchi et al. (2008). In the case of fault zones, correlations between surface and underground data are probable, where the projected surfaces intersect the underground target area within the EUV/EUA. Figure 10.11a additionally illustrates the maximum projection depth in this study. It is limited per individual lineament by three points representing topographic highs and topographic lows along its trace. Given the fact that they are located on a semicircle and therefore define a minimum 3D plane ($1/2\lambda$), this geometry can be expanded to depth, finally representing an extended 3D plane lying on a full circle. The point located on the full circle with the lowermost 3D spatial location defines the maximum projection distance for this assumption. Note that this assumption does not exclude that the projected fault continues to greater depths, but it definitely requires that the structures are not dissected and offset by other faults within the projection interval (see the discussion below). The latter point has to be evaluated on the base of surface information in a probabilistic manner.

Correlation Uncertainty Surface data are extrapolated to depth and visualized as central extrapolation surfaces (CES), accompanied by the EUV/EUA. Possible correlations between surface and underground data can be visualized by combining the EUA, CES, and the fault zone mapping at the tunnel level (Schneider, 1974). It has to be pointed out that Schneider (1974) did not consider areas of schistose granite, strongly schistose granite, or gneisses (e.g. Biotite gneiss, etc.) as zones of ductile deformation. As recognized in past decades, changing foliation intensities in originally isotropic plutonic rocks like the CAGr represent strain gradients related to deformation under mainly ductile deformation conditions (e.g. Wehrens, 2015; Baumberger, 2015; Oliot et al., 2010; Challandes et al., 2008; Choukroune et al., 1987; Marquer et al., 1985; Choukroune and Gapais, 1983). Since in the study area the temperatures required for ductile deformation of granites were high enough only during Alpine orogeny, gneissic and parallel textured rocks in the CAGr of the mapping of the Transitgas tunnel (Schneider, 1974) have to be reinterpreted as ductile shear zones of Alpine age. In this light, both gneissic and parallel textured domains of the Alplistock and Gärstenegg area represent former shear zones, which correlate one to one with the respective CES (Figure 10.12a, b). The faults in these two domains are interpreted as brittle reactivation of the ductile shear zones under retrograde deformation conditions (see Wehrens, 2015; Wehrens et al., 2017). While the contact between the mylonites (gneiss) and the coarse-grained CAGr was brittely reactivated in the case of the Alplistock shear zone, several tens of meters wide brittle damage zone evolved for the Gärstenegg shear zone (Figure 10.12a, b), as evident from the densely spaced 13 faults in the EUA (width: 130 m). Here the CES precisely hits one of the fault zones also indicating a brittle reactivation of the contact between mylonite (parallel textured granite) and massive granite. Owing to aforementioned problems in detecting phenomena of ductile strain localization in the strongly foliated biotite-rich GrGr on the base of the Transitgas tunnel mapping only, it is not possible to discriminate a potential ductile precursor shear zone in the case of the Geisschälen CES (Figure 10.12c). However, two fault zones occur within its EUA (width: 25 m), which might result out of ductile shear zones. Based on these three examples we conclude that ductile localization along mechanically weak zones and their reactivated brittle successor structures mapped at the surface (e.g. Wehrens, 2015; Oliot et al., 2010; Vouillamoz, 2009; Challandes et al., 2008; Baumberger, 2008; Marquer et al., 1985; Choukroune and Gapais, 1983; Steck, 1968; and Figure 10.1) are also abundant in underground facilities. As particularly examples (A) and (B) in Section 10.4.2.2 demonstrate, fault zone projections from the surface intersect the tunnel level at prominent transitions between zones of different mechanical strengths. Since they are located within the EUA, they represent the most likely continuations of the surface incisions.

10.4.4 Validation of Proposed Extrapolation Approach

Extrapolation to depth focused only on subsets of the surface and underground data, which are oriented subparallel to each other. Only a few prominent fault zones with widths ≥1 m detected at the surface ($n = 30/120$) are projected to depth and correlated with a subset of fault zones mapped ($n = 450/1800$) at the level of the Transitgas tunnel (see Figure 10.1 for location details). This discrepancy and the above findings indicate that the erodibility required to generate large-scale surface incisions strongly depend on the width of the originally ductile structures and the degree and spacing of subsequent overprinting by brittle faulting.

Figure 10.13a illustrates the general situation of the tunnel relative to the lithological sequence, its 2D trace and the occurrence of ductile deformation zones along this N–S transect. Special attention has to be paid to the red areas within the CAGr in Figure 10.13a, which are reinterpreted as ductile shear zones based on the argumentation presented above. Figure 10.13a visualizes the frequency of brittle (blue) and ductile (red) shear zones along the tunnel trace. The clustered occurrence of ductile shear zones clearly points out the localized style of strain accommodation within the CAGr, where domains of weak deformation are surrounded by high strain zones (e.g. Baumberger, 2015; Wehrens, 2015; Wehrens et al., 2017; Choukroune and Gapais, 1983). On the contrary, brittle fault zones are more frequent but still occur in a clustered manner. The correlation between clusters of ductile shear zones and those of brittle faults (Figure 10.13b) supports the brittle reactivation of former ductile shear zones (e.g. Wehrens, 2015; Vouillamoz, 2009; Baumberger, 2008; Keusen et al., 1989; Labhart, 1977; Steck, 1968). Figure 10.13c illustrates such possible correlations. The assignments proposed are truly geometrical and are based on the previously presented workflow (EUV/EUA, CES) and geological considerations, whereas kinematic arguments were not taken into account. Due to the fact that only shear zones running subparallel to the orientation of the Aar massif (055 +/−15°) are investigated, not all fault zones at tunnel level as well as not all prominent surface incisions are used for our correlation (Figure 10.13c). This restriction explains the missing allocations between surface incision and underground structures for tunnel meters 300 to 2,300, 2,500 to 3,000, and 3,300 to 4,200, as well as 5,800 to 8,400. Here, E–W or NW–SE trending shear zones would have to be correlated additionally (Figure 10.14a). Despite this aforementioned discrepancy between the large numbers of underground structures compared to fewer morphologically expressed surface structures, our three key examples suggest that a reasonable correlation between surface and underground data is possible. Requirements for this correlation are careful inspection of the type and dimensions of underground structures, knowledge on the reason for their location (e.g. reactivation of pre-existing mechanical anisotropies like lithological contacts), and quantification of the local density of multiple structures like faults. In this sense, wide ductile shear zones (≥1 m), overprinted by multiple brittle faults or the occurrence of fault gouges, represent the most suitable candidates for an enhanced surface erodibility and therefore a pronounced morphological incision at the surface.

Based on our correlation test (Appendix C, Figure C.1), we can state that all mapped surface faults can be successfully correlated with underground structures within their EUA. Depending on the wedge uncertainty and the depth of the Transitgas tunnel, minimum and maximum EUV/EUAs are 7 m and 135 m wide, respectively. On average, EUV/EUA are 36 m in width, meaning that any prediction smaller than +/−18 m for the localization of a fault zone along the tunnel trace is not possible at the corresponding depth levels based on surface information only.

10.4.5 Structural 3D Model and Shear Zone Map

The final shear zone map in Figure 10.14a is based initially on 2D remote sensing data (Baumberger, 2015; Appendix B, Figure B.1), which is subsequently evaluated for data accuracy and uncertainty during data processing along aforementioned procedures and requirements. In addition, field data is used for the verification of the regional-scale remote sensing data. Based on this combined approach, a verified and valid shear zone map is derived for the entire Haslital (Figure 10.14a). It illustrates the spatial orientation of the lithological boundaries (in gray) and five sets of shear zones with a minimum width of 1 m: NE–SW main (red), NE–SW secondary (red), NW–SE (green), E–W (yellow), and the large-scale Grimselpass shear zone (Wehrens, 2015; Wehrens et al., 2017). The largest, 100 m wide and therefore most prominent structure is the Grimselpass shear zone (see also Herwegh et al., 2020). Due to inheritance of compositional variations (paragneisses, orthogneisses, and post-Variscan intrusives), the Grimselpass shear zone is (i) slightly oblique to the NE–SW shear zones, (ii) was used repetitively under ductile and brittle deformation conditions during different stages of Alpine deformation (Egli et al., 2018; Wehrens et al., 2017; Belgrano et al., 2016; Wehrens, 2015), resulting in (iii) an enhanced fluvio-glacial erosion in this region (Baumberger, 2015). The latter is documented by the morphological carving along this structure.

Despite showing smaller widths (1 to 10 m) compared to the Grimselpass shear zone, all other NE–SW striking major shear zones also extend over several kilometers. In map view (Figure 10.14a), their spacing as well as the degree of apparent bending of the 2D fault traces decreases from N to S. The latter trend is attributed to the topography effect, since the change in apparent bending relates directly to the change in dip of the fault planes, which increases from moderate to vertical fault orientations from N to S (compare Figure 10.14a with b and c). Note also that the long-range bends of the NW–SW (green) and E–W (yellow) faults are mainly due to the topography effect induced by the crossing of the fault traces of the Aar valley (Figure 10.14a; see also Schneeberger et al., 2017).

In a geodynamic perspective, the decreasing spacing of the NE–SW striking shear zones towards the S manifests directly increasing deformation along the N–S temperature gradient and an increasing content in rheological weak zones either consisting of ultrafine-grained polymineralic mylonites or sheet-silicate rich lithologies like the GrGr or the ortho-/paragneisses of the Grimselpass shear zone (Wehrens et al., 2016, 2017). Here, a complex interplay of reverse faulting (Handegg phase) and strike-slip faulting (Oberaar phase) resulted in the present day steeply oriented fault zones (Wehrens et al., 2017). In contrast, the moderately south dipping fault zones in the N are related to a late north directed thrusting episode (Pfaffenkopf phase). In this sense, the approach followed provides a valuable tool for the investigations of the large-scale 3D deformation pattern. It is this complex 3D pattern of fault zones, together with geodynamic forces, such as lower crustal decoupling, buoyancy-/erosion-driven uplift within an overall transpressional regime, which provoked the exhumation of the Aar massif shaping its recent 3D geometry (Nibourel et al., 2021; Herwegh et al., 2017, 2020). Similar fault/shear zone patterns occur in the other External Crystalline Massifs of the Alps (Herwegh et al., 2022), making a 3D analysis with the presented workflow a suitable tool for further exploration of their Massif internal geometries.

10.5 Summary Discussion and Conclusions

In order to project fault traces to depth, we suggest a workflow consisting of the following steps: (i) the detection of fault traces on the base of 2D input data, (ii) their digitization with the highest resolution followed by (iii) an optimization of fault segments in light of their angular resolution, allowing (iv) the generation of 3D planes for sets of fault segments from which then (v) the 3D plane of the entire fault structure can be generated. Each of these steps has its intrinsic uncertainties, which have to be evaluated in order to gain knowledge of the final uncertainty of the projected fault planes at depth. In 2D, (i) the spatial resolution of image data and its implication for digitization, (ii) the ability of the operator regarding knowledge and experience, (iii) the loss of digitization accuracy due to the representation of natural structures in a pixel matrix on DEM or orthophotos, and (iv) the discrepancy between spatial and angular resolution of vector data account for the most important parameters for uncertainty assessment of a digitized line segment. Towards the third dimension, the calculation of the 3D geometry of a plane based on surface information (Fernandez, 2005) relies on such input data, where the projection wedge and projection corridor uncertainties yield, in the Central Extrapolation Surface and Extrapolation Uncertainty Area, both predicting the underground domain in which a projected fault plane should be found.

As a great advantage of the presented approach, large areas can efficiently be investigated by means of remote sensing and focused field investigation, allowing the generation of a reliable lineament map for large-scale shear/fault zones (Figure 10.14a). Among the strengths of the remote sensing technique is the uniformity and consistency of the lineament surface information at the scale of kilometers down to tens of meters. As a consequence, the derived shear zone map (Figure 10.14a) reliably represents all lineaments of similar relevance (width and extension) within an entire region under study at the level of given spatial resolution. Below the resolution limit of the remote sensing approach, field investigations can provide complementary information on location and orientation of primary and secondary structures. Of special importance is their ability to deliver detailed information about (1) the mechanical/structural characteristics and (2) the local dip of the shear zone. The first allows for better understanding of the various conditions and geologic settings that may be represented by a lineament observed by remote sensing (compare with Figure 10.2). The second allows extrapolation of local dip variations to scales of remote sensing derived lineaments and thus to generate reasonable constraints for regional 3D plane dip variations. In our study, the strengths of the two approaches are ideally combined to allow for 3D extrapolation of surface data to depth by giving corresponding uncertainty estimates (Figures 10.11 and 10.14b and c). For example, azimuth and dip angles can be generated with a higher precision on the base of the remote sensing approach (Figure 10.10), while the maximum uncertainty in dip results from field measurements. In this light, the proposed concepts of the Extrapolation Uncertainty Volume/Area (EUV/EUA) and the Central Extrapolation Surface (CES) provide an important base to evaluate the influence of different input datasets (Figure 10.11; see also Wellmann et al., 2010). The careful and detailed assessment of different uncertainty types of both datasets (Caumon et al, 2009; Bond et al., 2007, Kocal et al., 2007; Wang et al., 2005) provides valid input data for surface data extrapolation to depth with subsequent evaluation of the projection accuracy at tunnel level (Figures 10.12 to 10.14). As pointed out by Pakyuz-Charrier et al. (2018), adding additional information may locally decrease uncertainty (resulting in higher precision) in favor of a global increase (leading to less accurate models). In these terms, our approach accounts for the estimation of quantitative uncertainty, as demanded, for example, by Wellmann et al. (2010). Using the EUV/EUA and CES to visualize inherent uncertainty of shear zones, we are in line with the ideas of Pang et al. (1997), which postulate the uncertainty visualization as a tool to aid in data analysis and decision making.

Note that the correlations in this study are truly geometrical. In this context, inherited structures as discussed by, for example, Balestra et al. (2019) or structural behavior as introduced by, for example, Wellmann et al. (2018) are not taken into account. The correlation between particular surface and underground fault zones is limited, without accounting for detailed kinematics and structural considerations (Faulkner et al., 2010) as well as field characterizations, both at the surface and at the tunnel level. In this sense, a combination of the present study with the work of Wehrens et al. (2016, 2017) would allow enhanced constraints to be obtained for more precise extrapolations. Also note that our semiautomated approach differs from Jessell et al. (2021), as they provide a fully automated approach for the initial modelling steps.

Last but not least, a general geological uncertainty remains due to the heterogeneous appearance of structures in nature and their 3D continuation in space. Although the lateral continuations of major fault zones are very consistent in our study area, secondary structures connecting the major ones (e.g. splays, relay structures, etc.) are very difficult to handle due to their limited dimensions and irregular appearance (Faulkner et al., 2010). The prediction of the occurrence of such secondary structures at depth is rather challenging, requiring probabilistic/stochastic approaches (e.g. Schneeberger et al., 2017), which are out of the scope of the present study. Up to now, very little detailed structural information is available about the geometry and 3D propagation of deep-seated structures in the field area, as, for example, large-scale ductile shear zones that may act as thrust planes. Therefore, any depth projection of surface information introduces itself to a new degree of scale-dependent uncertainty, which must be accounted for.

Acknowledgments

The research resulting in this article was partly funded by the Swiss National Science Foundation (project 200021-132196) and additionally supported by the Swiss Geological Survey, a department of the Swiss Federal Office of Topography swisstopo. Both institutions are acknowledged for their support. We further acknowledge both Kraftwerke Oberhasli KWO and Nagra (Grimsel Test Site) for providing logistic support as well as giving access to their archives. Midland Valley (now Petroleum Experts) is thanked for providing the software package Move^{TM} within their Academic Software Initiative. The authors also thank Philip Wehrens for supporting fieldwork and the reviewers Clare Bond and Luca Penasa, whose comments and suggestions greatly improved the manuscript.

Appendix A: Topography Effect

Spatial analysis of 3D structures acquired by remote sensing only describes the orientations of the 3D fault zones in the 2D map space. When applying such a space reduction, one has to be aware that artifacts related to topography effects might arise. The apparent strike of a digitized 2D intersection line with topography deviates in the case of a non-planar surface topography from the true 3D strike of the plane. Therefore, in a 2D analysis of 3D trace lines, valid 3D strike orientations can only be obtained at those fault segments, which are oriented subparallel to the strike of the entire fault zone. All other orientations along the fault zone trace show apparent strike directions and do therefore not represent the true orientation of the plane (Figure A.1). This is called the topography effect. In order to evaluate the potential influence of such effects, we used a synthetic DEM and calculated its intersections with five artificial planes with a common strike direction of 180°, but each with an individual inclination (15, 30, 45, 60, and 90°, Figure A.1). In line with data processing in this study and Baumberger (2015), the intersection lines were cut into segments of 25 m length and their orientations are plotted as rose diagrams (see Figure A.1a–e). It becomes clear that the more parallel the strike of an individual segment is oriented to the one of the main orientations of a structure, the more valid its representation in the rose diagram. Figure A.1 indicates that the steeper a fault zone is oriented, the more the orientations of the single segments approximate the strike of the main structure and the more valid are its results. In contrast, with decreasing dip angles, the resulting segments form two branches in the rose diagrams, whose apparent opening angles increase. These are severe artifacts, where structural planes with dip values between 60° and 15° result in apparent opening angles of 40° to >90° (Figure A.2). This

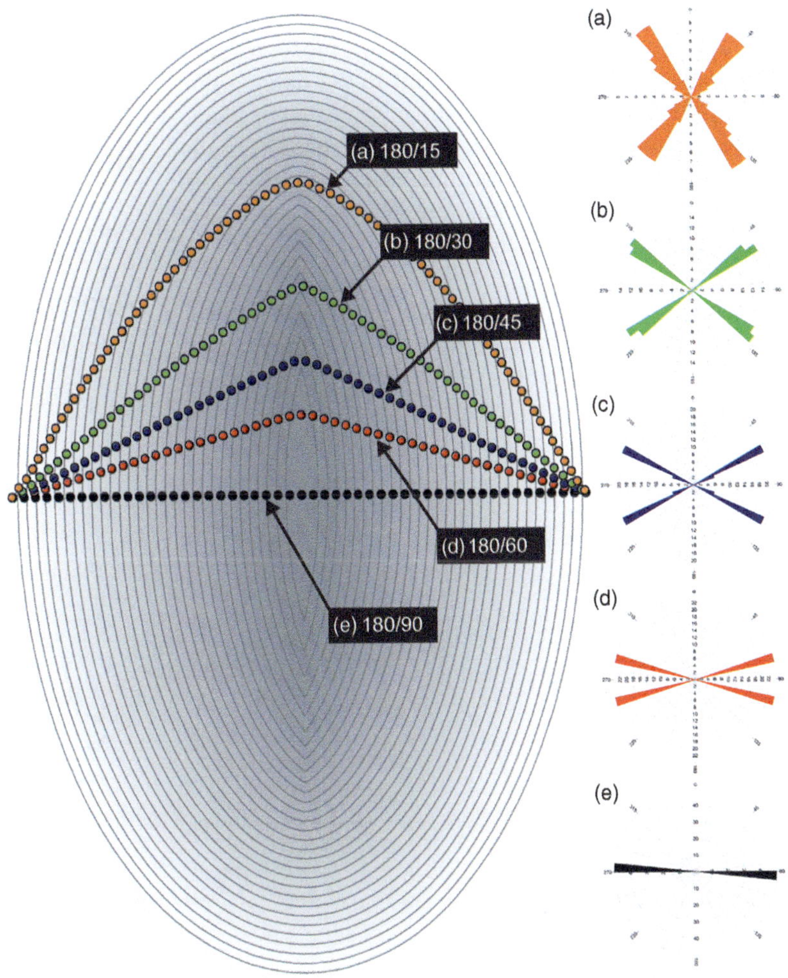

Figure A.1 Artificial digital elevation model, cross-cut by five artificial planes with inclinations of 15, 30, 45, 60, and 90°. Intersection between lineament and topography gives an intersection line that is cut into different segments (lineaments) of 25 m length. The 2D strike of the individual segments is shown as rose diagrams for the individual dip angles (a–e). The more the orientation of a single lineament is deviated from ideal strike orientation, the more the result is biased by the topography. The steeper the lineaments are dipping, the more valid their orientations appear.

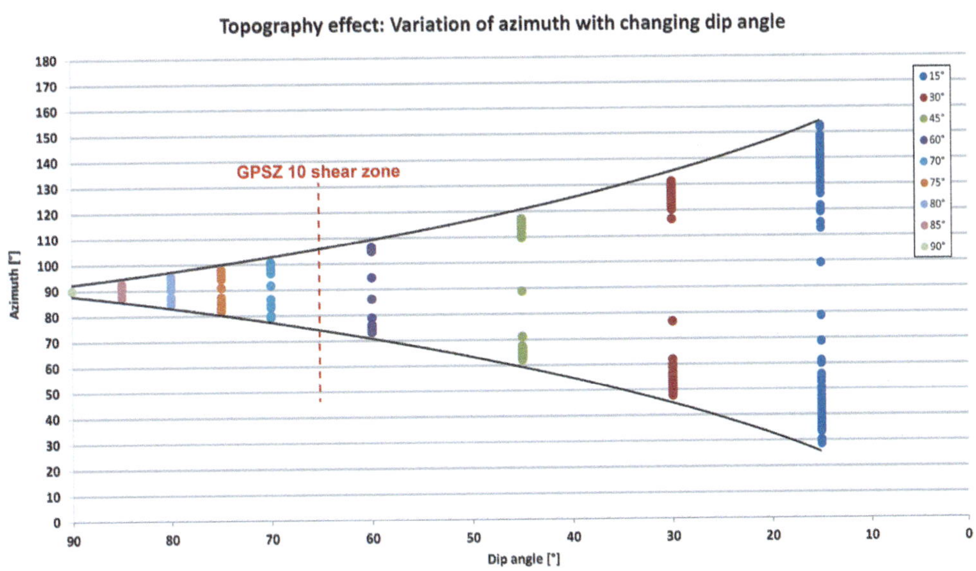

Figure A.2 Topography effect demonstrates the influence of dip angle on the scatter of the strike orientations for 25 m segments. The steeper the plane, the less scattered the strike orientations, the narrower the confidence interval. Vertical segments do not scatter at all. For example, for structure GPSZ 10, which shows a minimum dip of ~65°, the topography effect results in a 25° scatter for the strike orientations along a trace in a 2D map view.

unpleasant effect will broaden the distribution pattern of a rose diagram. As presented in Figure A.1, the topography effect is smallest, where the tangent to the outcrop line is subparallel oriented to the strike of the main plane. Therefore, such an effect is frequently apparent in gently and moderately dipping structural planes, and becomes scarcer the steeper the planes are oriented and, hence, the better they represent the real orientation of the main plane.

Figure A.2 demonstrates the influence of the inclination angle on the scattering of the orientation of each surface. In general, it can be stated that the steeper the orientation, the smaller the variation in strike. The influence of the topography effect on GPSZ10 is given as an example.

Appendix B: Lineament Map from Remote Sensing Data Acquisition

See Figure B.1.

Appendix C : Intersection Analysis at Tunnel Level

See Figure C.1.

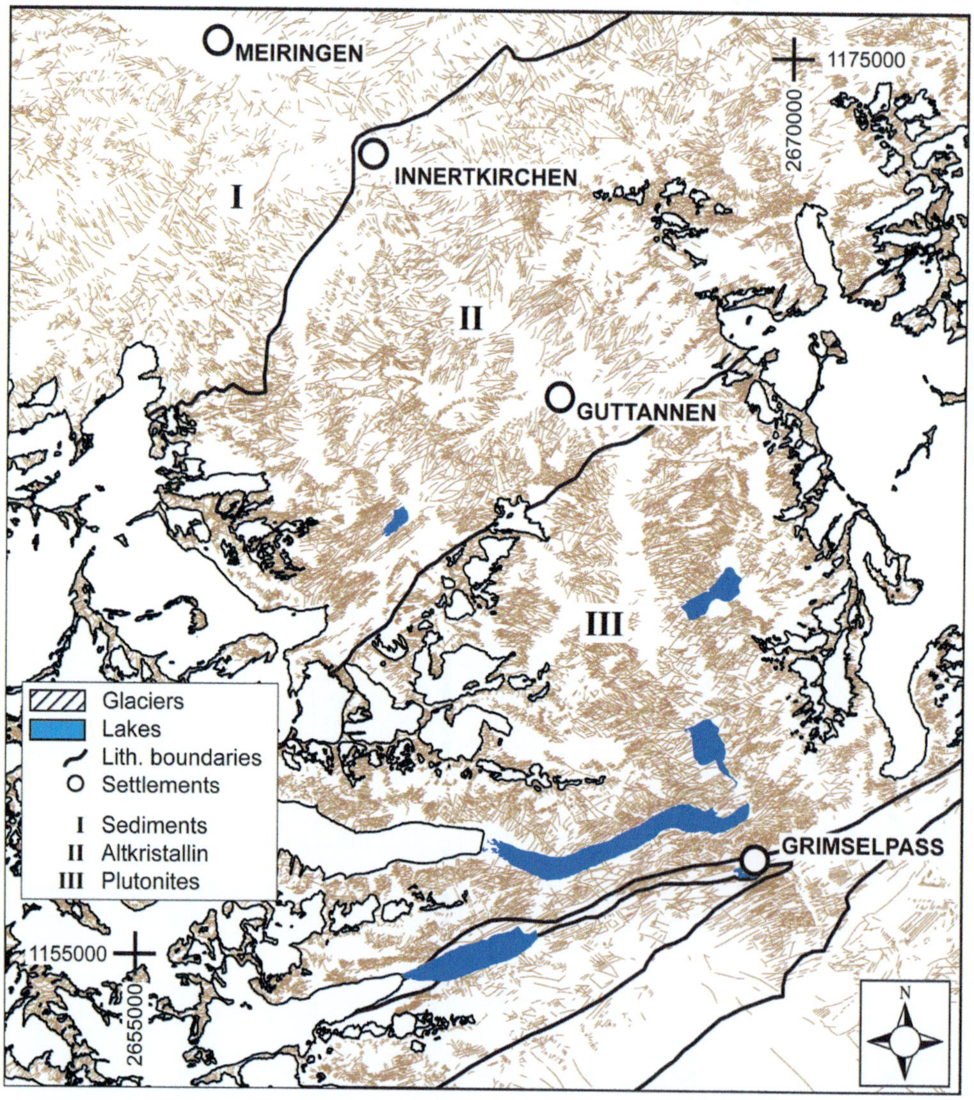

Figure B.1 Lineament map consisting of 178,000 standardized segments of 25 m length each (= 32,000 original lineaments). The map focuses on three lithological domains: (I) the Mesozoic sediments in the NW, (II) the pre-Variscan gneisses in the middle, and (III) the post-Variscan plutonites in the SE. White areas denote additional regions with no lineaments at all (beside glaciers), due to quaternary coverage or alluvial cones (Baumberger, 2015).

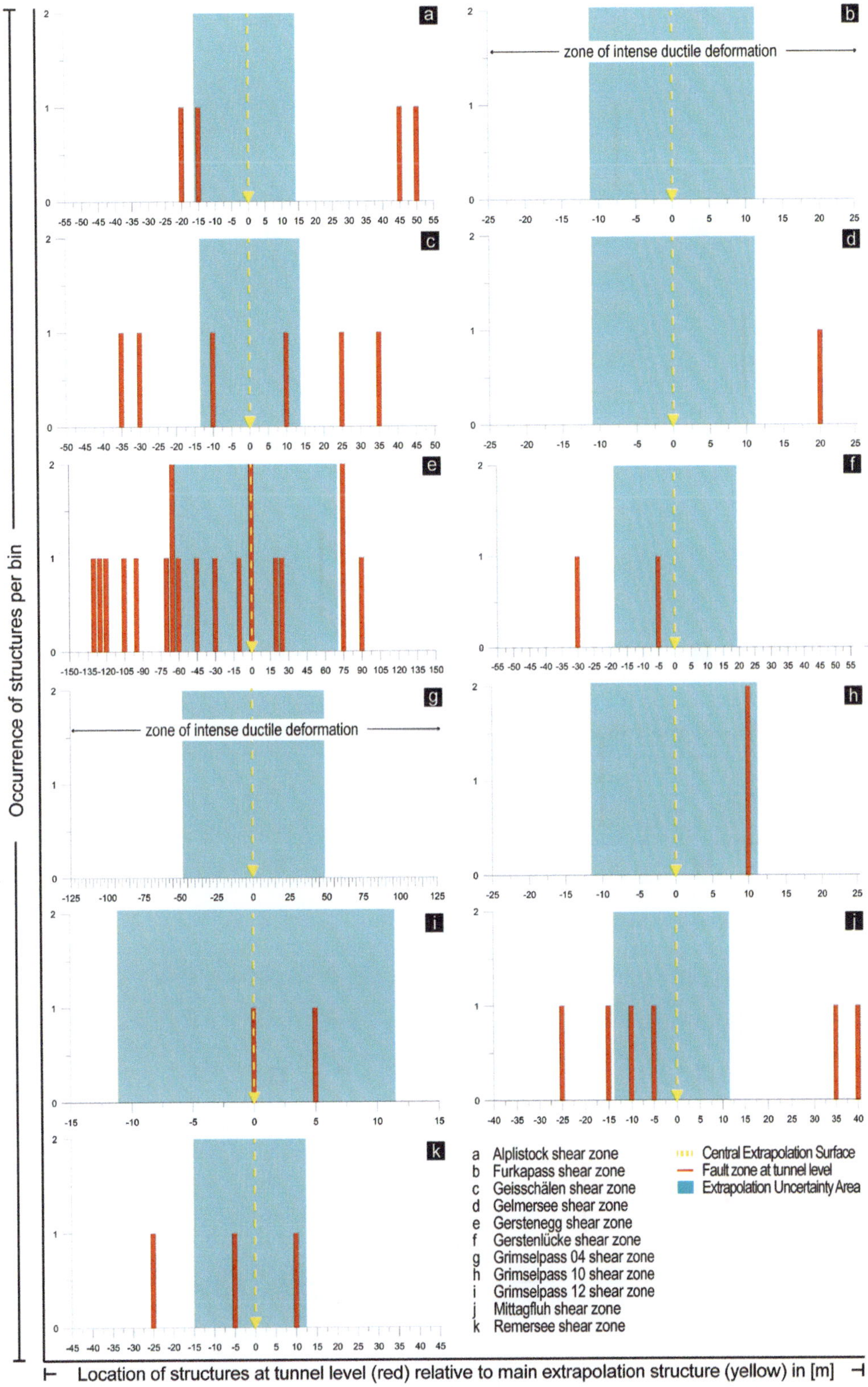

Figure C.1 Histograms for extrapolation of all surface structures, based on the method presented in Figure 10.11. (A) Main NE–SW shear zones. For most of the extrapolations, a correlation is likely within the maximum area of interest (see Figure 10.11 for details). Images (b) and (g) are located in zones of intense deformation, which are the Grimsel zone and the Grimsel Granodiorite respectively. Due to this reason, no discrete structures can be discriminated (Baumberger, 2015). (B) Secondary NE–SW shear zones. Within the area of interest (see Figure 10.11), a possible correlation can be postulated for most of the extrapolations. Structures (e) and (g) are located in areas of intense deformation (as in A), which are the Grimsel Granodiorite and the transition of the Central Aaregranite to the Grimsel Granodiorite. These strongly deformed domains do not allow discriminating single structures within the deformation zones (Baumberger, 2015).

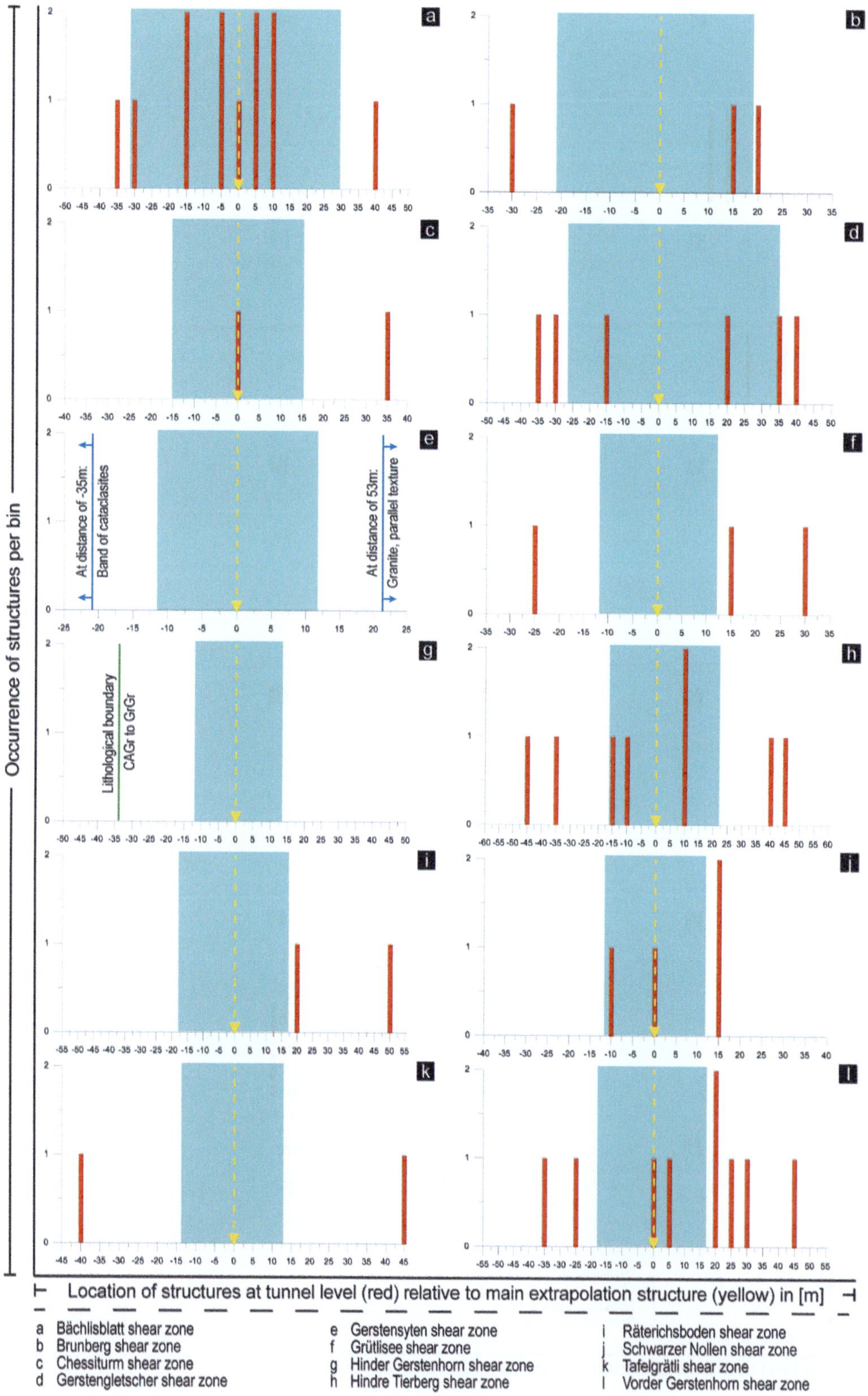

Figure C.1 (*Continued*)

References

Abrecht, J. (1994). Geologic units of the Aar massif and their pre-Alpine rock associations: A critical review. *Schweiz. Mineral. Petrogr. Mitt.* 74: 5–27.

Arboyela, M.L. and Engelder, T. (1995). Concentrated slip zones with subsidiary shears: their development on three scales in Cerro Brass fault zone, Appalachian valley and ridge. *Journal of Structural Geology* 17: 519–532.

Aydin, A. and Johnson, A.M. (1978). Development of faults as zones of deformation bands and as slip surfaces in sandstone. *Pure and Applied Geophysics* 116 (4–5): 931–942.

Balestra, M., Corrado, S., Aldega, L. et al. (2019). 3D structural modeling and restoration of the Apennine-Maghrebian chain in Sicily: Application for non-cylindrical fold-and-thrust belts. *Tectonophysics* 761: 86–107.

Bambauer, H.U., Bernotat, W., Breit, U., and Kroll, H. (2005). Perthitic alkali feldspar as indicator mineral in the Central Swiss Alps. Dip and extension of the surface of the microcline/sanidine transition isograd. *Eur. J. Mineral.* 17: 69–80.

Banerjee, S. and Mitra, S. (2004). Remote surface mapping using orthophotos and geologic maps draped over digital elevation models: Application to the Sheep Mountain anticline. *Wyoming. AAPG Bulletin* 88 (9): 1227–1237.

Baumberger, R. (2008). Reactivated shear zones in granites and their role as water conducting features. Master's thesis. University of Bern.

Baumberger, R. (2015). Quantification of lineaments: Link between internal 3D structure and surface evolution of the Hasli valley (Aar massif, central Alps, Switzerland). Doctoral thesis. University of Bern.

Belgrano, T., Herwegh, M., and Berger, A. (2016). Inherited structural controls on fault geometry, architecture and hydrothermal activity: An example from Grimsel Pass, Switzerland. *Swiss J. Geosciences* 109 (3): doi: 10.1007/s00015-016-0212-9.

Bergemann, C., Gnos, E., Berger, A. et al. (2014). Shear zone activity of the Grimsel area (Aar massif): Th–Pb data in hydrothermal cleft monazite. In: *Paper presented at the 12th Swiss Geoscience Meeting 2014*. Switzerland: Fribourg.

Berger, A., Mercolli, I., Herwegh, M., and Gnos, E. (2017a). Explanatory notes accompanying the geological map of the Aar massif and the Tavetsch- and Gotthard nappes, Geological Special Map 129, Federal Office of Topography, Wabern, Switzerland.

Berger, A., Wehrens, P., Lanari, P. et al. (2017b). Microstructures, mineral chemistry and geochronology of white micas along a retrograde evolution: An example from the Aar massif (Central Alps, Switzerland). *Tectonophysics* 721: 179–195.

Bistacchi, A. (2010). Three-dimensional characterization of a crustal-scale fault zone: the Pusteria and Sprechenstein fault system (Eastern Alps). *Journal of Structural Geology* 32: 2022–2041.

Bistacchi, A., Massironi, M., Dal Piaz, G.V. et al. (2008). 3D fold and fault reconstruction with an uncertainty model: An example from an Alpine tunnel case study. *Computers and Geosciences* 34: 351–372.

Blaschke, T. (2010). Object based image analysis for remote sensing. *ISPRS Journal of Photogrammetry and Remote Sensing* 65 (1): 2–16.

Bond, C.E., Shipton, Z.K., Jones, R.R. et al. (2007). Knowledge transfer in a digital world: Field data acquisition, uncertainty, visualization, and data management. *Geosphere* 3 (6): 568–576. doi: 10.1130/GES00094.1.

Bond, C.E., Johnson, G., and Ellis, J.F. (2015). Structural model creation: the impact of data type and creative space on geological reasoning and interpretation. *London, Geological Society, Special Publications* 421: 83–97.

Burkhard, M. (1988). L'Hélvétique de la bordure occidentale du massif de l'Aar (evolution tectonique et métamorphique). *Eclogae Geol. Helv.* 81 (1): 63–114.

Caumon, G., Collon-Drouaillet, P., Carlier, L. et al. (2009). Surface-Based 3D Modeling of Geological Structures. *Math. Geosci.* 41: 927–945.

Challandes, N., Marquer, D., and Villa, I.M. (2008). P-T-t modelling, fluid circulation, and ^{39}Ar-^{40}Ar and Rb–Sr mica ages in the Aar Massif shear zones (Swiss Alps). *Swiss J. Geosci.* 101 (2): 269–288. doi: 10.1007/s00015-008-1260-6.

Chernick, M.R. (2008). *Bootstrap Methods: A Guide for Practitioners and Researchers*, 2nde. Hoboken, NJ: John Wiley.

Choukroune, P. and Gapais, D. (1983). Strain pattern in the Aar Granite (Central Alps): Orthogneiss developed by bulk inhomogeneous flattening. *Journal of Structural Geology* 5 (3–4): 411–418.

Choukroune, P., Gapais, D., and Merle, O. (1987). Shear criteria and structural symmetry. *Journal of Structural Geology* 9 (5–6): 635–646. doi: 10.1016/0191-8141(87)90137-4.

Cowan, E.J., Beatson, R.K., Ross, H.J. et al. (2003). *Practical implicit geological modelling. Paper presented at the 5th International Mineral Geological Conference* 8: 89–99.

Crosetto, M., Tarantola, S., and Saltelli, A. (2000). Sensitivity and uncertainty analysis in spatial modelling based on GIS, Agriculture. *Ecosystems and Environment* 81: 71–79.

Dag, A. and Ozdemir, A.C. (2013). A comparative study for 3D surface modeling of coal deposit by spatial interpolation approaches. *Resource Geology* 63 (4): 394–403.

Dar, I.A., Sankar, K., and Dar, M.A. (2010). Remote sensing technology and geographic information system modeling: An integrated approach towards the mapping of groundwater potential zones in Hardrock terrain, Mamundiyar basin. *Journal of Hydrology* 394: 285–295.

Davis, T.J. and Keller, P. (1997). Modelling and visualizing multiple spatial uncertainties. *Computers and Geosciences* 23 (4): 397–408.

Dhont, D., Luxey, P., and Chorowicz, J. (2005). 3-D modeling of geologic maps from surface data. *AAPG Bulletin* 89 (11): 1465–1474.

Diamond, L.W., Wanner, C., and Waber, H.N. (2018). Penetration depth of meteoric water in orogenic geothermal systems. *Geology* 46: 1063–1066.

Efron, B. (1979). Bootstrap methods: Another look at the Jackknife. *The Annals of Statistics* 7 (1): 1–26.

Egli, D., Herwegh, M., Berger, A. et al. (2018). Structural characteristics and evolution of an exhumed long-lived hydrothermal reservoir. *Tectonophysics* 747–748: 239–258.

Engelder, T.J. (1987). Joints and shear fractures in rock. In: *Fracture Mechanics of Rock*, 27–69. London: Academic Press.

Engelder, T.J. (1989). Analysis of pinnate joints in the Mount Desert Island granite: Implications for postintrusion kinematics in the coastal volcanic belt. *Maine. Geology* 17 (6): 564–567.

Faulkner, D.R., Jackson, C.A.L., Lunn, R.J. et al. (2010). A review of recent developments concerning the structure, mechanics and fluid flow properties of fault zones. *Journal of Structural Geology* 32: 1557–1575.

Ferńandez, O. (2005). Obtaining a best fitting plane through 3D georeferenced data. *Journal of Structural Geology* 27: 855–858.

Ferńandez, O., Jones, S., Armstrong, N. et al. (2009). Automated tools within workflows for 3D structural construction from surface and subsurface data. *Geoinformatica* 13: 291–304.

Gahegan, M. and Ehlers, M. (2000). A framework for the modelling of uncertainty between remote sensing and geographic information systems. *ISPRS Journal of Photogrammetry and Remote Sensing* 55: 176–188.

Gajski, D. (2004). Rasterbasierte Geländeoberflächeanalysen. Doctoral thesis, unpublished. Technische Universität Wien.

Gallerini, G. and De Donatis, M. (2009). 3D modeling using geognostic data: The case of the low valley of Foglia river (Italy*)*. *Computers and Geosciences* 35: 146–164.

Gapais, D. (1989). Shear structures within deformed granites: Mechanical and thermal indicators. *Geology* 17 (12): 1144–1147.

Gapais, D., Bale, P., Choukroune, P. et al. (1987). Bulk kinematics from shear zone patterns: Some field examples. *Journal of Structural Geology* 9 (5–6): 525–530.

Goncalves, P., Oliot, E., Marquer, D., and Connolly, J.A.D. (2012). Role of chemical processes on shear zone formation: An example from the Grimsel metagranodiorite (Aar massif, Central Alps). *J. Metamorphic Geol.* 30: 703–722.

Gunnink, J., Maljers, D., and Hummelmann, J. (2010). Quantifying uncertainty of geological 3D layer models, constructed with a-priori geological expertise. Paper presented at the International Association of Mathematical Geoscience General Assembly 2010.

Heilbronner, R. and Barrett, S. (2014). *Image Analysis in Earth Sciences – Microstructures and Textures of Earth Materials.* Heidelberg: Springer.

Herwegh, M., Berger, A., Baumberger, R. et al. (2017). Large-scale crustal-block-extrusion during Late Alpine Collision. *Sci. Rep.* 7 (413): doi: 10.1038/s41598-017-00440-0.

Herwegh, M., Berger, A., Glotzbach, C. et al. (2020). Late stages of Continent–Continent collision: Timing, kinematic evolution, and exhumation of the Northern Rim (Aar massif) of the Alps. *Earth-Science Reviews* 102959.

Herwegh, M., Berger, A., Bellahsen, N. et al. (2022). Evolution of the external crystalline massifs of the European Alps: From massif to lithosphere scale. In: *Geodynamics of the Alps* (ed. C. Rosenberg and N. Bellahsen). ISTE Book Series.

Hofmann, B.A., Helfer, M., Diamond, L.W. et al. (2004). Topography-driven hydrothermal breccia mineralization of Pliocene age at Grimsel Pass, Aar massif, Central Swiss Alps. *Schweiz. Mineral. Petrogr. Mitt.* 84: 271–302.

Hopkinson, C., Hayashi, M., and Peddle, D. (2009). Comparing alpine watershed attributes from LiDAR, photogrammetric, and contour-based digital elevation models. *Hydrol. Process* 23: 451–463.

Jessell, M., Ogarko, V., de Rose, Y. et al. (2021). Automated geological map deconstruction for 3D model construction using map2loop 1.0 and map2model 1.0. *Geosci. Model Dev.* 14: 5063–5092. https://doi.org/10.5194/gmd-14-5063-2021.

Kaufmann, O. and Martin, T. (2008). 3D geological modelling from boreholes, cross-sections and geological maps, application over former natural gas storages in coal mines. *Computers and Geosciences* 34: 278–290.

Kenward, T., Lettenmaier, D.P., Wood, E.F., and Fielding, E. (2000). Effects of digital elevation model accuracy on hydrologic predictions. *Remote Sensing of Environment* 74 (3): 432–444.

Keusen, H.R., Ganguin, J., Schuler, P., and Buletti, M. (1989). *Felslabor Grimsel – Geologie, Technischer Bericht 87–14*. Baden, Switzerland: Nagra.

Kienzle, S. (2004). The effect of DEM Raster resolution on first order, second order and compound terrain derivatives. *Transactions in GIS* 8 (1): 83–111.

Kocal, A., Duzgun, H.S., and Karpuz, C. (2007). An accuracy assessment methodology for the remotely sensed discontinuities: A case study in Andesite Quarry area. *Turkey. International Journal of Remote Sensing* 28 (17): 3915–3936.

Labhart, T. (1977). *Aarmassiv und Gotthardmassiv, Sammlung Geologischer Führer*, vol. Bd. 63. Berlin, Stuttgart: Gebrüder Bornträger.

Lemon, A.M. and Jones, N.L. (2003). Building solid models from boreholes and user-defined cross-sections. *Computers and Geosciences* 29: 547–555.

MacCormack, K.E., Berg, R.C., Kessler, H. et al. (ed.) (2019). 2019 synopsis of current three-dimensional geological mapping and modelling in geological survey organizations. In: *Alberta Energy Regulator/Alberta Geological Survey, AER/AGS Special Report 112*, vol. 307.

Mair, D., Lechmann, A., Herwegh, M. et al. (2018). Linking Alpine deformation in the Aar Massif basement and its cover units – the case of the Jungfrau–Eiger mountains (Central Alps, Switzerland). *Solid Earth* 9: 1099–1122.

Mancktelow, N.S. and Pennacchioni, G. (2005). The control of precursor brittle fracture and fluid–rock interaction on the development of single and paired ductile shear zones. *Journal of Structural Geology* 27: 645–661.

Marquer, D., Gapais, D., and Capdevila, R. (1985). Comportement chimique et orthogneissification d'une granodiorite en faciès schistes verts (Massif de l'Aar, Alpes Centrales). *Bull. Minéral.* 108: 209–221.

Moore, I.D., Grayson, R.B., and Ladson, A.R. (1991). Digital terrain modelling: A review of hydrological, geomorphological and biological applications. *Hydrological Processes* 5: 3–30.

Morelli, M. and Piana, F. (2006). Comparison between remote sensed lineaments and geological structures in intensively cultivated hills (Monferrato and Langhe domains, NW Italy). *International Journal of Remote Sensing* 27 (20): 4471–4493.

Nibourel, L., Berger, A., Egli, D. et al. (2021). Structural and thermal evolution of the eastern Aar Massif: Insights from structural field work and Raman thermometry. *Swiss Journal of Geoscience* 114: 9. https://doi.org/10.1186/s00015-020-00381-3.

Niggli, C. (1965). *Petrographie und Petrogenesis der Migmatite und Gneise im südlichen Aarmassiv zwischen Obergesteln und Furkapass*. Doctoral thesis, unpublished. University of Bern.

O'Leary, D.W., Friedman, J.D., and Pohn, H.A. (1976). Lineament, linear, lineation: Some proposed new standards for old terms. *Geol. Soc. Am. Bull.* 87 (10): 1463–1469. doi: 10.1130/0016-7606(1976)87<1463.

Oliot, E., Goncalves, P., and Marquer, D. (2010). Role of plagioclase and reaction softening in a metagranite shear zone at mid-crustal conditions (Gotthard massif, Swiss Central Alps). *J. Metamorphic Geol.* 28: 849–871. doi: 10.1111/j.1525-1314.2010.00897.x.

Olsen, S.N., Johnson, C.M., Beard, B.L., and Baumgartner, L.P. (2000). New U-Pb zircon data and constraints on the age and mode of migmatization in the Aar massif, Central Alps. *European Journal of Mineralogy* 12: 1245–1260.

Pakyuz-Charrier, E., Giraud, J., Ogarko, V. et al. (2018). Drillhole uncertainty propagation for three-dimensional geological modeling using Monte Carlo. *Tectonophysics* 747: 16–39.

Pang, A.T., Wittenbrink, C.M., and Lodha, S.K. (1997). Approaches to uncertainty visualization. *The Visual Computer* 13: 370–390.

Pirotti, F. and Tarolli, P. (2010). Suitability of LiDAR point density and derived landform curvature maps for channel network extraction. *Hydrol. Process* 24: 1187–1197. doi: 10.1002/hyp.7582.

Raharimahefa, T. and Kusky, T.M. (2006). Structural and remote sensing studies of the southern Betsimisaraka Suture, Madagascar. *Gondwana Research* 10: 186–197.

Rahiman, T. and Pettinga, J.R. (2008). Analysis of lineaments and their relationship to Neogene fracturing, SE Viti Levu. *Fiji. GSA Bulletin* 120 (11–12): 1544–1555. doi: 10.1130/B26264.1.

Rauch, A. (2021). Dealing with uncertainty and ambiguity in geological maps, with focus on bedrock geology. In: *Rep. Swiss Geol. Surv*, 14.

Rolland, Y., Cox, S., and Corsini, M. (2009). Constraining deformation stages in brittle-ductile shear zones from combined field mapping and 40Ar/39Ar dating: The structural evolution of the Grimsel Pass area (Aar Massif, Swiss Alps). *Journal of Structural Geology* 31: 1377–1394.

Sala, P., Pfiffner, O.A., and Frehner, M. (2014). The Alpstein in three dimensions: fold-and-thrust belt visualization in the Helvetic zone, eastern Switzerland. *Swiss Journal of Geoscience* 107: 177–195. doi: 10.1007/s00015-014-0168-6.

Schaltegger, U. (1989). *Geochemische und isotopengeologische Untersuchungen am Zentralen Aaregranit und seinen assoziierten Gesteinen zwischen Aare und Reuss (Aarmassiv, Schweiz)*. Doctoral thesis, unpublished. University of Bern.

Schaltegger, U. (1994). Unravelling the pre-Mesozoic history of Aar and Gotthard massifs (Central Alps) by isotopic dating – a review. *Schweiz. mineral. petro. Mitt.* 74: 41–51.

Scheiber, T., Fredin, O., Viola, G. et al. (2015). Manual extraction of bedrock lineaments from high resolution LiDAR data: Methodological bias and human perception. *GFF* 137: 362–372.

Schmid, S.M. and Handy, M.R. (1991). Towards a genetic classification in fault rocks: Geological usage and tectonophysical implications. *Controversies in Modern Geology* 1: 339–361.

Schneeberger, R., de La Varga, M., Egli, D. et al. (2017). Methods and uncertainty-estimations of 3D structural modelling in crystalline rocks: A case study. *Solid Earth* 8: 987–1002. https://doi.org/10.5194/se-8-987-2017.

Schneeberger, R., Egli, D., Lanyon, G.W. et al. (2018). Structural-permeability favorability in crystalline rocks and implications for groundwater flow paths: A case study from the Aar massif (central Switzerland). *Hydrogeology Journal* 1–14.

Schneider, T.R. (1974). *Geologische Stollenaufnahmen Transitgas AG: Unterstock-, Urweid-, Gstelli-, Grimsel- und Obergesteln-Stollen*. Zürich: Elektro-Watt Ingenieurunternehmung AG.

Schober, A. and Exner, U. (2011). 3D structural modelling of an outcrop-scale fold train using photogrammetry and GPS mapping. *Austrian Journal of Earth Sciences* 104 (2): 73–79.

Sprague, K.B. and de Kemp, E.A. (2005). Interpretive tools for 3-D structural geological modelling, Part II: Surface design from sparse spatial data. *GeoInformatica* 9 (1): 5–32.

Stalder, H.A. (1964). Petrographische und mineralogische Untersuchungen im Grimselgebiet (mittleres Aarmassiv). *Schweiz. Mineral. Petrogr. Mitt.* 44: 187–399.

Stampfli, G.M., Borel, G.D., Marchant, R., and Mosar, J. (2002). Western Alps geological constraints on western Tethyan reconstructions. *Journal of the Virtual Explorer* 8: 77–106.

Steck, A. (1968). Die alpidischen Strukturen in den Zentralen Aaregraniten des westlichen Aarmassivs. *Eclogae Geol. Helv* 61 (1): 19–48.

Tacher, L., Pomian-Srzednicki, I., and Parriaux, A. (2006). Geological uncertainties associated with 3-D subsurface models. *Computers and Geosciences* 32: 212–221.

Tonini, A., Guastaldi, E., Massa, G., and Conti, P. (2008). 3D geo-mapping based on surface data for preliminary study of underground works: A case study in Val Topina (Central Italy). *Engineering Geology* 99: 61–69.

Tonini, A., Guastaldi, E., and Meccheri, M. (2009). Three-dimensional reconstruction of the Carrara Syncline (Apuane Alps, Italy): An approach to reconstruct and control a geological model using only field survey data. *Computers and Geosciences* 35: 33–48.

Turner, A.K., Kessler, H., and van der Meulen, M.J. (2021). Applied multidimensional geological modeling: informing sustainable human interactions with the shallow subsurface, 672. Wiley-Blackwell.

Vollgger, S.A., and Cruden, A.R. (2013). 3D implicit geological modelling of a gold deposit from a structural geologist's point of view. Paper presented at the 12th SGA Biennial Meeting, Uppsala, Sweden.

Vouillamoz, J. (2009). Strain localization along shear zones in the Juchlistock area, Master's thesis, unpublished. University of Bern.

Wang, G., Gertner, G.Z., Fang, S., and Anderson, A.B. (2005). A methodology for spatial uncertainty analysis of remote sensing and GIS products. *Photogrammetric Engineering and Remote Sensing* 71 (12): 1423–1432.

Wanner, C., Diamond, L.W., and Alt-Epping, P. (2019). Quantification of 3-D thermal anomalies from surface observations of an orogenic geothermal system (Grimsel Pass, Swiss Alps). *Journal of Geophysical Research: Solid Earth* 124: 10839–10854.

Warren, S.D., Hohmann, M.G., Auerswald, K., and Mitasova, H. (2004). An evaluation of methods to determine slope using digital elevation data. *Catena* 58: 215–233.

Wehrens, P. (2015). Structural evolution in the Aar Massif (Haslital transect): Implications for midcrustal deformation. Doctoral thesis, unpublished. University of Bern.

Wehrens, P., Herwegh, M., Berger, A. et al. (2016). Deformation at the frictional-viscous transition: Evidence for cycles of fluid-assisted embrittlement and ductile interseismic deformation. *Tectonophysics* 693: 66–84. doi: 10.1016/j.tecto.2016.10.022.

Wehrens, P., Baumberger, R., Berger, A., and Herwegh, M. (2017). How is strain localized in a mid-crustal basement section? Spatial distribution of deformation in the Aar massif (Switzerland). *J. Struct. Geol.* 94: 47–67. doi: 10.1016/j.jsg.2016.11.004.

Wellmann, F. and Caumon, G. (2018). 3-D structural geological models: Concepts, methods, and uncertainties. *Advances in Geophysics* 59: 1–121. https://doi.org/10.1016/bs.agph.2018.09.001.

Wellmann, F. and Regenauer-Lieb, K. (2011). Uncertainties have a meaning: Information entropy as a quality measure for 3-D geological models. *Tectonophysics* 526–529: 207–216.

Wellmann, F., Horowitz, F., Schill, E., and Regenauer-Lieb, K. (2010). Towards incorporating uncertainty of structural data in 3D geological inversion. *Tectonophysics* 490: 141–151.

Wellmann, J.F., De La Varga, M., Murdie, R.E. et al. (2018). Uncertainty estimation for a geological model of the Sandstone greenstone belt, Western Australia – insights from integrated geological and geophysical inversion in a

Bayesian inference framework. *Geological Society, London, Special Publications* 453 (1): 41–56.

Wenzhong, S., Bin, W., and Yan, T. (2014). Accuracy analysis of digital elevation model relating to spatial resolution and terrain slope by bilinear interpolation. *Mathematical Geoscience* 46: 445–481. doi: 10.1007/s11004-013-9508-8.

Wolock, D.M. and McGabe, G.J. (2000). Differences in topographic characteristics computed from 100- and 1000-m resolution digital elevation model data. *Hydrol. Process* 14: 987–1002.

Woodcock, N.H. (1977). Specification of fabric shapes using an eigenvalue method. *GSA Bulletin* 88 (9): 1231–1236. doi: 10.1130/0016-7606(1977)88<1231:SOFSUA>2.0.CO;2.

Wüthrich, H. (1963). Rb-Sr-Altersbestimmungen an Gesteinen aus dem Aarmassiv. *Eclogae Geol. Helv* 56 (1): 103–112.

Wycisk, P., Hubert, T., Gossel, W., and Neumann, C. (2009). High-resolution 3D spatial modelling of complex geological structures for an environmental risk assessment of abundant mining and industrial megasites. *Computers and Geosciences* 35: 165–182.

Xiong, Z., Guo, J., Xia, Y. et al. (2018). A 3D multi-scale geology modeling method for tunnel engineering risk assessment. *Tunnelling and Underground Space Technology* 73: 71–81.

Yang, L., Achtziger-Zupančič, P., and Caers, J. (2021). 3D modeling of large-scale geological structures by linear combinations of implicit functions: Application to a large banded iron formation. *Natural Resources Research* 1–25.

Zanchi, A., Salvi, F., Zanchetta, S. et al. (2009). 3D reconstruction of complex geological bodies: Examples from the Alps. *Computers and Geosciences* 35: 49–69.

Zhu, L. and Zhuang, Z. (2010). Framework system and research flow of uncertainty in 3D geological structure models. *Mining Science and Technology* 20: 306–311.

11

Application of Implicit 3D Modelling to Reconstruct the Layered Structure of the Comet 67P

Luca Penasa[1], Matteo Massironi[2], Emanuele Simioni[3,4], Marco Franceschi[2], Giampiero Naletto[1,4,5], Sabrina Ferrari[1], Ivano Bertini[5], Pamela Cambianica[3], Elisa Frattin[3,5], Fiorangela La Forgia[5], Alice Lucchetti[3], Maurizio Pajola[3], Frank Preusker[6], Frank Scholten[6], Laurent Jorda[7], Robert Gaskell[8], and Holger Sierks[9]

[1] *University of Padova, Center of Studies and Activities for Space (CISAS) "G. Colombo," Via Venezia 15, 35131 Padova, Italy*
[2] *University of Padova, Department of Geosciences, Via G. Gradenigo 6, 35131 Padova, Italy*
[3] *INAF, Astronomical Observatory of Padova, Vicolo dell'Osservatorio 5, 35122 Padova, Italy*
[4] *CNR-IFN UOS Padova LUXOR, Via Trasea 7, 35131 Padova, Italy*
[5] *University of Padova, Department of Physics and Astronomy "Galileo Galilei," Via Marzolo 8, 35131 Padova, Italy*
[6] *Deutsches Zentrum für Luft- und Raumfahrt (DLR), Institut für Planetenforschung, Rutherfordstraße 23, 12489 Berlin, Germany*
[7] *Laboratoire d'Astrophysique de Marseille, UMR 7326, CNRS & Aix Marseille Université, 13388 Marseille Cedex 13, France*
[8] *Planetary Science Institute, 1700 East Fort Lowell, Suite 106, Tucson, AZ 85719, USA*
[9] *Max Planck Institute for Solar System Research, Justus-von-Liebig-Weg 3, 37077 Göttingen, Germany*

Abstract

We provide details about the procedure employed for the three-dimensional geological modelling of the lobes of comet Churyumov-Gerasimenko (67P). The two lobes of 67P are characterized by well visible terraces and elongated cliffs that revealed a pervasive layering arranged in an onion-like fashion. None of the layers can be traced for a continuity large enough to provide a means of stratigraphic correlation. Therefore, an explicit modelling approach of the layered structure is not easily applicable. We show how a very simple modelling strategy based on implicitly-defined analytical surfaces (spheres or ellipsoids), and requiring very limited operator decision-making, can be successfully applied to produce a geological model that easily fits the available scattered attitude observations. Our formulation has the advantage of providing a small set of parameters with a precise geometrical meaning that can be compared with other parameters of the lobes (i.e. center of ellipsoids and center of mass). The presented method originates outside commercial geological modelling software packages and required devising some ad-hoc solutions for the visualization of the resulting models in comparison with observations made on OSIRIS images. Some extracts of the visualization code, helpful for applying this procedure in similar operational contexts, are also presented.

11.1 Introduction

ESA's Rosetta spacecraft made it possible to observe the surface of the comet Churyumov-Gerasimenko (67P) at unprecedented resolution. The camera system onboard the spacecraft, the OSIRIS (Keller et al., 2007), captured tens of thousands of images during the two years of the mission, which were used for shape model reconstruction (Jorda et al., 2016; Preusker et al., 2017), spectrophotmetric studies (Fornasier et al., 2015; Oklay et al., 2016), geomorphological mapping (Giacomini et al., 2016; Lee et al., 2016), and many other investigations on surface morphology and geology (El-Maarry et al., 2015a, 2016; Thomas et al., 2015; Birch et al., 2017; Vincent et al., 2017).

Comet 67P has a bilobate shape: the *big lobe* and the *small lobe* are connected through a bridge of material, known as the *neck* (see, for example, Sierks et al., 2015). The comet's nucleus is extremely rich in geomorphological forms. Among the most prominent ones are small- to large-scale fractures (El-Maarry et al., 2015b; Auger et al., 2017), landslides (Pajola et al., 2016) and morphological terraces. Terraces are flat portions of terrain, bounded on one or more sides by almost vertical cliffs: terraces and cliffs appear to be strongly associated and produce the whole range of intermediate morphologies between staircase-patterns of terraces to flat-topped *mesas*.

Terraces and cliffs are spatially organized, outlying a large-scale pervasive pattern of concentric and

3D Digital Geological Models: From Terrestrial Outcrops to Planetary Surfaces, First Edition.
Edited by Andrea Bistacchi, Matteo Massironi, and Sophie Viseur.

semi-parallel layers (Thomas et al., 2015; Massironi et al., 2015; Rickman et al., 2015). This arrangement strongly recalls exposures of sedimentary and, less frequently, volcanic rocks observed on Earth and other rocky planets where the action of *differential erosion* enhances the original differences in the physical properties of the rocks. On comets the more prominent erosive process is expected to be sublimation (e.g. Keller et al., 2015; Guilbert-Lepoutre et al., 2015). Thus a process of *differential sublimation* might be a good candidate for explaining the terrace formation. The processes controlling terrace formation have produced a topographic profile of non-equilibrium, which alternates flat terrains with gravitationally failure-prone cliffs (Pajola et al., 2016). Figure 11.1, modified from Penasa et al. (2017a), shows the evidence for a tight alternation of terraces and cliffs on a typical topographic profile of 67P.

Terraces seem to form in correspondence of pre-existent extended surfaces, or bedding planes, which continue deep inside the nucleus. Terraces are observed for at least 1.2 and 1.38 km of stratigraphic thickness for the big lobe and the small lobe respectively, suggesting that a large portion of the cometary body is layered (Penasa et al., 2017a).

The same surfaces controlling the terraces can also be observed on vertical walls and inside pits, as linear traces produced by the intersection of the bedding planes with the surface (see, for example, Thomas et al., 2015; Massironi et al., 2015; Basilevsky et al., 2017). These traces are clearly aligned with nearby terraces and therefore represent the continuation of the terraces within the subsurface. No element contradicting this evidence has been found so far. While these traces are locally visible they cannot be followed for large distances, making the identification of key horizons difficult, and precluding any lateral correlation at the scale of the entire lobes. Their elusive appearance, which seems to be highly sensitive to the illumination condition (the same outcrop might show or not show well-defined bedding planes, depending on the shadowing), suggests that these traces might actually be the result of small-scale terrace-like morphologies, at the limit with the image resolution.

In the next sections we will describe the process that led from the observation of terraces to a three-dimensional model of the layering. We will discuss the main problems related with such an uncommon case of a three-dimensional layered structure and how a working solution was obtained. The analytical solution described in this chapter was also used as a basis to create a more adaptive 3D model by Franceschi et al. (2020), where more details on this topic can be found.

11.1.1 From Terraces to Geological Cross-sections

The OSIRIS system is composed by two cameras: the Wide Angle Camera (WAC) camera with a larger field of view (FoV) of 12° and the Narrow Angle Camera (NAC) with a smaller FoV of 2.35°. Thanks to the large number of images collected during the Rosetta mission, three-dimensional meshes of the surface were produced by using either the stereophotogrammetry (SPG; Preusker et al., 2017) or the stereophotoclinometry method (SPC; Jorda et al., 2016). The models were produced incrementally by including new images in the reconstruction pipeline as they became available, so that the first generation of models comprised only the northern hemisphere of the comet.

From the analysis of the orientation of the terraces, Massironi et al. (2015) drew the first geological cross-sections of the inner layering, also showing that the terraces are aligned with the gravity equipotentials of each lobe taken individually, rather than being coherent with the gravity field of the entire bi-lobated nucleus. That evidence demonstrated the independent nature of the layering of the two lobes. To derive geological cross-sections

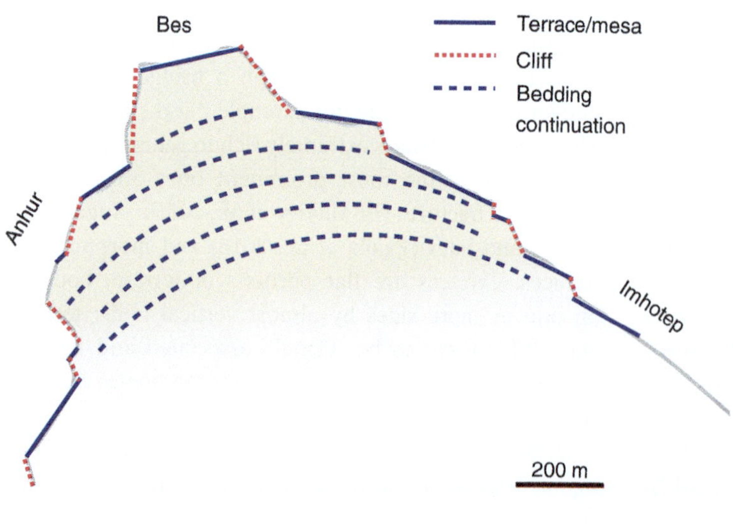

Figure 11.1 A topographic profile obtained by cutting the shape model of Churyumov-Gerasimenko (67P) with a plane. Terraces and cliffs alternate, producing *staircase pattern* morphologies. Modified from Penasa et al. (2017a).

of the cometary nucleus, Massironi et al. (2015) estimated the orientation of the terrace on the 3D shape model that was available at the time, hence limited to the northern hemisphere of the Churyumov-Gerasimenko (67P) comet (Jorda et al., 2016) and characterized by a lower resolution. In particular, to estimate a best-fitting normal for each terrace surface, portions of the 3D shape model were extracted by using the meshlab software (Cignoni et al., 2008). The extracted vertices of the triangular mesh were then employed to obtain a best-fitting plane estimate of the terrace by means of a least squares method (e.g. Fernández, 2005).

From the three-dimensional shape model a set of cross-sections were extracted providing topographic profiles of the comet nucleus. A subset of the obtained terraces, located within a buffer from the section plane, was then plotted on the same cross-section to represent the local orientation of the layering. From this information, Massironi et al. (2015) interpreted the inner layering as formed by extended surfaces, whose intersections with the two-dimensional section were traced by respecting the orientation imposed by the terraces. A similar methodology was employed for drawing geological cross-sections used during planning of the Brenner basis tunnel (Bistacchi et al., 2008; Schiavo et al., 2015).

This approach provided geological sections like the one in Figure 11.2, in which the terraces, interpreted as proxies for the underlying bedding planes, are extended inside

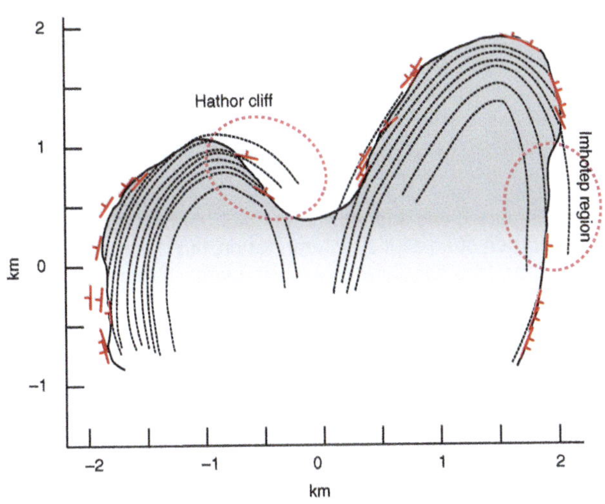

Figure 11.2 One of the geological cross-sections realized by Massironi et al. (2015), redrawn and simplified. Red marks represent the measured orientations at the section plane of the observed terraces and their normals. Dashed lines are the interpreted continuation of the terraces as extended bedding planes inside the nucleus, which were traced in accordance with the observed attitudes at the surface. On some regions the stratigraphic series is well exposed for some hundreds of meters of the total thickness (dotted ellipses).

the nucleus. This provided an indication not only about the expected orientation of the layers below the surface but also about the overall layered sequence thickness, which was shown to be of the order of hundreds of meters. Although being limited to the northern hemisphere, this analysis showed that large portions of the layered sequence were actually missing, for example on the Hathor cliff, where the layers meet the topographic surface at a high angle (Figure 11.2).

To better constrain the geometry of the layers, the creation of a three-dimensional model was of primary importance. A typical workflow for producing three-dimensional models of sedimentary sequences involves the creation of a set of cross-sections on which the subsurface continuation of layers/horizons is tracked and/or interpreted. As layers are intersected by each cross-section in depth, it is possible to create surfaces fitting the traces of each layer in three dimensions. This approach is defined as *explicit modelling* (Caumon et al., 2009). Although cross-sections of Churyumov-Gerasimenko (67P) were readily available (at least for the northern hemisphere) from Massironi et al. (2015), the creation of a explicit model turned out to be a difficult task. Using 67P it is not possible to clearly identify any stratigraphic key horizon that could be identified and traced on all the cross-sections. This makes it difficult to establish a sure lateral continuity of the strata over a couple of hundred meters. Indeed, on 67P only scattered attitude information exists, and hence each geological section can be internally self-consistent with the nearby strata attitudes but there is no objective way to establish the stratigraphic correlation among different sections (see Figure 11.3). Although it would in principle be possible to solve this problem by explicitly correlating the layers on different sections, any resulting model would be highly interpretative and subject to the choices enforced by the operator performing the modelling.

The 67P case indeed highlights the severe limitations of generating 3D models from two-dimensional cross-sections when little or no information about the stratigraphic sequence is available.

11.2 A Modelling Strategy for Onion-Like Layers

Three-dimensional modelling strategies in geoscience can be grouped into two categories: *implicit* or *explicit* (e.g. Rolo et al., 2017). In the *explicit* approach, the surfaces are constructed as meshes and modified through direct interaction with their geometrical parameters. In the *implicit* approach, instead, surfaces are constructed as iso-potential surfaces of a scalar field. The first method

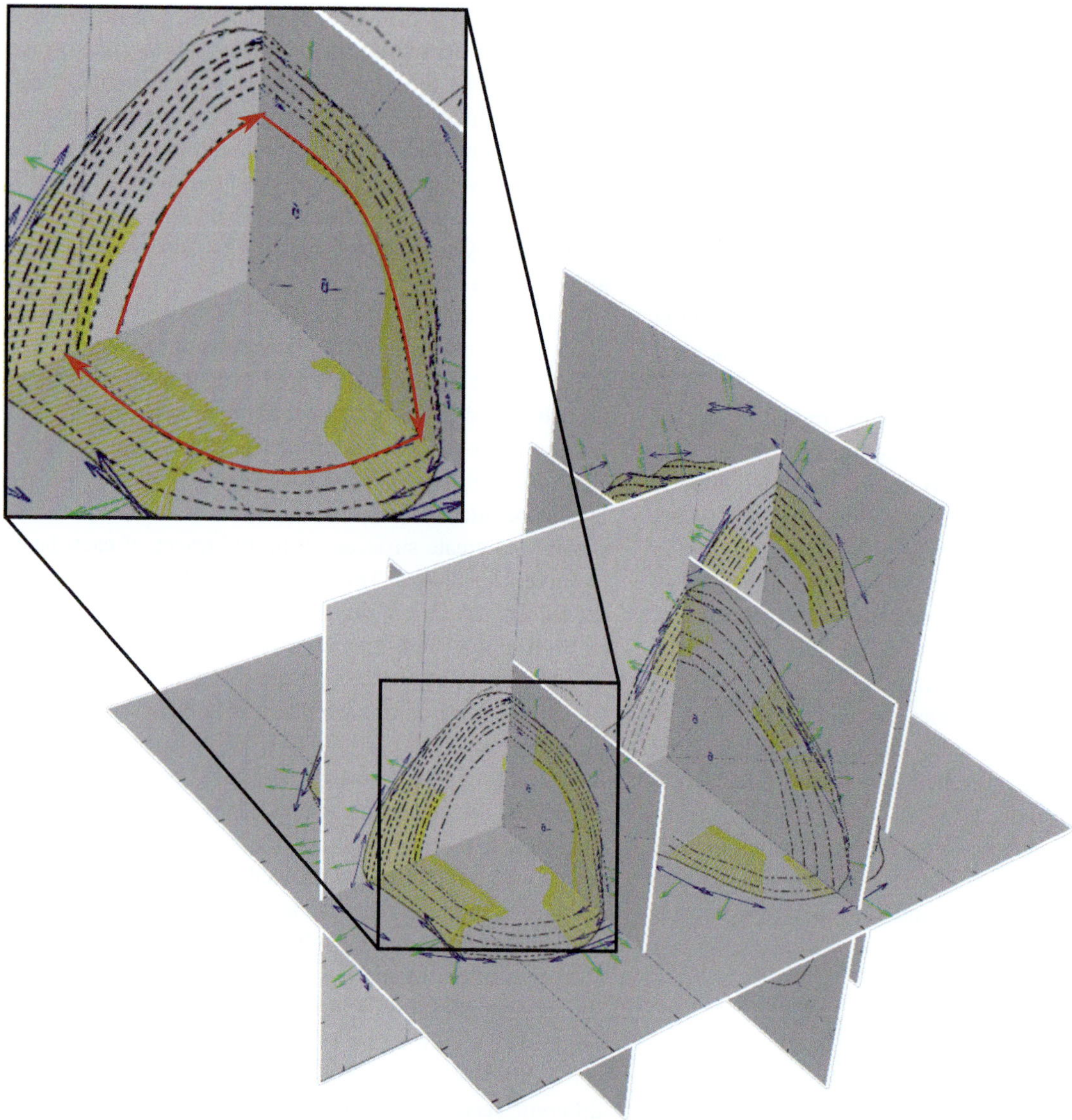

Figure 11.3 The geological cross-sections realized by Massironi et al. (2015) imported and aligned within *GoCAD/SKUA©* 3D geological modelling software: the sections provide a framework for the generation of an explicit interpretation, but the absence of any clear stratigraphic marker that could be extended throughout all the sections does not allow an explicit approach to be strictly followed. The top frame highlights the inconsistencies that are introduced when trying to correlate layers on different sections that were independently realized.

works better when good constraints (e.g. in the form of key horizons) are available but it proved not to be well suitable for comet 67P as the explicit creation of bedding planes would have required the identification of lines and terraces well correlated at the scale of the entire comet (i.e. terraces belonging to an original continuous or semicontinuous bedding plane of global extension). Such correlation, as already noted, could not be done on 67P without involving a series of assumptions not supported by clear evidence.

An *implicit* modelling approach (see, for example, Hillier et al., 2014; Calcagno et al., 2008; Grose et al., 2021) was thus pursued to obtain a simplified numerical representation of the layered sequences of Churyumov-Gerasimenko (67P) lobes. Implicit models seek a representation of a specific layer as a contour surface of a three-dimensional, real-valued function f:

$$f(\mathbf{p}) = f_1 \tag{11.1}$$

where f_1 is a scalar associated with a specific layer and $\mathbf{p} = [x, y, z]^T$ is a point part of it. Function f should ideally represent the *stratigraphic position* (a value representing the position of a point in the stratigraphic pile) of any given three-dimensional point. The modelling function

f is normally an interpolator that satisfies some a-priori constraints, typically related to the smoothness of the generated surfaces. Several different interpolation algorithms have been proposed, mostly based on Radial Basis Functions (RBFs) (Hardy, 1990), kriging (Hardy, 1990), or *discrete* interpolants (Caumon et al., 2013). Although interpolation-based modelling schemes are extremely adaptive and can be used to represent highly complex geological bodies (e.g. mineralization ores), the parameters of the models lack a concise physical meaning, making it more difficult to interpret the results of the modelling.

For these reasons we made use of the knowledge about the layering on 67P lobes, to formulate an ad-hoc modelling function. The terraces, and hence the layering, appear to be geometrically aligned to the gravity equipotentials of each lobe (Massironi et al., 2015), thus suggesting that a roundish shape could be well representative of the layers.

A simple spherical modelling (SP) function in three dimensions can be constructed as follows (see Figure 11.4):

$$f(\mathbf{p}) = \|\mathbf{p} - \mathbf{c}\| \tag{11.2}$$

This kind of implicit function predicts the stratigraphic position of any point $\mathbf{p} = [x, y, z]^T$ as a distance from a center $\mathbf{c} = [x_0, y_0, z_0]^T$. Such a simple model represents each layer by a spherical surface, centered on the point $\mathbf{c}$ (see the two-dimensional example in Figure 11.4). Notice that in this case the gradient of the function at any given point is always unitary:

$$\|\nabla f(\mathbf{p})\| = 1 \tag{11.3}$$

Hence this kind of model represents a real metric of distance (being defined as such) in the depth space: if we consider two different layers at f_1 and f_2 we can compute the thickness of the comprised layering as the difference of the scalar field between the two $(f_1 - f_2)$.

The modelling strategy based on spherical shells can also be adapted to other analytical shapes. To derive a modelling function that could account for the evident non-sphericity of the lobes of 67P we made use of ellipsoids: Figure 11.4 illustrates this concept in two dimensions. A first possibility consists in using a family of concentric ellipsoids sharing the same axial ratios (Fixed axial Ratios model, hereafter FR). The obtained modelling function (see also Penasa et al., 2017a) is

$$f(\mathbf{p}) = g(\mathbf{R}^T \cdot (\mathbf{p} - \mathbf{c})) \tag{11.4}$$

and the relative gradient is

$$\nabla f(\mathbf{p}) = \mathbf{R} \cdot \nabla g(\mathbf{R}^T \cdot (\mathbf{p} - \mathbf{c})) \tag{11.5}$$

where the function $g(\mathbf{p})$ is defined as

$$g(\mathbf{p}) = \frac{1}{bc}\sqrt{b^2c^2x^2 + b^2z^2 + c^2y^2} \tag{11.6}$$

The rotation matrix $\mathbf{R}$ represents the orientation of the ellipsoidal shells in space, and can also be parametrized as three angles α, β, γ (yaw–pitch–roll), while $\mathbf{c}$ is the common center of the ellipsoidal shells. The parameters b and c are the axial ratio of the ellipsoidal family and x, y, z are the coordinates of any point $\mathbf{p}$. This choice provides a modelling function that has a simple analytical expression but has the drawback of producing ellipsoidal shells that are not parallel to each other; thus the thickness of the layers is variable (see Figure 11.4, FR model).

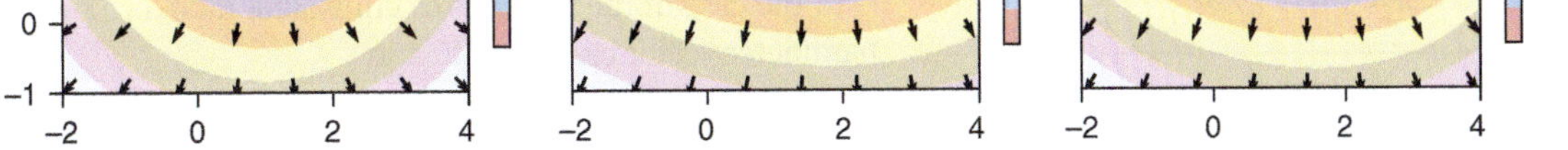

Figure 11.4 Three different modelling functions for roundish concentric layers. A simple modelling function (SP) can be defined as the distance from a point. Contour lines (or surfaces in 3D) can then be traced and used to represent concentric shells. The gradient at any given point (arrows) provides a means of estimating the local orientation of the layers, which can be compared with measured orientations (the observed orientations of terraces). The layers for an FR model (concentric ellipsoids with fixed axial ratios) are not parallel to each other, which is a common requirement in geological modelling. This issue can be solved with the Parent Ellipsoid (PE) model (distance field from a parent ellipsoid), but this would produce unrealistic lens-shaped layers in the interior of the parent ellipse (or ellipsoid in 3D). In this example units are arbitrary.

On the other hand, a model in which thickness is preserved (Parallel Ellipsoids model, hereafter PE) can be obtained by defining the function $f(\mathbf{p})$ as the distance from a *parent ellipsoid*, which can be defined by its a,b,c parameters, **R** and **c** (see also Laurent, 2016, for a discussion on thickness preservation when dealing with discrete implicit models). This kind of model thus requires the *distance* from an ellipsoid to be computed, a problem that does not have an analytical solution and has to be solved numerically by finding the zeroes of a sixth degree polynomial. A *robust* algorithm is described in Eberly (2013) and its implementation can be found in the *Geometric Tools* repository (www.geometrictools.com). This type of model, which would have the desirable property of having a unitary gradient everywhere, has the serious drawback of producing unrealistic lenticular shapes inside the parent ellipsoid (see the two-dimensional representation of Figure 11.4, PE model).

Table 11.1 summarizes some simple models that can be used for modelling onion-like layered bodies. All these models can find application when modelling roundish layered bodies, although an SP model can be too simple to approximate complex shapes well. For solving this issue an FR or PE model can be used to obtain a more adaptive shape, although the FR model has better properties in term of the resulting shapes (no lens-shaped shells inside the parent ellipsoid) and it is simpler to implement, having an analytical solution for the function value and its gradient. This function has the only drawback of not being thickness-preserving, but it might be acceptable if there are not strict reasons to impose this constraint, as in the case of 67P, where continuous layers of constant thickness have not been observed.

As a side note, we highlight the fact that the implicit formulation of these models simplifies their use as "base models" for other implicit modelling techniques: if the object to be modelled deviates locally from the base shape (e.g. the ellipsoid), additional implicit functions (e.g. RBF) can be used to account for residuals by just summing the contribution of the base-shape implicit model with a more adaptive one. For example, an extended model $f'(\mathbf{p})$ could be obtained as

$$f'(\mathbf{p}) = f(\mathbf{p}) + r(\mathbf{p}) \tag{11.7}$$

and the gradient as

$$\nabla f'(\mathbf{p}) = \nabla f(\mathbf{p}) + \nabla r(\mathbf{p}) \tag{11.8}$$

where $f(\mathbf{p})$ is any of the SP, FR, or PE models, and $r(\mathbf{p})$ is another, more flexible, interpolator (e.g. RBF or kriging) that could be used to take into account the deviations from the base shape.

Table 11.1 Some of the functions that were tested as a modelling function for the layers of 67P lobes.

Abbreviation	Function	Number of parameters	Analytical solution
SP	Spherical	3 (x,y,z)	Yes – very simple
FR	Fixed ratios	3 (x,y,z) + 2 (b,c) + 3 (α, β, γ)	Yes – very simple
PE	Parallel Ellipsoid	3 (x,y,z) + 3 (a,b,c) + 3 (α, β, γ)	None

11.3 Model Fitting

The parameters of the model must be determined by maximizing the agreement between predictions and observations. In this case, each observation is a point in space $\mathbf{p}_i = [x, y, z]^T$ associated with a unit vector $\boldsymbol{n}_i = [nx, ny, nz]^T$ representing the direction perpendicular to the layers at that location, as depicted in Figure 11.5. The available data does not provide any direct observation of the scalar field to be modelled (values of the function f), but only of its normalized gradient $\boldsymbol{n}_i$, which can be compared with the predicted gradient as an angular measure:

$$\theta_i = \arccos\left(\frac{\nabla f(\mathbf{p}_i)}{\|\nabla f(\mathbf{p}_i)\|} \cdot \boldsymbol{n}_i\right) \tag{11.9}$$

By minimizing these angles via a non-linear least-squares solver, the best-fitting parameters for any selected f model can be found. We made use of the CERES C++ package (Agarwal et al., 2016) to develop a simple toolkit for fitting implicit models. Each model requires the definition of two methods: one for computing the value of $f(\mathbf{p})$ and one for its gradient $\nabla f(\mathbf{p})$, and can then be easily fitted to any dataset of observations. Alternatively to unit normals, polylines describing the edges of the terraces could also be used to constrain the model, with just minor changes to the fitting strategy of Equation (11.9). More information on this approach can be found in Ruzicka et al. (2019).

In the case of the fixed-ratios ellipsoidal model (FR) the parameter vector can be written as

$$\boldsymbol{x} = [c_x, c_y, c_z, b, c, \alpha, \beta, \gamma]^T \tag{11.10}$$

Given a set of initial values, the solver finds the vector of parameters that minimizes the angular residuals of Equation (11.9).

Figure 11.6 shows the histograms of the best-fitting model residuals in the case of the spherical model (SP), the fixed ratios model (FR), and for a model based on shells

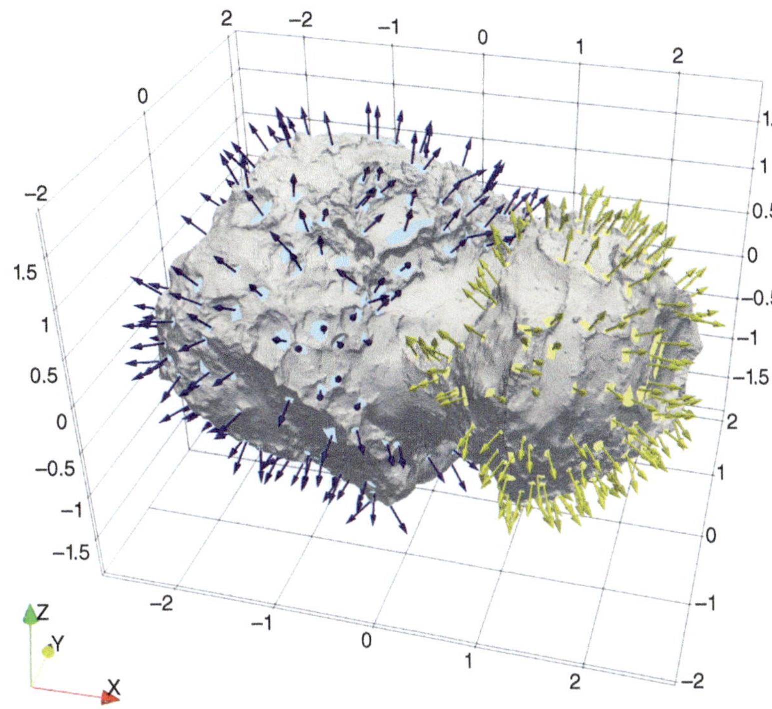

Figure 11.5 The 3D shape model of Churyumov-Gerasimenko (67P) (Preusker et al., 2017; Jorda et al., 2016). Each arrow represents an observation used as a constraint to model the layers. These attitudes were obtained by selecting a portion of the 3D model representative of terraces. A best-fitting procedure on selected vertices provided a unit normal perpendicular to each terrace. Colored patches correspond to the portions of the shape model that were used for the plane fitting.

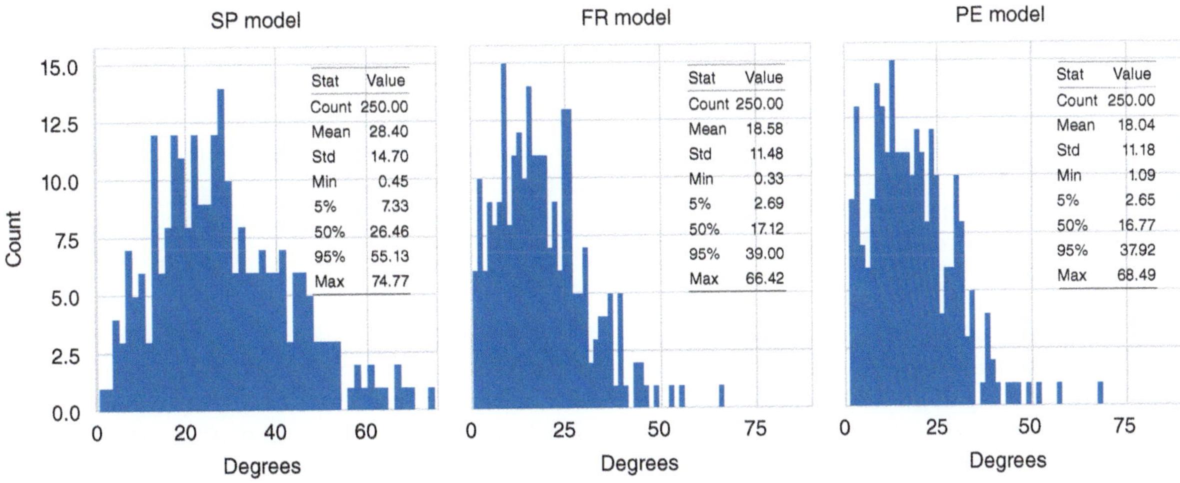

Figure 11.6 Histograms and statistics for the angular residuals for three different best-fitting models for the Big Lobe of comet 67P. In the case of the spherical model (SP), each layer is approximated by a concentric sphere. A family of ellipsoids with fixed axial ratios (FR) provide similar average residuals as the one based on parallel shells (PE). The FR model is simpler to implement and avoids the creation of lens-shaped shells within the parent ellipsoid. The average residuals is of ~10° smaller using the PE ellipsoidal model, with respect to the simpler spherical one, where almost no residuals below 10° are present.

parallel to a parent ellipsoid (PE). The FR model was used in Penasa et al. (2017a) and was chosen to be the *reference* model for the layers on 67P because it provided a residual comparable to the PE model with the additional benefit of not producing the inner lenticular shapes that can be observed in Figure 11.4. The residuals can be also evaluated for their location on the cometary body. Figure 11.7 shows each observation as a vector in space, colored by the angular residual of the model at that location. Observations with higher residuals does not cluster in specific locations, but are rather well-scattered on the whole body, suggesting that they can mostly be considered as outliers, rather than deviations of the layers from the ellipsoidal shape that was employed.

In the case of an FR model, three parameters are needed to define the center of the ellipsoid, while b and c are the axial ratios of the ellipsoidal shells and α, β, γ are *yaw*, *pitch*, and *roll* angles, defining their orientation in space.

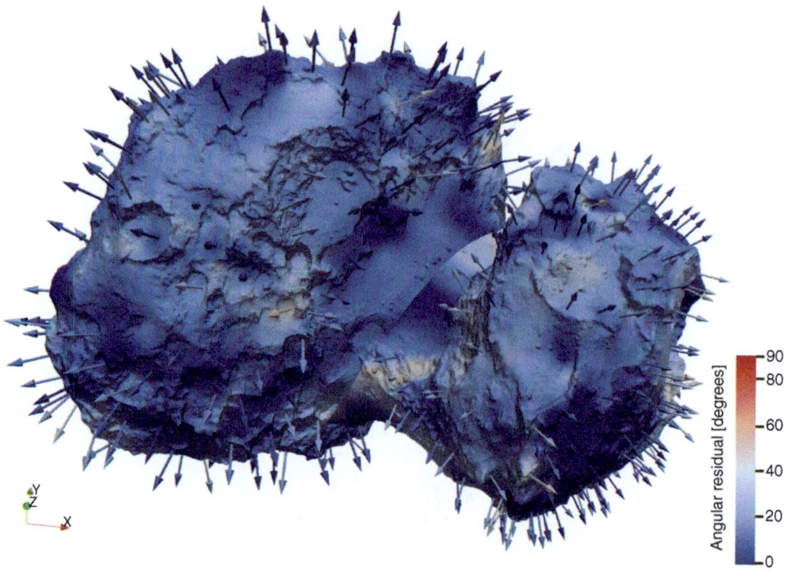

Figure 11.7 The angular residuals resulting from the model can be visualized by mapping the angles to a color scale (0° to 90°) and using them to color the vectors representing the observations. In this case (the best-fitting FR model), the residuals have been also interpolated on the surface for visualization purposes.

The management of the angular parameters α, β, and γ requires some special attention during implementation because an ellipsoid has three symmetry axes and thus different triplets of angles might actually represent ellipsoidal models with an equivalent orientation in space. As an example, consider a generic triaxial ellipsoid: a rotation of 180° around any axis produces an ellipsoidal model that is equivalent for the scope of our modelling to the one in the non-rotated state.

Also, the a, b, and c parameters require a similar care. These parameters are often constrained in this way:

$$a \geq b \geq c > 0 \tag{11.11}$$

Although several non-linear optimization packages allow the solution to be constrained by defining the bounds for each parameter, we found that this strategy could be counter-productive, slowing down or preventing convergence. A working solution consists in letting the solver reach convergence through the more convenient path in the parameters space, for example by providing a solution of the kind $b \geq c \geq a > 0$, which can be easily reformulated in the expected form by reordering the columns of the orientation matrix **R** and the respective values of the semi-axis.

When determining parameters by nonlinear methods, it is important to initialize the vector of parameters with values that are *near* the global minimum, to have faster convergence and possibly avoid local minima. The overall topography of the object can be used as a first approximation of the shape of the layers and thus the center of the model can be set to match the centroid of the observations. Initializing the orientation and the axial parameters might be more complex and can be, for example, obtained by Principal Component Analysis (PCA) applied on the locations of the observations.

11.3.1 Errors Determination

To associate an error to each model's parameter, a *bootstrap* method based on a Monte Carlo resampling (Efron, 1979; Chernick, 2008) has been applied. The method can be implemented in the following iterative way:

1. A new dataset of observations is generated by resampling with replacement of the original dataset. This simulates the effect that might be introduced by a new data-collection campaign.
2. The best-fitting solution is determined as described above, and the vector of the best-fitting parameters is stored.

These two steps are repeated for a large number of times (i.e. 10,000), thus providing a new best-fitting solution at each iteration. We found that in the specific case, when fitting an FR ellipsoidal model, even starting with a naive initialization of the parameters (i.e. the center located in the centroid of the observations, randomly chosen axial ratios in between 0.6 and 1.0, and axes aligned with the Cartesian axes of the reference frame), the rate of convergence resulted in 100% (all iterations converged properly).

The bootstrap method provides a large number of parameter vectors for the chosen model, which can be summarized as histograms. As an example, Figure 11.8 shows three histograms for each lobe representing the coordinates of the FR ellipsoidal model center. The histograms exhibit well-shaped Gaussian distributions, suggesting that the parameters converge toward the same solution: this makes the hypothesis of multiple solutions for the considered model unlikely. From these distributions it is possible to obtain the maximum likelihood solution, while the associated standard errors can be obtained by statistical

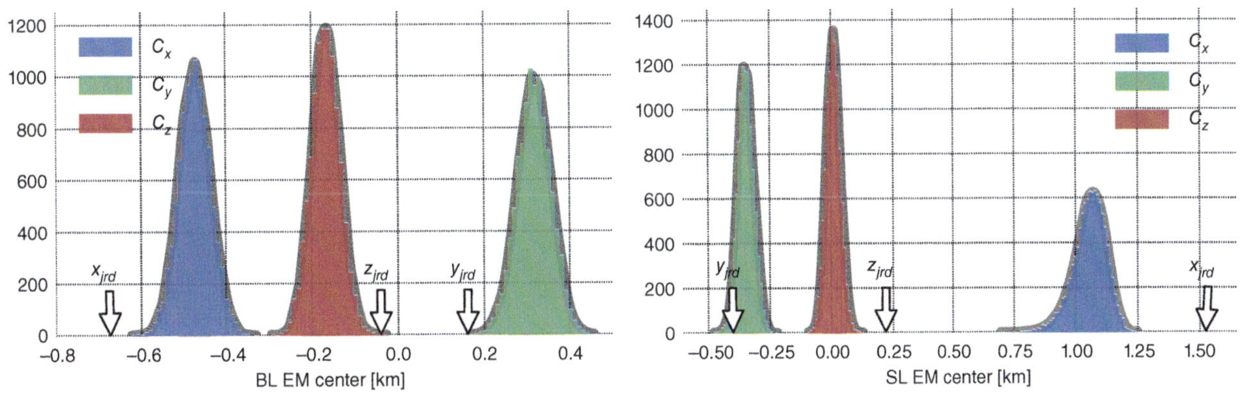

Figure 11.8 Distributions of the ellipsoidal model center $[c_x, c_y, c_z]$ resulting from the bootstrap iterations. The coordinates for the center of mass computed by Jorda et al. (2016) are reported with the *jrd* subscript.

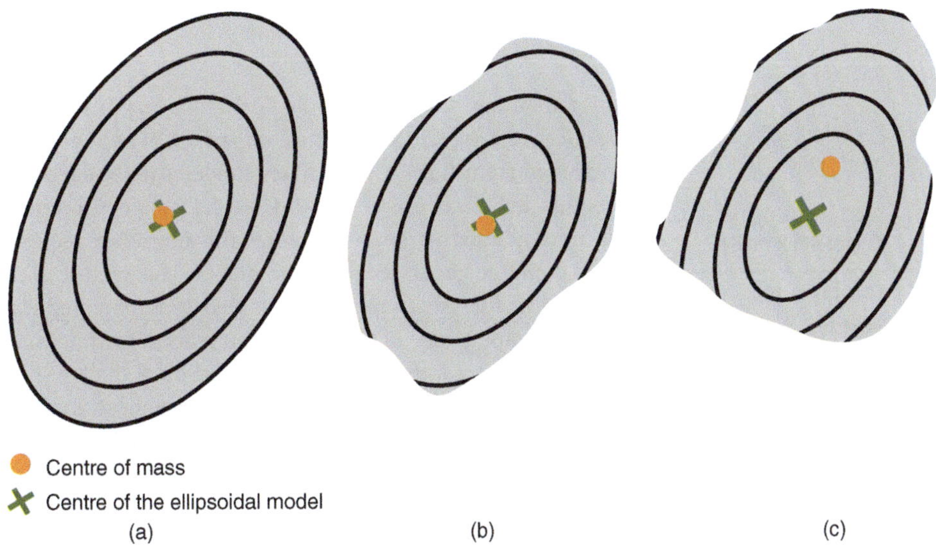

Figure 11.9 The center of mass of the modelled lobe would be expected to correspond to the center of the ellipsoidal model whenever the object is (a) intact or (b) when it underwent an approximately homogeneous erosional process. In the case of localized material removals, the center of mass is expected to be significantly shifted (c), as was in fact observed for comet 67P (Based on Penasa et al., 2017a).

means, taking care to use circular statistics for the angles (Fisher, 1993). A complete table of the parameters and their errors, as determined by the described procedure assuming an FR ellipsoidal model, can be found in Penasa et al. (2017a). We will limit the discussion here to the coordinates of the center of the ellipsoidal models.

The proposed modelling strategy has the advantage of providing a set of parameters with an unambiguous geometrical meaning and may help to provide quantitative/semi-quantitative measurements useful to the study of the comet. For instance, the *center* $[c_x, c_y, c_z]$ of the ellipsoids can be compared with the center of mass of the respective lobe. Assuming a homogeneous density of the nucleus, the center of the ellipsoidal model and the center of mass of the lobe are expected to coincide. Any deviation from this configuration might suggest that the lobe underwent a localized erosional process that removed more material in specific locations. Figure 11.9 illustrates this concept.

In the case of 67P this analysis revealed that the center of mass in both cases does not coincide with the center of the respective ellipsoidal models. Table 11.2 reports on the center of mass for the two lobes of 67P, which are also displayed in Figure 11.8 as white arrows.

We found that the two centers of mass are significantly shifted with respect to the centers of the FR ellipsoidal models, by about 300 and 500 meters for the big and the small lobe respectively. To evaluate whether these values are significant or not, the estimated errors must be considered. The center of mass was computed by considering a homogeneous density for the material and, under this assumption, the error can be considered as negligible when

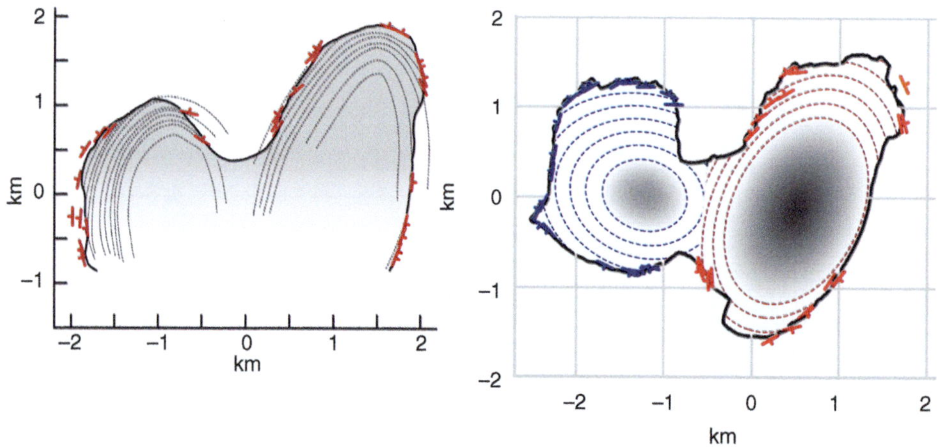

Figure 11.10 Comparison of a cross-section. On the left, the section number 4, from Extended Data Figure 8 by Massironi et al. (2015), was obtained by interpretation of orientation data. The one on the right was obtained from the ellipsoids-based FR model by Penasa et al. (2017b).

Table 11.2 Center of mass of the lobes, as computed by Jorda et al. (2016), compared to the center of the FR model from Penasa et al. (2017a).

	Center of mass	Center of FR model
Big lobe	[−0.673, 0.147, −0.049] km	[−0.473, 0.320, −0.167] km
Small lobe	[1.524, −0.391, 0.208] km	[1.060, −0.346, 0.010] km

compared to the standard errors for the centers of the FR ellipsoidal models. When these values are compared with the distributions of the ellipsoidal model parameters (see Figure 11.8), it appears evident that there is a consistent shift of the coordinates (well above the significance level), especially along some directions. For example, the x coordinate of the center of the FR ellipsoidal model for the small lobe is shifted by about 500 meters. This shift can be ascribed to the large volumes of cometary material, which are missing on the Hathor cliff (see Figure 11.2). The statistically significant differences between the centers of mass and the center of the respective ellipsoidal models thus support the idea that the nucleus underwent significant processes of uneven material removal (see Figure 11.9).

11.4 Visualization and Validation of the Models

Once the most likely ellipsoidal model is found through the described bootstratp method, it is necessary to convey the model so that is can be used to validate the geological interpretation. The simplest approach consists in generating a series of geological cross-sections of the model. Figure 11.10 shows the comparison of two sections obtained from the FR most-likely model and the original section by Massironi et al. (2015). The obtained section is now granted to be consistent with any other section that could be generated from the model. This picture also highlights the great improvements that were obtained for the 3D shape model of 67P during the years of the mission.

Alternatively, a certain number of ellipsoidal shells of the model can then be overlaid in any 3D visualization software. Although this approach gives the precise idea of the geometry of the used model, it does not allow the model to be easily compared with other information because the shells tend to occlude the view, even if some transparency rendering is employed. In particular, the obtained model is expected to produce shells that, once intersected with the local topography, should be parallel to nearby terraces following the related scarps.

For this reason we used the modelling equation (i.e. Equation (11.4)) to compute the scalar value of the ellipsoidal model function $f(\boldsymbol{p})$ for each vertex of the shape model, thus providing a scalar field that can be easily visualized thanks to the ParaView software (Squillacote et al., 2007) by mapping the values to a color scale (Figure 11.11a). The scalar values computed using Equation (11.4), in the case of an ellipsoidal model with fixed axial ratios, correspond to the value of the semimajor axis of the ellipsoidal shell passing for that location (hence its name, Reference Ellipsoid Semimajor axis, or simply RES). ParaView is an open source software for scientific visualization built on top of the VTK library (Schroeder et al., 2006) and can be used to visualize three-dimensional mesh models and the associated scalar fields (scalar values associated with the vertices or the facets of the models).

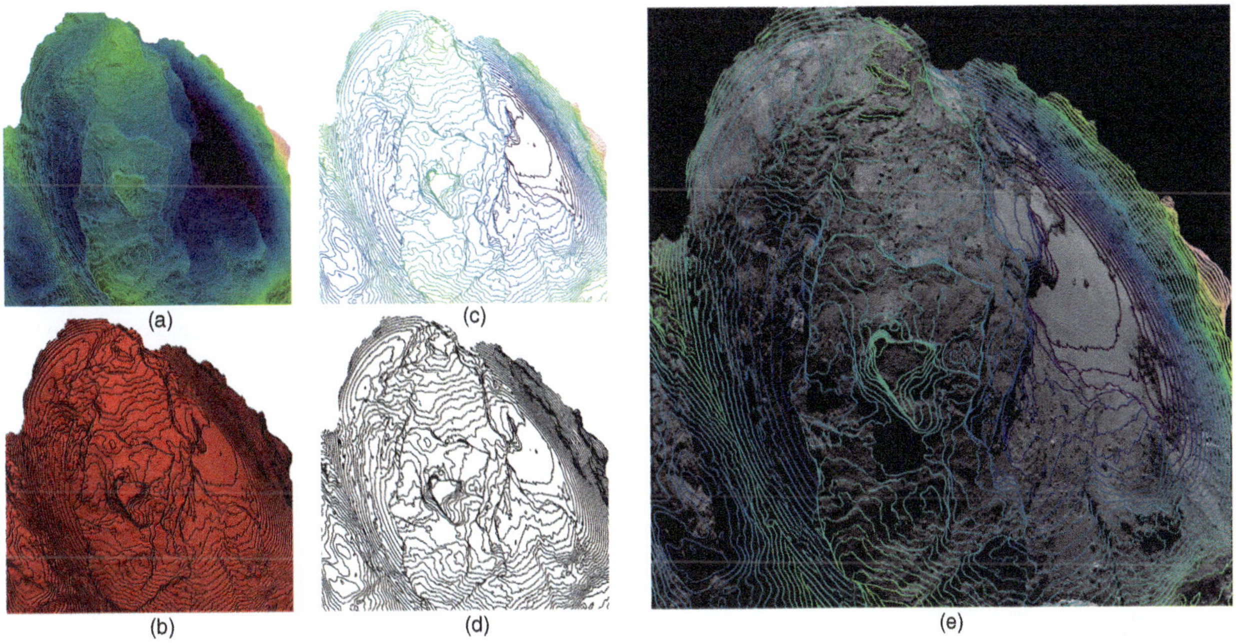

Figure 11.11 To overlay the RES contour lines on a real OSIRIS image, the geometry of the shape model must be taken into account. This can be done by (d) creating a mask and (b) using an ad-hoc color mapping. The mask can then be used to extract the contour lines from (a) to obtain the color contour-lines image, as shown in (c). These can then be easily overlaid to any NAC or WAC image (e). (ImageID reference `NAC_2016-01-27T16.27.58.970`.)

The primary source of information used in this work is the collection of images generated by osiris, thus comparing the modelling results with any NAC or WAC image (Figure 11.11a) is of primary importance. To correctly overlay ParaView-generated images with OSIRIS images, we created a Python script to align the camera to the NAC or the WAC cameras, by using the information provided by the mission's SPICE kernels (https://www.cosmos.esa.int/web/spice/spice-for-rosetta)(Acton et al., 2018). Retrieving the camera's orientation and position with respect to 67P can be done in Python by using the spiceypy toolkit (Annex et al., 2017):

```
import os
import spiceypy

# load the spice kernels
os.chdir("cgkernels/mk/")
spiceypy.furnsh("ROS_V111.TM")

# the time at which the camera position is needed
t = "2016-01-27T16:27:58.970"
# transform the string to a time
t = spiceypy.str2et(t)

# get the position of Rosetta spacecraft in 67p
# refence frame,
# centered in 67P, no light time correction
center = spiceypy.spkpos("ROSETTA", t,
"67P/C-G_CK", "NONE", "67P/C-G")[0]
```

```
# rotation matrix of the NAC camera,
# in respect to 67P reference frame
Rot = spiceypy.pxform("ROS_OSIRIS_NAC",
"67P/C-G_CK", t);
```

The variables center and Rot correspond to the optical center and the rotation matrix of the camera, and can be used to align the view of ParaView with the NAC or WAC camera at the time t. The code first obtains the rotation matrix and the position vector as above, then resets the current camera position to a known one, and finally applies the transformation matrix build by using the variables center and Rot. This strategy can be easily scripted to process a large number of camera's positions and to automatically obtain matching screenshots for any number of NAC or WAC images.

```
# and the active camera
camera = GetActiveCamera()
view = GetActiveView()
view.CameraParallelProjection = 0

# get center and rotation of camera
center, Rot = getNacCenterAndRotationAtTime(t)
# as above
# we create a homogeneous transformation matrix,
# embedding center and rotation
Rt = np.zeros((4,4))
Rt[3,3] = 1
Rt[:3,:3] = Rot
Rt[:3,3] = center
```

```
# we do a total reset to start in a known
  position
camera.SetFocalPoint(0, 0, 1)
camera.SetPosition(0, 0, 0)
camera.SetViewUp((0, 1, 0))
camera.SetRoll(-90)
camera.SetWindowCenter(0, 0)

# create a transform
t = vtk.vtkTransform()

# finally set to the transform
t.SetMatrix(Rt.ravel().tolist())

# and apply to the camera
camera.ApplyTransform(t)

# set the FoV depending on the camera
camera.SetViewAngle(2.35) # 12 for WAC

Render() # update render

# we might save a screenshot
SaveScreenshot(/path/to/screenshot.png,
ImageQuality=100,
viewOrLayout=view,
TransparentBackground=True,
ImageResolution=(2048, 2048))
```

By computing the contours of the RES field at evenly spaced intervals, which can be easily done in ParaView, it is also possible to obtain three-dimensional traces of the predicted layers (see, as an example, Figure 11.11). These traces do not correspond to any specific layer but can convey the orientation of the layering in different regions of the body in a fairly efficient way. ParaView displays the ellipsoidal model contours on top of the three-dimensional shape model, but producing contour lines overlaid to the actual osiris image (e.g. Figure 11.11e) requires the occlusions generated by the shape of 67P to be correctly taken care of. This operation can be scripted in the following way:

1. The paraview camera orientation, position, and field of view is set as the osiris camera was when the original image was acquired.
2. The screenshot of Figure 11.11a is then saved.
3. The colors of the shape model are changed to red, while the contour lines are colored with a unique black color and a screenshot is saved (Figure 11.11b).
4. The screenshot is processed to create a mask of black pixels (Figure 11.11d).
5. This mask is applied to the first screenshot taken, in order to extract only the contour lines (Figure 11.11c).
6. The extracted image can then be overlaid on the original osiris image (Figure 11.11e).

The 3D shape model (Preusker et al., 2017; Jorda et al., 2016) is the result of processing hundreds of OSIRIS images to produce a three-dimensional mesh of the surface, and can be used to create additional views of the same scene that can be helpful in correctly understanding the resolved morphological features on the comet surface. Figure 11.12b was created in Blender (e.g. Kent, 2013) to reflect the alternation of terraces and cliffs. Gravitational slopes greater than 35° were used to artificially color the cliffs, while flatter terrains were left in gray by creating an ad-hoc shader. Also, Blender can be easily scripted to align its own camera with the view of any OSIRIS image, as shown in the next script:

```
import bpy
# using transform3d package convert rot mat to
euler angles import transforms3d as tr

render = bpy.context.scene.render
render.resolution_x = 2048
render.resolution_y = 2048
render.resolution_percentage = 100

cc = bpy.data.cameras[0]
cc.sensor_width = 1.35e-02*2048 # pixel size
times number of pixels
cc.sensor_height = 1.35e-02*2048
cc.angle = 2.35 / 180 * np.pi # would be 12 for
WAC

center, Rot = getNacCenterAndRotationAtTime(t)
# as explained above

camera = bpy.context.scene.camera
camera.location = center

# this transform is needed because the camera in
# blender is differently defined in respect to
# NAC/WAC inside spice
Rb = np.array([[0,1,0],[1,0,0],[0,0,-1]])

bpy.context.object.rotation_mode = 'XYZ'
eulers = np.array(tr.euler.mat2euler(Rot.dot(Rb),
'sxyz'))
camera.rotation_euler = eulers

# render the scene to a file
render.filepath = "/path/to/render.png"
bpy.ops.render.render(write_still=True)
```

This script requires the user to correctly set up Blender beforehand by importing the three-dimensional model together with per-vertices gravitational slopes and setting the lighting and the materials as desired. The code can be easily extended to correctly orient the light source to reproduce the observed shadows, as was done in Figure 11.12b, where the shadowing effect was slightly reduced. The complete scripts can be obtained from the author's github page (https://github.com/luca-penasa) or by contacting the author.

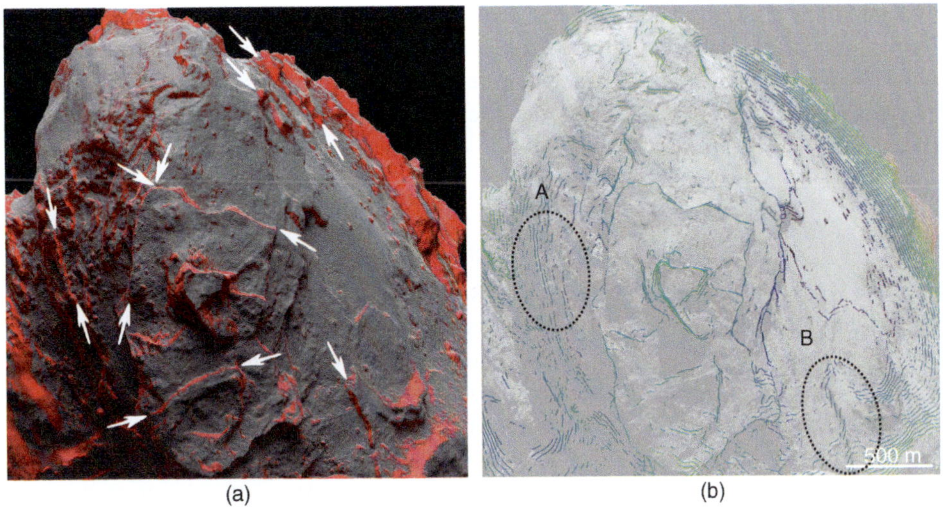

Figure 11.12 (a) Render of the shape-model made with Blender: the cliffs (regions with gravitational slopes greater than 35°) were shaded in red to highlight the alternation of cliffs with terraces. White arrows highlight some of the more extended cliffs. The same point of view of the NAC camera was obtained by means of the Pythons scripts reported in this work. (b) RES contour lines traced only on cliffs to highlight locations where layers are expected to be well exposed. The two ellipses show regions where the model fits the observed cliffs perfectly (A) and where the predicted lines are instead at a slight angle (B).

Figure 11.12 demonstrates the use of the proposed methods as a visual aid for validating the modelling results. The layering of comet Churyumov-Gerasimenko (67P) is documented by the presence of terraces separated by steep cliffs. The cliffs themselves can be easily identified by isolating the regions with a gravitational slope greater than 35°, as shown in Figure 11.12a: cliffs are clearly aligned in strips of terrain displaying lateral continuity. Some of these extended cliffs are highlighted by arrows in Figure 11.12a in iorder to help the interpretation. The contour lines obtained from the model (Figure 11.12b) are expected to be parallel to the orientation of the cliffs. More comparisons of this kind can be found in Penasa et al. (2017a) and its additional material.

11.5 Conclusions

The attempt to reconstruct a 3D model of the layered structure of the lobes of comet 67P is challenged by the scattered observations of the bedding planes. This issue was tackled by a very simple modelling strategy based on the approximation of the layers as shapes derived from analytical functions (spheres or ellipsoids). The implementation of the solver was required to take into account the rotational symmetry of the ellipsoids and to correctly devise a strategy for the initialization of the model parameters. We found that convergence between model and observations for 67P was robust to random initialization of the parameters. The described approach has the advantage of providing a small set of parameters with a precise physical meaning whose standard errors can be easily determined by using a bootstrap strategy. These errors made it possible to compare the *center* of the concentric layering (the center of the ellipsoidal model) with the center of mass of the respective lobe from a statistical point of view, demonstrating that the two points do not correspond.

The proposed modelling strategy was implemented *outside* commercial 3D geomodelling platforms and required the development of ad-hoc visualization strategies useful for comparing the results with the morphological features that could be observed from osiris images.

We stress that the modelling of the layers of 67P lobes through ellipsoids is a simplification of a more complex shape whose further investigation will require improvements able to take into account higher complexity.

Acknowledgments

OSIRIS was built by a consortium of the Max-Planck-Institut für Sonnensystemforschung in Göttingen, Germany, CISAS-University of Padova, Italy, the Laboratoire d'Astrophysique de Marseille, France, the Instituto de Astrofisica de Andalucia, CSIC, Granada, Spain, the Research and Scientific Support Department of the European Space Agency, Noordwijk, The Netherlands, the Instituto Nacional de Tecnica Aeroespacial, Madrid, Spain, the Universidad Politechnica de Madrid, Spain, the Department of Physics and Astronomy of Uppsala University, Sweden, and the Institut für Datentechnik und Kommunikationsnetze der Technischen Universität Braunschweig, Germany. The support of the national funding agencies of Germany (DLR), Italy (ASI), France (CNES), Spain (MEC), Sweden (SNSB), and the ESA Technical Directorate is gratefully acknowledged. We thank the ESA teams at ESAC, ESOC, and ESTEC for their work in support of the Rosetta mission.

References

Acton, C., Bachman, N., Semenov, B., and Wright, E. (2018). A look towards the future in the handling of space science mission geometry. *Planetary and Space Science* 150: 9–12. doi: 10.1016/j.pss.2017.02.013.

Agarwal, S., Mierle, K., and Others (2016). Ceres solver, http://ceres-solver.org.

Annex, A., Carcich, B., McAuliffe, J. et al. (2017). *Andrewannex/Spiceypy: Spiceypy* 2.1.0. doi: 10.5281/zenodo.1044745.

Auger, A.-T., Groussin, O., Jorda, L. et al. (2017). Meter-scale thermal contraction crack polygons on the nucleus of comet 67P/Churyumov-Gerasimenko. *Icarus*. doi: 10.1016/j.icarus.2017.09.037.

Basilevsky, A., Mall, U., Keller, H. et al. (2017). Geologic analysis of the Rosetta NavCam, Osiris and ROLIS images of the comet 67P/Churyumov-Gerasimenko nucleus. *Planetary and Space Science* 137: 1–19. doi:10.1016/j.pss.2017.01.002.

Birch, S.P.D., Tang, Y., Hayes, A.G. et al. (2017). Geomorphology of comet 67P/Churyumov–Gerasimenko. *Monthly Notices of the Royal Astronomical Society* 469 (Suppl. 2): S50–S67. doi: 10.1093/mnras/stx1096.

Bistacchi, A., Massironi, M., Dal Piaz, G.V. et al. (2008). 3D fold and fault reconstruction with an uncertainty model: An example from an Alpine tunnel case study. *Computers and Geosciences* 34 (4): 351–372. doi: 10.1016/j.cageo.2007.04.002.

Calcagno, P., Chilès, J., Courrioux, G., and Guillen, A. (2008). Geological modelling from field data and geological knowledge. *Physics of the Earth and Planetary Interiors* 171 (1–4): 147–157. doi: 10.1016/j.pepi.2008.06.013.

Caumon, G., Collon-Drouaillet, P., Carlier, L. et al. (2009). Surface-based 3D modeling of geological structures. *Mathematical Geosciences* 41 (8): 927–945. doi: 10.1007/s11004-009-9244-2.

Caumon, G., Gray, G., Antoine, C., and Titeux, M.-O. (2013). Three-dimensional implicit stratigraphic model building from remote sensing data on tetrahedral meshes: Theory and application to a regional model of La Popa Basin, NE Mexico. *IEEE Transactions on Geoscience and Remote Sensing* 51 (3): 1613–1621. doi: 10.1109/TGRS.2012.2207727.

Chernick, M.R. (2008). *Bootstrap Methods: A Guide for Practitioners and Researchers, Wiley Series in Probability and Statistics*, 2nd edn., Wiley-Interscience, Hoboken, N.J.

Cignoni, P., Callieri, M., Corsini, M. et al. (2008). MeshLab: An open-source mesh processing tool. In: *Eurographics Italian Chapter Conference* (ed. V. Scarano, R.D. Chiara and U. Erra). The Eurographics Association doi: 10.2312/LocalChapterEvents/ItalChap/ItalianChapConf2008/129-136.

Eberly, D. (2013). Distance from a point to an ellipse, an ellipsoid, or a hyperellipsoid. https://www.geometrictools.com/Documentation/DistancePointEllipseEllipsoid.pdf.

Efron, B. (1979). Bootstrap methods: Another look at the jackknife. *The Annals of Statistics* 7 (1): 1–26. doi: 10.2307/2958830.

El-Maarry, M.R., Thomas, N., Giacomini, L. et al. (2015a). Regional surface morphology of comet 67P/Churyumov-Gerasimenko from Rosetta/OSIRIS images. *Astronomy and Astrophysics* 583: A26. doi: 10.1051/0004-6361/201525723.

El-Maarry, M.R., Thomas, N., Gracia-Berná, A. et al. (2015b). Fractures on comet 67P/Churyumov-Gerasimenko observed by Rosetta/OSIRIS. *Geophysical Research Letters* 42 (13): 5170–5178. doi: 10.1002/2015GL064500.

El-Maarry, M.R., Thomas, N., Gracia-Berná, A. et al. (2016). Regional surface morphology of comet 67P/Churyumov-Gerasimenko from Rosetta/OSIRIS images: The southern hemisphere. *Astronomy and Astrophysics* 593: A110. doi: 10.1051/0004-6361/201628634.

Fernández, O. (2005). Obtaining a best fitting plane through 3D georeferenced data. *Journal of Structural Geology* 27 (5): 855–858.

Fisher, N.I. (1993). *Statistical Analysis of Circular Data*. Cambridge University Press, Cambridge, England and New York, USA.

Fornasier, S., Hasselmann, P.H., Barucci, M.A. et al. (2015). Spectrophotometric properties of the nucleus of comet 67P/Churyumov-Gerasimenko from the OSIRIS instrument onboard the ROSETTA spacecraft. *Astronomy and Astrophysics* 583: A30. doi: 10.1051/0004-6361/201525901.

Franceschi, M., Penasa, L., Massironi, M. et al. (2020). Global-scale brittle plastic rheology at the cometesimals merging of comet 67P/Churyumov–Gerasimenko. In:

Proceedings of the National Academy of Sciences, 201914552. doi: 10.1073/pnas.1914552117.

Giacomini, L., Massironi, M., El-Maarry, M.R. et al. (2016). Geologic mapping of the Comet 67P/Churyumov–Gerasimenko's Northern hemisphere. *Monthly Notices of the Royal Astronomical Society* 462 (Suppl. 1): S352–S367. doi: 10.1093/mnras/stw2848.

Grose, L., Ailleres, L., Laurent, G., and Jessell, M. (2021). LoopStructural 1.0: Time-aware geological modelling. *Geoscientific Model Development* 14 (6): 3915–3937. doi: 10 .5194/gmd-14-3915-2021.

Guilbert-Lepoutre, A., Besse, S., Mousis, O. et al. (2015). On the evolution of comets. *Space Science Reviews* 197 (1–4): 271–296. doi:1\ignorespaces0.1007/s11214-015-0148-9.

Hardy, R.L. (1990). Theory and applications of the multiquadric-biharmonic method 20 years of discovery 1968–1988. *Computers and Mathematics with Applications* 19 (8–9): 163–208.

Hillier, M.J., Schetselaar, E.M., de Kemp, E.A., and Perron, G. (2014). Three-dimensional modelling of geological surfaces using generalized interpolation with radial basis functions. *Mathematical Geosciences* 46 (8): 931–953. doi: 10.1007/s11004-014-9540-3.

Jorda, L., Gaskell, R., Capanna, C. et al. (2016). The global shape, density and rotation of Comet 67P/Churyumov-Gerasimenko from preperihelion Rosetta/OSIRIS observations. *Icarus* 277: 257–278. doi: 10 .1016/j.icarus.2016.05.002.

Keller, H.U., Barbieri, C., Lamy, P. et al. (2007). OSIRIS – The Scientific Camera System Onboard Rosetta. *Space Science Reviews* 128 (1–4): 433–506. doi: 10.1007/s11214-006-9128-4.

Keller, H.U., Mottola, S., Davidsson, B. et al. (2015). Insolation, erosion, and morphology of comet 67P/Churyumov–Gerasimenko. *Astronomy and Astrophysics* 583: A34. doi: 10.1051/0004-6361/201525964.

Kent, B.R. (2013). Visualizing astronomical data with Blender. *Publications of the Astronomical Society of the Pacific* 125 (928): 731.

Laurent, G. (2016). Iterative thickness regularization of stratigraphic layers in discrete implicit modeling. *Mathematical Geosciences* 48 (7): 811–833. doi: 10.1007/s11004-016-9637-y.

Lee, J.-C., Massironi, M., Ip, W.-H. et al. (2016). Geomorphological mapping of comet 67P/Churyumov–Gerasimenko's Southern Hemisphere. *Monthly Notices of the Royal Astronomical Society* 462 (Suppl. 1): S573–S592. doi: 10.1093/mnras/stx450.

Massironi, M., Simioni, E., Marzari, F. et al. (2015). Two independent and primitive envelopes of the bilobate nucleus of comet 67P. *Nature* 526 (7573): 402–405. doi: 10 .1038/nature15511.

Oklay, N., Vincent, J.-B., Fornasier, S. et al. (2016). Variegation of comet 67P/Churyumov-Gerasimenko in regions showing activity. *Astronomy & Astrophysics* 586: A80. doi:10.1051/0004-6361/201527369.

Pajola, M., Oklay, N., La Forgia, F. et al. (2016). Aswan site on comet 67P/Churyumov–Gerasimenko: Morphology, boulder evolution, and spectrophotometry. *Astronomy and Astrophysics* 592: A69. doi: 10.1051/0004-6361/201527865.

Penasa, L., Massironi, M., Naletto, G. et al. (2017a). A three-dimensional modelling of the layered structure of comet 67P/Churyumov–Gerasimenko. *Monthly Notices of the Royal Astronomical Society* 469 (Suppl. 2): S741–S754. doi: 10.1093/mnras/stx2899.

Penasa, L., Massironi, M., Ferrari, S. et al. (2017b). The layered structure of the nucleus of the comet 67P: Implications on missing volumes and lobes orientations. *In European Planetary Science Congress* 11: pp. EPSC2017–485.

Preusker, F., Scholten, F., Matz, K.-D. et al. (2017). The global meter-level shape model of comet 67P/Churyumov–Gerasimenko. *Astronomy and Astrophysics* 607: L1. http://doi: 10.1051/0004-6361/201731798.

Rickman, H., Marchi, S., A'Hearn, M.F. et al. (2015). Comet 67P/Churyumov–Gerasimenko: Constraints on its origin from OSIRIS observations. *Astronomy and Astrophysics* 583: A44. doi: 10.1051/0004-6361/201526093.

Rolo, R.M., Radtke, R., and Costa, J.F.C.L. (2017). Signed distance function implicit geologic modeling, *REM – International. Engineering Journal* 70 (2): 221–229. doi: 10 .1590/0370-44672016700146.

Ruzicka, B.-K., Penasa, L., Boehnhardt, H. et al. (2019). Analysis of layering-related linear features on comet 67P/Churyumov–Gerasimenko. *Monthly Notices of the Royal Astronomical Society* 482 (4): 5007–5011. doi: 10 .1093/mnras/sty3079.

Schiavo, A., Dal Piaz, G.V., Monopoli, B. et al. (2015). Geology of the Brenner Pass–Fortezza Transect, Italian Eastern Alps. *Journal of Maps* 11 (1): 201–215. doi: 10.1080/17445647.2014.980337.

Schroeder, W., Martin, K., and Lorensen, B. (2006). *The Visualization Toolkit: An Object-Oriented Approach to 3D Graphics; Visualize Data in 3D – Medical, Engineering or Scientific; Build Your Own Applications with C++, Tcl, Java or Python; Includes Source Code for VTK (Supports Unix, Windows and Mac)*, 4th edn. Kitware, Inc., Clifton Park, NY.

Sierks, H., Barbieri, C., Lamy, P.L. et al. (2015). On the nucleus structure and activity of comet

67P/Churyumov–Gerasimenko. *Science* 347 (6220): aaa1044.

Squillacote, A.H., Ahrens, J., Law, C. et al. (2007). *The Paraview Guide* 366: Kitware.

Thomas, N., Sierks, H., Barbieri, C. et al. (2015). The morphological diversity of comet 67P/Churyumov–Gerasimenko. *Science* 347 (6220) aaa0440–aaa0440: doi: 10.1126/science.aaa0440.

Vincent, J.-B., Hviid, S.F., Mottola, S. et al. (2017). Constraints on cometary surface evolution derived from a statistical analysis of 67P's topography. *Monthly Notices of the Royal Astronomical Society* 469 (Suppl. 2): S329–S338. doi: 10.1093/mnras/stx1691.

Index

3D Digital Geological Models: From Terrestrial Outcrops to Planetary Surfaces, First Edition.
Edited by Andrea Bistacchi, Matteo Massironi, and Sophie Viseur.